# Biomedical Visualization

## Volume 9

**Series Editor**

Paul M. Rea, Anatomy Facility, School of Medicine, Dentistry and Nursing,
College of Medical, Veterinary and Life Sciences
University of Glasgow
Glasgow, UK

The Biomedical Visualization book series invites contributions related to visualization and imaging in the biomedical sciences and related fields such as medicine, dentistry, veterinary surgery, informatics, and the allied health professions.

This can encompass work in image analysis, workflow methodologies, photogrammetry in science, animations, digital reconstructions and applications, big data and visualizations, educational methodologies, usability and evaluations, augmented and virtual reality, and 3D and 4D technologies, including 3D printing, as well as informatics, e-tutorials, MOOCs, HCI and public engagement. It can be from macroscopic to microscopic but show how we can view data and information related to the biomedical field in a much more accessible, innovative and engaging way using technology.

This series is intended for researchers, clinicians, students, and teaching staff in biomedical and medical schools who use and develop visualization and imaging techniques in medical education, patient care, and biomedical and medical research.

Iain D. Keenan • Isabel Stabile
Asha Venkatesh
Editors

# International Anatomical Education

## The Trans-European Pedagogic Anatomy Research Group

Springer

*Editors*
Iain D. Keenan
School of Medicine
Newcastle University
Newcastle Upon Tyne, UK

Isabel Stabile
Faculty of Dental Surgery
University of Malta
Msida, Malta

Asha Venkatesh
University of Aberdeen
Aberdeen, UK

ISSN 2731-6130 ISSN 2731-6149 (electronic)
Biomedical Visualization
ISBN 978-3-031-91848-3 ISBN 978-3-031-91849-0 (eBook)
https://doi.org/10.1007/978-3-031-91849-0



Cover Illustration: Catherine MacRobbie.

This Springer imprint is published by the registered company Springer Nature Switzerland AG
The registered company address is: Gewerbestrasse 11, 6330 Cham, Switzerland

# Preface

The Trans-European Pedagogic Anatomy Research Group (TEPARG) was founded in 2003 by Professor Bernard Moxham. At that time, Professor Moxham was President of the European Federation for Experimental Morphology (EFEM) and was concerned that scholarly, research-informed and evidence-based approaches to the design and implementation of anatomical education were not being adequately pursued.

Consequently, Professor Moxham created TEPARG to promote and support research in anatomical education. Enthusiastic anatomy teachers and anatomy pedagogic researchers from across Europe were invited to meet to share good practice and to create important new projects for the development of anatomical education.

The first TEPARG conference was held in-person in Paris, France, in March 2005 and continued annually until the COVID-19 pandemic, with the first hybrid conference held in March 2020 and the first entirely virtual TEPARG conference delivered in 2021. Hybrid conferences have now become the norm, with meetings held physically (initially in Barcelona and now in Lisbon), with a large virtual element, allowing much larger numbers of TEPARG members to attend. TEPARG has also hosted symposia at renowned international meetings, including at the International Federation of Associations of Anatomists (IFAA) 2019 meeting in London, UK.

As of 2025, TEPARG has more than 100 active members based at academic institutions across Europe, including Austria, Basque Country, Cyprus, the Czech Republic, France, Greece, Ireland, Italy, Malta, Poland, Portugal, Serbia, Spain, Switzerland, Turkey and the nations of England, Scotland and Wales within the United Kingdom. TEPARG has also recently welcomed members and contributors based in other continents, from nations including Australia, Bahrain, Canada, India, Kuwait, Malaysia, Nepal, Nigeria, Saudi Arabia, South Africa, the USA and Zambia.

TEPARG conference abstracts have previously been published in the *European Journal of Anatomy* and in *Anatomia, Histologia, Embryologia* and articles arising from projects devised and presented at TEPARG meetings, including work on body donation and contributions to core anatomical syllabuses, have been published in the *Journal of Anatomy* and *Anatomical Sciences Education*. However, this is the first volume of work entirely authored by TEPARG members and demonstrates the scope, depth and impact of pedagogic projects conducted by the group.

The work presented in this volume demonstrates careful consideration by the authors of several key areas within the current complex landscape of international anatomical education. This volume is presented in two sub-themes, with the first concerning broad considerations of modern anatomy curricula (Chaps. 1, 2, 3 and 4) and the second involving discussion of pedagogic innovations for the delivery of anatomical education to learners and to the wider public (Chaps. 5, 6, 7 and 8).

It is fitting that TEPARG founder Bernard Moxham (Cardiff, Wales) has authored the first chapter in this first volume of work by TEPARG members. This chapter introduces the volume by addressing an important and broad area within the context of higher education and medical training in the United Kingdom, which is likely to be of concern for anatomical and medical educators in the United Kingdom, Europe and beyond. The second chapter has been authored by current (2024–present) TEPARG President Erich Brenner (Innsbruck, Austria), who provides an overview of the 'gold standard' anatomical learning resource and the privilege and value provided by donated human bodies for study of anatomy. The curriculum theme is then developed within the specific contexts of two UK institutions, from North and South of the Anglo-Scottish border, where an inclusive anatomy curriculum has been introduced (Aberdeen, Scotland) and challenges of learning anatomy within a case-based curriculum have been explored (Newcastle upon Tyne, England).

The second part of this volume describes innovations in anatomical education and, like part one, provides an overview of the topic. In this case, the author Ian Johnson (Sydney, Australia) takes us further afield internationally to describe an investigation of student perceptions of innovations that have been introduced to their anatomy learning experience. We then return to Europe to explore innovations in the integration of new digital approaches into the teaching of traditional cadaveric anatomy (Padua, Italy) and the development of digital simulations in dental anatomy (Basque Country). The volume concludes with a chapter describing a highly successful public engagement event (Newcastle upon Tyne, England), surrounding the recently introduced 'World Anatomy Day', now a widespread shared international event celebrated on 15th October each year.

The work presented in this volume will have implications for anatomical educators and pedagogic researchers in the anatomical sciences who are seeking to develop their own curricula and to implement effective, evidence-based, and research-informed innovations into their teaching.

Newcastle upon Tyne, UK — Iain D. Keenan
Msida, Malta — Isabel Stabile
Aberdeen, UK — Asha Venkatesh

# Contents

# Editors and Contributors

## About the Series Editor

**Paul M. Rea** is an internationally recognised leading expert in the applications of digital technologies to biomedical/medical visualisation in education, clinical settings and industry-linked research. Initially, he trained in medicine, undertaking roles in general medicine, paediatric surgery and emergency medicine, followed by a period as a pathologist. He then changed direction, pursuing a clinical anatomy educator role as Full Professor of Digital and Anatomical Education. It was in academia that he realised the potential digital technologies had for education, simulation, training and public/civic engagement. He led the strategic partnership with the School of Simulation and Technology (SIT), the Glasgow School of Art (GSA) and founded the MSc Medical Visualisation and Human Anatomy, following a significant grant-funded research. This degree improved graduate employability and had significant high-profile media coverage. He was the founding and now series editor (editor-in-chief equivalent) for the internationally acclaimed Biomedical Visualization book series with Springer Nature. He holds strategic leadership roles for innovation, engagement and enterprise; body donation; education; and previously project management for large-scale infrastructural developments. Professor Rea is a licensed teacher of anatomy, licensed by the Scottish Government, and has held this role for 19 years.

He is also the Director of Innovation, Engagement and Enterprise within the School of Medicine, Dentistry and Nursing, a role he has held since August 2022. He is responsible for leading, delivering and managing the three pillars of the strategy: driving innovation and enterprise, employability for careers and growing entrepreneurs, and regional economic and societal impact.

## About the Volume Editors

**Iain D. Keenan,** BSc (Hons.), PhD, MMedEd., SFHEA, NTF, FAS, is Full Professor of Anatomical Education at the School of Medicine, Newcastle University, UK, and has been teaching gross anatomy, histology and embryology for medical, medical sciences and healthcare profession degree programmes in the Faculty of Medical Sciences since 2012. Professor Keenan's research involves development and investigation of digital visualisation and art-based learning approaches in anatomical education, which has informed Professor Keenan's leadership of the global 'Exploring 3D Anatomy' and 'Exploring 3D Anatomy Plus' courses in collaboration with Leonard Shapiro at the University of Cape Town, South Africa. Professor Keenan has authored several high-profile publications in esteemed international anatomical education journals, has authored several chapters in Biomedical Visualization, and has edited a previous volume in the series. Professor Keenan's wide-ranging contributions to anatomical education have been recognised in the award of a National Teaching Fellowship (2020) and Fellowship of the Anatomical Society (2021). Professor Keenan has been a TEPARG member since 2018, winning the best presentation prize at the TEPARG conference that year. Professor Keenan was first appointed Honorary Secretary of TEPARG in 2021 and began a second term in 2024.

**Isabel Stabile** FRCOG, PhD, had a long career in academic medicine in the United Kingdom and the USA, where she worked for 20 years. She is the author of over 80 publications and six books. Professor Stabile is currently engaged as Full Professor in the Faculty of Dental Surgery at the University of Malta. She also has a part-time private practice as a gynaecologist.

During her tenure at Florida State University in the USA, she was responsible for generating over $10,000,000 USD in research funding, which has led to numerous publications as well as international awards. She was also responsible for redesigning the clinical curriculum in the Faculty of Medicine and Surgery and the introduction of new strategies to facilitate learning. The final product is student centred, vertically and horizontally integrated, and builds on the success of the biomedical science curriculum jointly developed previously. Professor Stabile is the immediate past president of TEPARG.

**Asha Venkatesh** MBBS, PGDip (Ophth), MSc (Clinical Education), SFHEA, FAS, following a brief career in clinical ophthalmology in India, now works as a Senior Lecturer in Anatomy at the University of Aberdeen, UK. In addition to teaching gross anatomy, histology and embryology, she is interested in anatomy assessment, having steered the move to platform-based assessment of anatomy at the University of Aberdeen. She has won three university-level excellence awards for her teaching and supervision. Dr. Asha Venkatesh is also the equality diversity and inclusion lead of her school and is passionate about inclusive anatomy education.

Externally, she is the Membership Officer of the Anatomical Society (UK and Ireland) participating in the charitable actions of this learned society. Since 2022, she has been representing the Anatomical Society at TEPARG.

## Contributors

**Jon Jatsu Azkue** Department of Neurosciences, School of Medicine and Nursing - University of the Basque Country UPV/EHU, Leioa, Spain

**Erich Brenner** Institute for Clinical and Functional Anatomy, Medical University of Innsbruck, Innsbruck, Austria

**Thomas Butts** School of Medicine, Faculty of Health Sciences and Wellbeing, University of Sunderland, Sunderland, UK

**Raffaele De Caro** Section of Human Anatomy, Department of Neuroscience, University of Padua, Padua, Italy

**Evie M. Donald** School of Medicine, Faculty of Health Sciences and Wellbeing, University of Sunderland, Sunderland, UK

**Kate Dulohery** School of Medicine, Faculty of Health Sciences and Wellbeing, University of Sunderland, Sunderland, UK

**Yasmin Elebessy** Newcastle University Medicine Malaysia, Iskandar Puteri, Johor, Malaysia

**Yousr Elebessy** Newcastle University Medicine Malaysia, Iskandar Puteri, Johor, Malaysia

**Emily Green** Faculty of Medical Sciences, School of Medicine, Newcastle University, Newcastle upon Tyne, UK

**Ian P. Johnson** Adelaide, South Australia

**Chloe Kelly** Faculty of Medical Sciences, School of Medicine, Newcastle University, Newcastle upon Tyne, UK

**Munesh P. Khamuani** School of Medicine, Faculty of Health Sciences and Wellbeing, University of Sunderland, Sunderland, UK

**Marko Konschake** Institute for Clinical and Functional Anatomy, Medical University of Innsbruck, Innsbruck, Austria

**Roberto Di Lenarda** Department of Medical, Surgical, and Health Sciences, University of Trieste, Trieste, Italy

**Eire Lopetegi** Centre dentaire Bondy Hôtel de Ville – Dentylis, Bondy, France

**Veronica Macchi** Section of Human Anatomy, Department of Neuroscience, University of Padua, Padua, Italy

**Joanna Matthan** Newcastle University Medicine Malaysia (NUMed), Johor, Malaysia

**Harry Miles** School of Medicine, Faculty of Health Sciences and Wellbeing, University of Sunderland, Sunderland, UK

**Bernard J. Moxham** Cardiff School of Biosciences, Cardiff University, Cardiff, Wales

St. George's University, Grenada, West Indies

**Vanessa Nicolin** Department of Medical, Surgical, and Health Sciences, University of Trieste, Trieste, Italy

**Diogo Pais** NOVA Medical School, NOVA University of Lisbon, Lisbon, Portugal

**Debs Patten** School of Medicine, Faculty of Health Sciences and Wellbeing, University of Sunderland, Sunderland, UK

**Elisabeth Pechriggl** Institute for Clinical and Functional Anatomy, Medical University of Innsbruck, Innsbruck, Austria

**Taisha Peplowska** Faculty of Medical Sciences, School of Medicine, Newcastle University, Newcastle upon Tyne, UK

**Odile Plaisant** University of Paris Citė, Paris, France

**Andrea Porzionato** Section of Human Anatomy, Department of Neuroscience, University of Padua, Padua, Italy

**Aydan Rahmanova** Department of Anatomy, School of Medicine, Medical Sciences and Nutrition, University of Aberdeen, Aberdeen, Scotland, Aberdeen, United Kingdom

**Alice E. M. Roberts** School of Medicine, Faculty of Health Sciences and Wellbeing, University of Sunderland, Sunderland, UK

**Kayleigh Scotcher** Faculty of Medical Sciences, School of Medicine, Newcastle University, Newcastle upon Tyne, UK

**Stefano Sivolella** School of Dentistry, Department of Neuroscience, University of Padua, Padua, Italy

**Edoardo Stellini** School of Dentistry, Department of Neuroscience, University of Padua, Padua, Italy

**Elena Stocco** Section of Human Anatomy, Department of Neuroscience, University of Padua, Padua, Italy

Department of Women's and Children's Health, University of Padua, Padua, Italy

Department of Surgery, Oncology and Gastroenterology, University of Padua, Padua, Italy

**Hannah Swainson** Faculty of Medical Sciences, School of Medicine, Newcastle University, Newcastle upon Tyne, UK

**Asha Venkatesh** Department of Anatomy, School of Medicine, Medical Sciences and Nutrition, University of Aberdeen, Aberdeen, Scotland

# Part I

# Developing Anatomy Curricula

# 1 Should Medicine Continue to Be Taught at a University? An Opinion Piece

Bernard J. Moxham, Odile Plaisant, and Diogo Pais

**Abstract**

This chapter alludes to the recent, and dramatic, changes in medical education that appear to go against the values and characteristics of a university education. Recent turn of events require critical debate, as education and training for medicine at university are being questioned and potentially undermined by the development of medical apprenticeships, which are thus far limited to the UK, and by the advent of artificial intelligence.

Following introductory comments concerning how changes to the medical curriculum and assessments and examinations have departed from past practice, the authors outline what the features of a university education should be and conclude that medical education has critically departed from this model. Questions are raised about who is responsible for this situation and how medical training might be arranged to be once more worthy of being labelled a 'university education'.

The points raised are wide-ranging and are deliberately provocative as debate has for too long been one-sided. Indeed, a practical synthesis can only emerge when thesis meets antithesis. In our view, change is inevitable, but change should occur in the context of what is appropriate for a university education and what best benefits patients. The issues raised have important implications for the esteem of the medical profession. Indeed, this has a considerable impact upon medical practitioners who play an important role in society and who are presently much appreciated by the public as being amongst its most learned members.

**Keywords**

Medical education · University education · Medical degree apprenticeships

## 1.1 Introduction

Medicine is a caring profession that is justifiably held in high esteem by the public; such esteem probably reaching a zenith following the heroic efforts made by healthcare workers during the COVID-19 pandemic. That esteem is translatable into the use of the honorary title 'Doctor' conferred upon medical practitioners ('Doctor' being

B. J. Moxham (✉)
Cardiff School of Biosciences, Cardiff University, Cardiff, Wales

St. George's University, Grenada, West Indies
e-mail: moxham@cardiff.ac.uk

O. Plaisant
University of Paris Cité, Paris, France

D. Pais
NOVA Medical School, NOVA University of Lisbon, Lisbon, Portugal


I. D. Keenan et al. (eds.), *International Anatomical Education*, Biomedical Visualization 9, https://doi.org/10.1007/978-3-031-91849-0_1

used to designate those who are deemed to be learned). Universities, the highest centres for research and education in contemporary societies, proudly boast the existence of their medical schools where, to obtain a medical degree, high academic achievers from schools study a great variety of scientific, sociological and clinical subjects over a timescale of 5–7 years (4 years for graduate entry).

The practice of medicine is probably as old as the existence of mankind, and this history extends to the teaching of anatomy (e.g. Persaud 1984, 1997; Porter 2002, 2003, 2006; Mandressi 2003; Strathern 2005; Ameisen et al. 2011; Klestinec 2011). The earliest reference in Europe to the presence of a medical school in a higher educational establishment is at Salerno in Southern Italy in the tenth century (Koehler et al. 2021). However, Montpellier in France (established in 1220) is the oldest medical school still functioning (Moxham and Plaisant 2014) (Fig. 1.1). Nowadays, globally medicine is practised by firstly obtaining a medical degree through a university. However, we will allude below to the imminent appearance in the UK of medical apprenticeships that might not involve significant university education.

Until the recent past, medical students began their education by intensive study of the basic sciences before gaining clinical experience on hospital wards to translate their knowledge of the normal human body into the effects, amongst other things, of pathology and trauma. The tuition of the basic sciences was often practical-based, involving dissection of the human body for the anatomical sciences and laboratory-based practicals for physiological and molecular sciences (e.g. Jones and Whitaker 2009). In the past, and certainly prior to the 1980s in the UK, many of the academics teaching the basic sciences were clinically qualified and could, in addition to taking students towards the frontiers of knowledge

**Fig. 1.1** University of Montpellier, Faculty of Medicine

from understanding of their own research endeavours, easily convey the clinical importance of what they were teaching. Furthermore, the students, knowing the clinical credentials of their teachers, would, it was presumed, have less motivational issues relating to the clinical relevance of their studies (Moxham et al. 2011; Riederer et al. 2018). Then, as now, examinations were important hurdles in the progression of students through their basic science and clinical years.

Presently, the long-established model outlined above has changed radically. In many countries, control of medical education has been transferred away from the teachers and academic departments to national authorities, deans and medical educationalists. Integration, both vertical and horizontal, is employed, both for teaching and examinations (Dahle et al. 2002). Subject matter has often been restricted to 'core material', although universally recognised 'core' syllabuses are lacking, and this has occurred alongside an educational policy of not teaching material 'just in case' but instead 'just in time'. Consequently, the frontiers of knowledge might only be hinted at. Additionally, practical classes are no longer the mainstay of basic science programmes and, in many of the anglophone countries, fewer and fewer clinically qualified basic science teachers are employed. The loss of practical classes seems particularly ill-advised. For example, dissection by students for anatomy is a project-related, experiential way of learning that develops a variety of skills and professional attitudes as well as encouraging independent learning. Similarly for physiology, although critical and data analytical skills can be developed alongside important aspects of independent learning, the use of live animals or human subjects for student practicals is increasingly regarded as unethical and more computer-based work is undertaken (e.g. Goyal et al. 2017).

Student learning has also changed. In the UK, it used to be said that one went up to university to *read* for a degree. Nowadays, the symbolic significance of this designation has been lost and instead students say they go to university to *study* for a degree. The presence of the Internet and digital learning has become prevalent and at some universities library stock is being sold off since students no longer rely on university libraries and reading books (Orlando Figes 2014; Becker 2023; Hoeft 2012; Beckers et al. 2016; St Clair-Thompson et al. 2017). The advent of artificial intelligence (AI) is likely to accelerate these developments and has implications for the ways in which students can be taught and examined.

One hears academics complain that medical students no longer have good study skills or even verbal and writing skills (Moxham et al. 2018), the presumption being that many assessments they had at school prior to their admission to university were of the 'tick-box' variety! Given that students appear to be less and less 'deep learners' and becoming more and more 'superficial or strategic' learners, this has fostered increased student reliance on assessment feedback to the detriment, it can be argued, to the development of independent learning so necessary for lifelong learning. Given that we live in a competitive age, ill-advised (or badly organised) feedback where students require frequent reassurance about their performance, might only loop back to feed their insecurity. Should this be the case, we are faced with a dilemma: how do we strike a balance between being developmental and being judgemental in providing feedback? Although presently evidence is lacking, it is not unreasonable to hypothesise that university students who perform well and are conscientious develop self-critical facilities, such that they can spur themselves on to greater endeavours. Spoon-feeding could be seen as being ultimately self-defeating.

Perhaps the most marked changes in medical education relate to assessments and examinations (e.g. Epstein 2007; Ferris and O'Flynn 2015). Before 1990, it was frequently the case that each subject would have its own examinations that students were required to pass to progress to the next stage of their medical education (although a limited degree of 'compensation' was sometimes permitted between related subjects). These days, examinations are more integrated such that, for example, all the basic science subjects would be examined together and with a clinical context. On paper, this seems a better approach since it can be argued that students should be able to integrate material and not compartmentalise their knowledge (e.g. Cooke et al. 2006; Kadirvelu and Gurtu 2015). However, there are downsides! We have noted that students are evaluating how much material in the range of different subjects is needed to be learned to pass the examinations. Accordingly, they are 'rationing' their studies and not revising each subject to the depth that they would have done previously (i.e. learning strategically). We have also noted from our own personal observations that at some universities, while students would have previously obtained high marks in the anatomical sciences, students are now able to progress with considerably lower marks when taking the same assessments. There is a danger, therefore, that integrative assessments can lead to an extraordinary high level of 'compensation' between subjects that would have formerly been unthinkable. We need further to investigate whether the lower marks being obtained relate to students favouring those subjects they perceive as being 'easy' and failing to study intensively those subjects and topics that they perceive as being 'hard' or 'demanding'. Indeed, if it is seen that there are few questions in the integrative examinations for the harder topics this would further demotivate study of the harder subjects. We are not here arguing against incorporating an integrative paper into the diet of examinations but warn about the possible consequences of completely disbanding subject-specific assessments.

In recent times, the types of questions employed in medical examinations have also changed radically. Many medical schools used to rely on written examinations where answers to sight-unseen essay questions would be required. Oral examinations (*viva voce*) would also often be an essential component of the diet of examinations (sometimes just for borderline or distinction level students) and practical examinations (e.g. using cadaveric specimens for anatomy) would be conducted. Although it is frequently argued that changes to these types of examinations were necessary to improve objectivity, more often the changes have arisen primarily because of the increased numbers of students entering medical school and the consequent logistical problems. As an example of increases in student numbers, at the University of Cardiff there were 120 medical students each year in 1990 but now there are closer to 500 students (a fourfold increase) and the UK government is proposing that each medical school should double its existing numbers of students. Furthermore, academics appear to be increasingly loathed to spend time marking essays, particularly if they are under pressure to improve their research output! Instead of essays and oral examinations, assessments are nowadays often largely based upon multiple-choice questions (MCQs), extended-matching questions (EMQs) and short answer questions (despite some dispute concerning the reliability of these types of question; e.g. see Haris Ali et al. 2016). The consequence of this change is the opposite to what medical educational authorities wished for when they criticised the heavy reliance on examinations that tested just factual recall (or, in the case of MCQs, factual recognition; e.g. the UK's *Tomorrow's Doctors* directives from the General Medical Council through 1993 to 2016 (General Medical Council n.d.); and now replaced by *Outcomes for Graduates* (General Medical Council n.d.). Moreover, they called for more student-centred teaching but the loss of essays and reliance on MCQs, etc., has meant that students are now just learning what they are told to learn in their classes, handouts and video presentations. (Indeed, the ability of students to repeatedly 'relive' lectures electronically because they are video-captured might encourage this.) Previously, and certainly for the better students, in their sight-unseen (closed book) examination essays, candidates were able to include information in their responses that had not been previously taught in class, because they had 'read around' the topic in the research literature. For example, for this essay question—'*Write an essay to describe the development of the female reproductive system'*, not only would the student have to logically order their material, prioritise the important pieces of information and write cogently and use diagrams to provide understanding of spatial issues, but could introduce material that was obtained from private independent study. These skills have now all but gone!

Concerning oral examinations, they have a bad reputation because of the stress students might be put under and because of biased, and bad, examiner technique. And yet, if such examinations were structured so that students would be gradually asked questions from the very simple to the more complex, students can be eased into the examination. The oral assessment also offers an opportunity to gauge the level of knowledge and understanding. For the excellent student, knowledge obtained beyond what has been delivered in class or in the texts can be elicited in the oral examination. The 'take-home message' from this is that closed book and oral examinations, for all their faults and logistical problems, did encourage independent and deep learning beyond the delivered class material, while more recent examination practices may not be so effective. We are aware that, to ameliorate these criticisms, students are sometimes provided with in-course assignments where they undertake literature searches or provide critical evaluations of clinical cases or research reports. However, these might be seen by the superficial or strategic learner as just exercises to get through before the main examinations take place that really count! Perhaps academia has lost interest and/or there is insufficient time available to encourage the deep learner! Indeed, the university's traditional role in recognizing, and encouraging, excellence is a time-consuming endeavour.

## 1.2 Medical Degree Apprenticeship

So far, we have assumed that medical education and training will continue within a university setting. However, recent pronouncements by politicians in the UK signal the advent of some form of medical degree apprenticeships largely outside of a university. Information about the apprenticeship scheme can be found on the National Health Service (NHS) website (NHS Health Education England (2023a, b) and NHS Long term workforce plan (2023). Skills for Health (2023), the Institute for Apprenticeships and Technical Education 2023, the UK Government (2023) and the British Medical Association (2023a, b).

The main motivation for this development relates to the dire workforce crisis in the UK's NHS and the restrictions placed upon the employment of medical professionals who have trained overseas, on the poor pay and conditions within the profession, on the increasing emigration of medical graduates to obtain better pay and conditions, on poor resourcing of the NHS as a result of the imposition of 'austerity', as well as increased patient demand. Furthermore, regardless of where politicians stand on the political spectrum, questions are asked about the ability (or effectiveness) of professions to self-regulate. On the NHS website it is stated that the medical apprenticeship scheme will enable the medical profession to become '*more accessible, more diverse and more representative of local communities*'. This presumably stems from the notion that apprentices will be recruited locally and will be more likely to live at home. It is to be hoped that these sentiments express a real communitarian commitment. It would be appalling if it reflects an appeasement to the illiberal who somehow resent being seen by a 'foreign' healthcare worker!

The need for dramatic action to support the UK's NHS is highlighted by reports from the UK's National Audit Office (2023), The Institute for Government and the King's Fund (2023) where the problems are all related to resourcing! Although the apprenticeship scheme is signalled as a 'historic and radical' change to solve the workforce crisis, in reality it is a return to medical training as existing in the nineteenth century and before a medical practitioner had to be university trained!

In addition to the apprenticeships, the UK government has announced that there will be a doubling of student numbers within university medical schools by 2031 (NHS Long Term Workforce Plan 2023), despite concern that adding large numbers of students to already overcrowded university medical programmes would be counterproductive to quality education because of likely lack of teaching infrastructure and insufficient numbers of highly qualified academics. Presumably use of digital technologies, AI and reduction in face-to-face tuition and practical tuition will be required! Moreover, to speed up the appearance of new medical graduates on the wards, there is the suggestion in the UK government's proposals that the medical degree could be reduced from 5 to 4 years. One can anticipate that this will be accomplished by further reductions in the teaching and learning of the biomedical sciences. Doubling university medical student numbers potentially also raises concerns about the quality of those admitted from school to medical school. Should it be clear that the pool of talent is insufficient, then there could be political pressure within, and without, the medical schools to drop standards to ensure minimal student withdrawals. Regardless, the pressure on teaching staff to get students to reach acceptable standards could be enormous, and, in the process, those students who have the potential for excellence might be overlooked or become frustrated by the lack of challenge!

If apprenticeships are established despite authorities in the UK laying down strict criteria for competence (Institute for Apprenticeships and Technical Education 2023), who on the wards will find the time to supervise the apprentices? Already, clinical staff are having difficulties in balancing time spent on patient care with training existing university medical students, junior graduate doctors and other healthcare practitioners. Financial support for the scheme is also an issue but there are other more concerning issues that relate to university education. First, the fact that the medical apprentices will receive payment during training, while full-time university undergraduates will not (while simultaneously building up huge debts because of student loans and university fees), will undoubtedly lead to potential conflict between the two cohorts. Second, there is the possibility of the development of a two-tier medical profession. Will those going through the apprenticeship scheme have the same career opportunities as those educated through the universities? Third, it is unsure how the public and patients will regard medical apprentices and how this might affect the esteem of the medical profession in general. Furthermore, how will the esteem of medical practitioners trained in the UK be viewed by the medical profession globally? Fourth, as set out in the information provided by the Institute for Apprenticeships and Technical Education, a skills training body created by the Department for Education of the UK government (Institute for Apprenticeships and Technical Education 2023), there are in total 181 'KSBs' (Knowledge, Skills and Behaviour statements) but only 3 or 4 that relate to the biomedical sciences and to science in general. There are several possible consequences to this. Will the apprentices regard learning normal anatomy, biochemistry and physiology as a priority? Will there be sufficiently skilled staff to teach the apprentices normal biomedical sciences and the understanding of science? Will critical and analytical skills linked to science be prioritised? Will the normal biomedical sciences be also further undermined in medical university education as it is perceived that it is not particularly highlighted in the apprenticeship scheme? In response to the third and fourth points, a study by Moxham et al. (2016) suggests that failure to provide a significant education in anatomy (and with the use of human cadaveric material) severely diminishes the esteem of the medical profession in the eyes of the general public. Finally, the British Medical Association has pointed out that '*there must be no illusions about the limited impact medical degree apprenticeships will have in solving the dire NHS workforce crisis*', having estimated that the medical workforce in England alone is presently short of over 43,000 doctors when a comparison is made with the average doctor to population ratio in EU countries (British Medical Association 2023a). Overall, the government's enthusiasm for the apprenticeship scheme, combined with the doubling of university medical student numbers and the possible cutting of the medical course to 4 years, comes almost inevitably at the expense of a university-type education, as clearly the changes are viewed as 'instrumental' in solving one of the government's more vulnerable vote-losing areas. No one could argue against the need to solve the workforce crisis but where is the concern for the learned doctor?

## 1.3 Questions Needing Answers

Having outlined medical education in the recent past and the changes that are increasingly being put in place everywhere in the world, several questions remain to be answered.

Firstly: **what should be the characteristics of a university education?**

Universities have the responsibility of providing their students with those subject-specific cognitive, psychomotor and attitudinal tools that overall we consider to be essential for underpinning ***professionalism***. Knowledge acquisition is an important aspect of the cognitive domain, including learning terminologies and descriptive frameworks that are associated with communication skills. For the psychomotor domain, students must acquire appropriate skills, including observational, descriptive, structuring and integrative skills, technical gestures or professional abilities, teamwork and research abilities that encourage the development of a critical and creative mind. For the attitudinal domain, the student must acquire ethical values and a deontological behaviour acceptable in the practice of their profession. Just as the spokes of a wheel must be of the equal dimension for the wheel to run smoothly, the three domains must also be the target of a balanced development so that the university student is, in their own area of expertise, a competent, valued and respected junior professional.

Consequently, aside from gaining essential knowledge and understanding of the discipline, students should develop the skills of independent study that are crucial for lifelong learning, acquire critical acumen to question both old and new ideas, cultivate self-critical evaluation skills, develop where appropriate a mature attitude and behaviour related to their chosen profession, pursue excellence and develop towards becoming a deep learner, be able to communicate effectively orally and in the written form, gain knowledge of the limits of their discipline (including where the frontiers of knowledge are), develop an understanding of important and essential research methodologies relating to their discipline and understand procedures and philosophies underpinning ethical issues within their discipline/profession. Less essential, but nevertheless of benefit, would be an understanding of the history of their discipline/profession and the advantages of interdisciplinary work (e.g. the humanities within medicine). One cannot expect excellence in all domains but at least competence would be something to aim for. As already indicated in the first part of this chapter, we seem to be moving further and further away from medical education fulfilling this university ethos. Indeed, it seems to us that medical education has become too often fixated on factual delivery and retainment and on examination procedures, particularly following the Covid pandemic (e.g. Potu et al. 2021). It is as if teachers and educationalists have themselves become superficial and strategic! We conclude that in many aspects medical education unfortunately has presently strayed from being a university education.

With the appearance of medical apprenticeships in the UK, and the possible reduction of the UK university medical course to 4 years, we have little confidence in the maintenance of the essential characteristics and standards associated with a university education.

The second question that needs to be addressed is: **who is to blame for this situation?**

We should first acknowledge that blame should not rest with the students. They are frequently recipients of an educational system dictated by political concerns and by examination processes before entering university that, being often more ‘tick-box’ in form, encourage superficial and strategic rather than deep learning and where teachers and schools vie against each other in ‘league tables’ that have no real educational purpose. Indeed, historically academics often claim that students enter university having insufficient basic knowledge or skills to adapt readily to university education (Moxham et al. 2018). This applies across the board, including medical students. It has become fashionable for sociologists, psychologists and educationalists to label and characterise different generations (from the 'lost generation', 1883–1900, to 'baby boomers', 1946–1964, to 'Generation Alpha', 2013–2025; see Center for Generational Kinetics (2023) and to characterise how they learn and how they should be taught. Furthermore, members of contemporary generations are often labelled as 'digital natives' who have grown up in the informational age (see Selwyn 2009. 'The digital native – myth and reality'). All of this presupposes uniformity within each generation, but what is too frequently forgotten is that medical students are, in most parts of the world, a 'select', highly capable, group. Why then should we shirk from providing them with challenges (not just for their own benefit but for the benefit of future patients)? So, who then is to blame?

We can apportion some blame on the political establishment. In many democratic countries, governments are constantly looking to see where votes are, and pupils and students failing is not a vote winner! Furthermore, they seem to perceive education, and particularly university education, as being merely drivers of the economy and a conveyer belt for employment. Moreover, they seem not to invite, or welcome, objective criticism! This predicament was perceptively thought about as long ago as 1933–1943 by Hermann Hesse (1943) in his novel *The Glass Bead Game* (*Das Glasperlenspiel*). Hesse imagined a country where, in almost a monastic setting, a game that is an abstract synthesis of all sciences and arts was played that maintained the intellectual and cultural credibility of the country. However, being considered an ivory tower institution and endeavour, it was seen by the politicians not to provide societal/economic benefits and was consequently abandoned. The master game player (Magister Ludi) leaves the institution and, towards the end of the novel, takes a swim in a lake where, symbolically, he drowns. The message is that intellectual life is precious and once lost is in danger of being forever lost to political expediency. Academics, it seems, can be their own worst enemy in being complicit in losing sight of their intellectual, as well as their societal, responsibilities!

The UK government, in its 'instrumentalist' approach to solving the workforce crisis in the NHS, appears to care little about the need to develop learning in the medical curriculum. Instead, in consort with the administrative leaders of the NHS, the UK government has devised a reduced 4-year university medical course and medical apprenticeships to attempt to ease societal concerns while also seeming to care little about the consequences to the reputation of the medical profession internationally and without adequate consultation with the profession.

Universities should also take some responsibility. They have complied with political pressure such that the concept of them being 'communities of scholars' is being (has been) lost. They have instead become 'businesses' where students, being essentially just 'bums on seats', are fee-paying clients. Furthermore, student failures and drop-outs are not to be countenanced, academic attainments are artificially elevated (grade inflation) and student opinion trumps academic discipline requirements (Universities UK 2024, Times Higher Education n.d. see also Stefan Collini: Speaking of Universities (2018)). In the context of medical education, medical authorities, deans and medical educationalists are not blameless. They have in recent times taken control over the courses in place of that previously exerted by the specialised academics within subject departments. This has supposedly, and understandably, happened to ensure standardisation of educational approach and examination reliability; for the curriculum to be integrated; and for the teaching to be vocationally appropriate. However, it was also a political 'power-play' that has wrested control (including financial control) away from departments.

Educational theory and practice, away from the traditional model, have also produced a rather anarchic organisation of medical education across the world such that, without international (let alone national) standards being agreed, there is little consistency or transparency. Many new pedagogic approaches have been instituted without adequate and rigorous assessment so that the approaches sometimes seem more ideological than practical. The latest assault on university-level medicine might come from the introduction of 'competency-based' medical courses (e.g. Carraccio et al. 2002). If, as happened with too many problem-based courses, such courses are improperly organised and resourced, ill-thought-out, or conceptually and practically misunderstood, the pursuit of excellence could be further downgraded, and the basic sciences could become marginalised and merely 'tools' in the training process rather than an important characteristic of the learned medical practitioner. The situation is compounded by a misunderstanding as to who exactly are the 'clients' of medical education. This issue is particularly important given that it is said frequently that we live in a 'Consumerist Society'. In this regard, the consumer of medical education is not the political governing authorities, nor indeed the medical profession (let alone the deans or medical educationalists), and, unlike many other university courses, it is not the medical student either! The consumer of medical education is the future patient, and their views as to how medical students should be educated are at best not considered and at worst dismissed.

Academics teaching medical students are additionally at fault. Given that careers and promotions are at stake, it is perhaps not surprising that there has been so little effective and public opposition to the trends in contemporary medical education and training. Furthermore, careers are being built upon research into medical education that often is little more than an advertisement for how one particular medical school has reorganised its curriculum and how 'content' the students are with the new course, despite the student having experienced nothing different!

For the third and final question: **what can be done to ameliorate issues concerning the validity of medicine being a university education?**

Tinkering with the present system will not solve the problem, and anyway 'the horse has bolted and it's now little use shutting the barn door'. Assuming the notion that medical apprenticeships do not gain acceptance, and that most members of the medical profession remain university educated and trained, the simplest solution within present constraints is for *all* students to complete an extra year of advanced studies that includes a substantial research project. This should be accompanied by significant changes to the early parts of the medical course, including more self-directed learning (perhaps supplemented by near-peer teaching), library assignments and learning from practical work instead of a reliance on didactic (including computer-assisted) teaching. At the same time, fewer 'tick-box' assessments should be employed. As argued earlier, the nature of the examinations employed dictates both how medicine is learned and taught. However, examinations should be the 'effect' (or consequence) of appropriate teaching and learning and not the 'cause' of how learners study. Furthermore, the examination diet should (re) introduce subject-specific requirements. There is a place for an integrative examination, but this should not dominate the assessments and progression at the early stages of the curriculum. Alternatively, but hardly likely, medical schools should cease trying to teach every element of medicine within the curriculum and should be more specialised. For example, there could be some schools that concentrate on general practice, while others specialised in surgery, and others on community medicine and psychiatry. This might seem to some to be a regressive policy, but does it make economic sense, let alone academic sense, to try to teach and learn the whole range of medical disciplines when this can only be accomplished by increasingly superficial coverage?

## 1.4 Conclusions

In answer to the question posed by the title of this article, it is our firm opinion that medicine should be taught at a university and that, regardless of political pressures, medical apprenticeships essentially outside of a university should not gain professional approval. If the teaching of medicine remains university-based, a change in direction from present educational trends will be required. Should the authorities in charge of medical schools not wish to shift direction, then perhaps medical schools will have to become self-standing and independent of universities and financed from ministries of health and not education. At this point the development of medical apprenticeships becomes more politically attractive and achievable. We would regret such a development, but the matter is of concern, not merely to the medical profession, but to the public who are the real clients of medical education and who would still wish for their doctors to be learned and who would also wish for the medical profession to continue to enjoy high esteem (Moxham et al. 2016).

What are the possible implications of these matters to visualisation in biomedical and related areas? As anatomists, we are of the opinion that our discipline is 'iconophilic' (displaying a love of images), a notion that no doubt could apply to many other disciplines in medicine and recent research has highlighted the importance of drawing in learning anatomy (Backhouse et al. 2017; Keenan et al. 2017; Keenan and Powell 2020; Shapiro et al. 2020, 2023; Horne et al. 2024). In the past, students in, for example, anatomy and histology practicals used to have to draw what was observed. With the loss of time for practicals this activity, and the commensurate learned skill, has become diminished in importance. Additionally, lectures often required the teacher to draw on a board and the students would copy the drawing. As a result, the student would not be presented with the fully formed image but would learn how the image was built up and how structures could be related to each other (additionally for anatomy helping with an appreciation of the 3D arrangement). Nowadays, images are projected fully formed in programmes such as PowerPoint or are provided in handouts or referred to in textbooks or websites or provided in 3D digital visualisation resources (e.g. Complete Anatomy, Primal Pictures, Sectra, Anatomage).

The advent of artificial intelligence (AI) potentially could exacerbate this situation as students can call up with simple instructions images that appear to be novel and previously unpublished. As AI is capable of devising realistic imagery (Palou 2023), an extensive literature on this topic is likely to be developed over the coming years (Palou, 2023). While some authors already claim that AI will lead to a better future for the teaching and learning of the biomedical sciences (e.g. Abdellatif et al. 2022; Totlis et al. 2023), others draw attention to the pitfalls amongst the promise (e.g. Buzzaccarini et al. 2024). In the expectation of ameliorating any deficiencies resulting from the absence of the need to draw, various techniques have been suggested, including getting students to draw a variety of non-medical objects to improve skills and powers of observation (e.g. Shapiro et al. 2020; Horne et al. 2024. We are also aware that in some medical schools (including one of those associated with an author of this chapter), students participated in life-drawing sessions with nude models.

The decline in the need to draw also feeds in to changes in the methods of assessment and examinations. For sight-unseen examinations where essays were written, students could draw diagrams to illustrate their answers. The introduction of multiple-choice question examinations instead of essays often dispenses with diagrams or, if used, most frequently only require the student to label diagrams. That this matters comes from evidence reported by Keenan and colleagues who have demonstrated that anatomical knowledge and learning are improved by drawing (Keenan et al. 2017; Backhouse et al. 2017), alongside better 3-D appreciation (Shapiro et al. 2020; Keenan and Powell 2020). Some might argue that drawing is a skill not given to everyone. However, Horne et al. (2024) argue, and present evidence, that drawing is for everyone and that it is an inclusive activity.

Whether or not learning styles or learning preferences (e.g. Reichman and Grasha 1974; Riding and Cheema 1991; Allinson and Hayes 1996; Rayner and Riding 1997; Price 2004; Honey and Mumford 2006) are reliable indicators of the ways students study and of academic performance remains debatable (e.g. Wilkinson et al. 2013; Gurpinar et al. 2011; Fleming et al. 2011; Lodge et al. 2015). While it might be presumed that visual learners, especially 'pragmatists' who enjoy educational methods with practical value and learning from demonstrations

(Honey and Mumford 2006), are those who appreciate learning via images and from anatomical dissection, Wilkinson et al. (2013) reported that medical students are more likely to be 'reflectors' (who stand back and think before acting) or 'theorists' (who enjoy conceptual and theoretical thinking). Furthermore, they found that different learning preferences have no influence on academic achievement, although only multiple-choice questions and short-answer questions were included in their study! It would be instructive to assess how different learning preferences affect performance at anatomy 'spotter' examinations where medical students are required to identify structures on anatomical prosections.

Overall, therefore, it seems that presently the importance of visualisation in student learning and assessment is generally, and regrettably, not considered as important as in the past.

Finally, we must state that we are not 'dyed in the wool' conservatives or reactionaries. Indeed, we have been involved in, and supported, curricular changes that have introduced problem-based teaching, more student-centred tuition (e.g. project-based learning, near-peer teaching, practical and ethical approaches to anatomy), the development of 'core' syllabuses, and the inclusion of medical humanities and issues relating to equality and diversity in the medical curriculum. We are therefore not arguing against change, but we strongly believe that change should occur in the context of what is appropriate for a university education and what best benefits patients from having learned medical practitioners. We are neither traditionalists nor modernists but postmodernists, in that we see what is good in the old as well as in the new. What, in our eyes, is most reprehensible is that medical education appears to be drifting into a situation without rigorous discussion about what is being lost, and politicians are stepping in to fill 'the void' and thus undermining university medical education. We appreciate that we run the risk of sounding like the councillors to King Canute who believed he could command the sea tides to turn back! Should it therefore not be possible to retrench and return medical education to its proper place within the ethos of a university, the future could be to go backwards in history to apprenticeships where training occurs mainly outside of the university. These, in our view, are potentially to the detriment of the discipline, and to the esteem of the medical profession. Merely increasing the number of persons with stethoscopes on the wards or in the offices of general practitioners may cover short-term workforce problems but will it necessarily be commensurate with improving patient care?

## References


Abdellatif H, Mushaiqri MA, Albalushi H, Al-Zaabi AA, Roychoudhury S, Das S (2022) Teaching, learning and assessing anatomy with artificial intelligence: the road to a better future. Int J Environ Res Public Health 19:14209–14221

Allinson J, Hayes C (1996) The cognitive style index: a measure of intuition-analysis for organisational research. J Manage Stud 33:119–135

Ameisen JC et al (2011) Une histoire de la médecine. Éditions de la Martinière, Paris

Backhouse M, Fitzpatrick M, Hutchinson J, Thandi CS, Keenan ID (2017) Improvements in anatomy knowledge when utilizing a novel cyclical "observe-reflect-draw-edit-repeat" learning process. Anat Sci Edu 10:7–22

Becker B (2023) The pandemic's impact on the academic library: adapting to the new normal. J Acad Librarianship. https://www.sciencedirect.com/journal/the-journal-of-academic-librarianship/special-issue/10TC8F2PXJC. Accessed 26 Apr 2024

Beckers R, van der Voordt T, Dewulf G (2016) Why do they study there? Diary research into students learning space choices in higher education. High Educ Res Dev 35:142–157

British Medical Association (2023a) Medical apprenticeships. https://www.bma.org.uk/news-and-opinion/medical-degree-apprenticeships. Accessed 8 Aug 2023

British Medical Association (2023b) Pre-ARM briefing—Medical degree apprenticeships. https://www.bma.org.uk/media/4162/medical-degree-apprenticeships_bma-briefing.pdf. Accessed 8 Aug 2023

Buzzaccarini G, Degliuomini RS, Borin M, Fidanza A, Salmeri N, Schiraldi L, Di Summa PG, Vercesi VS, Candiani M, Pagliardini L (2024) The promise and pitfalls of AI-generated anatomical images: evaluating midjourney for aesthetic surgery applications. Aesthetic Plast Surg 48:1874–1883

Carraccio C, Wolfsthal SD, Englander R, Ferentz K, Martin C (2002) Shifting paradigms: from flexner to competencies. Acad Med 77:361–367. https://doi.org/10.1097/00001888-200205000-00003. Accessed 30 July 2023

Center for Generational Kinetics (2023) Generational breakdown: info about all of the generations. https://genhq.com/the-generations-hub/generational-faqs/. Accessed 8 Aug 2023

Cooke M, Irby DM, Sullivan W, Ludmerer KM (2006) American medical education 100 years after the Flexner report. N Engl J Med 355:1339–1344

Dahle LO, Brynhildsen J, Behrbohm Fallsberg M, Rundquist I, Hammar M (2002) Pros and cons of vertical integration between clinical medicine and basic science within a problem-based undergraduate medical curriculum: examples and experiences from Linköping, Sweden. Med Teach 24:280–285

Epstein R (2007) Assessment is medical education. NEJM 356:387–396

Ferris H, O'Flynn D (2015) Assessment in medical education: what are we trying to achieve? Int J Higher Educ 4:139–144

Fleming S, McKee G, Huntley-Moore S (2011) Undergraduate nursing students' learning styles: a longitudinal study. Nurse Educ Today 31:444–449

General Medical Council (n.d.). https://www.gmc-uk.org/-/media/documents/Outcomes_for_graduates_Jul_15_1216.pdf_61408029.pdf, https://www.gmc-uk.org/-/media/documents/outcomes-for-graduates-2020_pdf-84622587.pdf. Accessed 31 May 2024

Goyal R, Garg R, Goyal PR (2017) Need for changes in the practical physiology curriculum of medical undergraduates. J Clin Diagn Res 11:CC06–CC08

Gurpinar E, Bati H, Tetik C (2011) Learning styles of medical students change in relation to time. Adv Physiol Educ 35:307–311

Haris Ali S, Carr PA, Ruit KG (2016) Validity and reliability of scores obtained on multiple-choice questions: why functioning distractors matter. J Schol Teach Learn 16:1–14

Hesse H (1943) The glass bead game (Das Glasperlenspiel). Holt, Rinehart and Winston. isbn:080501246X, 0312278497

Hoeft ME (2012) Why university students don't read: what professors can do to increase compliance. Int J Schol Teach Learn 6:12. https://doi.org/10.20429/ijsotl.2012.060212

Honey P, Mumford A (2006) The learning styles helper's guide. Peter Honey Publications, Maidenhead

Horne CA, Hepworth D, Saunders E, Keenan ID (2024) Everyone can draw: an inclusive and transformative activity for conceptualization of topographic anatomy. Anat Sci Educ. https://anatomypubs.onlinelibrary.wiley.com/doi/epdf/10.1002/ase.2460. Accessed 3 June 2024

Institute for Apprenticeships and Technical Education. Doctor (Degree) (2023). https://www.instituteforapprenticeships.org/apprenticeship-standards/doctor-degree-v1-0. Accessed 8 Aug 2023

Jones DG, Whitaker MI (2009) Speaking for the dead—The human body in biology and medicine. Ashgate, Aldershot

Kadirvelu A, Gurtu S (2015) Integrative learning in medical education: are students ready? Sci Educ 25:549–555

Keenan ID, Powell M (2020) Interdimensional travel: visualisation of 3D-2D transitions in anatomy learning. Adv Exp Med Biol 1235:103–116

Keenan ID, Hutchinson J, Bell K (2017) Twelve tips for implementing artistic learning approaches in anatomy education. MedEdPublish 6(2):106

Klestinec C (2011) Theaters of anatomy. The John Hopkins University Press, Baltimore

Koehler U, Beutel B, Hildebrandt W (2021) Salerno, the first medical school in Europe—birthplace of university medicine. Dtsch Med Wochenschr 146:1593–1597

Lodge J, Hansen L, Cottrell D (2015) Modality preference and learning style theories: rethinking the role of sensory modality in learning. Learn Res Pract 2:4–17

Mandressi R (2003) Le regard de l'anatomiste—dissections et invention du corps en Occident. Seuill, Paris, p 2

Moxham BJ, Plaisant O (2014) The history of the teaching of gross anatomy. How we got to where we are! Eur J Anat 18:219–244

Moxham BJ, Shaw H, Crowson R, Plaisant O (2011) The future of clinical anatomy. Eur J Anat 15:29–46

Moxham BJ, Hennon H, Lignier B, Plaisant O (2016) An assessment of the anatomical knowledge of laypersons and their attitudes towards the clinical importance of gross anatomy in medicine. Ann Anat 208:194–203

Moxham BJ, Plaisant O, Lignier B, Brahim F (2018) The perceptions of anatomists in the US and Europe of the skills and attributes required of newly-recruited medical students. Ann Anat 217:103–110

National Audit Office (UK) (2023) Report—Value for money. Access to unplanned or urgent care. https://www.nao.org.uk/reports/access-to-unplanned-or-urgent-care/. Accessed 8 Aug 2023

NHS Health Education England (2023a) Medical doctor degree apprenticeship. https://www.hee.nhs.uk/our-work/talent-care-widening-participation/apprenticeships/medical-doctor-degree-apprenticeship. Accessed 8 Aug 2023

NHS Health Education England (2023b) Medical doctor degree. https://www.hee.nhs.uk/search/new%20medical%20doctor%20degree. Accessed 8 Aug 2023

NHS Long Term Workforce Plan (2023). https://www.england.nhs.uk/ltwp/about-the-plan/. Accessed 26 Apr 2024

Orlando Figes (2014). https://www.tes.com/magazine/archive/students-dont-read-books-any-more-why. Accessed 26 Apr 2024

Palou N (2023) How to create realistic images using generative AI. https://telefonicatech.com/en/blog/the-influence-of-chemical-photography-to-produce-realistic-images-with-generative-ai#:~:text=Generative%20AI%20models%20for%20image,and%20aesthetic%20quality%20in%20seconds. Accessed 30 May 2024

Persaud TVN (1984) Early history of human anatomy. Charles C Thomas, Springfield
Persaud TVN (1997) A history of anatomy—The post-Vesalian era. Charles C Thomas, Springfield
Porter R (2002) Blood and guts. Allen Lane, London
Porter R (2003) Flesh in the age of reason. Allen Lane, London
Porter R (ed) (2006) The Cambridge history of medicine. Cambridge University Press, Cambridge
Potu BK, Atwa H, Nasr El-Din WA, Othman MA, Sarwani NA, Fatima A, Deifalla A, Fadel RA (2021) Learning anatomy before and during COVID-19 pandemic: students' perceptions and exam performance. Morphologie 106:188–194
Price L (2004) Individual differences in learning: cognitive control, cognitive style and learning style. Educ Psychol 24:681–698
Rayner S, Riding RJ (1997) Towards a categorisation of cognitive styles and learning styles. Educ Psychol 17:5–27
Reichman SW, Grasha AF (1974) A rational approach to developing and assessing the construct validity of a study learning style scales investment. J Psychol 87:213–223
Riding RJ, Cheema I (1991) Cognitive styles: an overview and integration. Educ Psychol 11:193–215
Riederer BM, Spinosa J-P, Macchi V, Porzionato A, De Caro R, McHanwell S, Moxham BJ (2018) Clinically-orientated anatomy: five exemplars to portray the concept. Eur J Anat 22:157–172
Selwyn N (2009) The digital native—myth and reality. ASLIB Proc 61:364–379
Shapiro L, Bell K, Dhas K, Branson T, Louw G, Keenan ID (2020) Focused multisensory anatomy observation and drawing for enhancing social learning and three-dimensional spatial understanding. Anat Sci Educ 13:488–503
Shapiro L, Hobbs E, Keenan ID (2023) Transforming musculoskeletal anatomy learning with haptic surface painting. Anat Sci Educ 16:677–693
Skills for Health (2023) Healthcare apprenticeships: apprenticeship standards. https://haso.skillsforhealth.org.uk/standards/#standard-22949. Accessed 8 Aug 2023
St Clair-Thompson H, Marsham AG, Marsham S (2017) Exploring the reading practices of undergraduate students. Edu Inq 9:284–298
Strathern P (2005) A brief history of medicine. Robinson, London
The King's Fund (UK) (2023) How does the NHS compare to the health care systems of other countries? https://www.kingsfund.org.uk/publications/nhs-compare-health-care-systems-other-countries. Accessed 8 Aug 2023
Times Higher Education (n.d.). https://www.timeshighereducation.com/news/sharp-increase-first-class-degrees-triggers-standards-debate. Accessed 26 Apr 2024
Totlis T, Natsis K, Filos D, Ediaroglou V, Mantzou N, Duparc P, Piagkou M (2023) The potential role of ChatGPT and artificial intelligence in anatomy education: a conversation with ChatGPT. Surg Radiol Anat 45:1321–1329
UK Government (2023) NHS doctor apprenticeships: everything you need to know. https://educationhub.blog.gov.uk/2023/06/30/nhs-doctor-apprenticeships-everything-you-need-to-know/. Accessed 8 Aug 2023
Universities UK (2024) How universities are turning the corner on grade inflation. https://www.universitiesuk.ac.uk/what-we-do/policy-and-research/publications/how-universities-are-turning-corner. Accessed 26 Apr 2024
Wilkinson T, Boohan M, Stevenson M (2013) Does learning style influence academic performance in different forms of assessment? J Anat 224:304–308

**Bernard J. Moxham** B.Sc. (Hons), B.D.S., Ph.D. (Bristol), F.S.B., F.H.E.A., F.A.S., is a craniofacial biologist and embryologist and is the Emeritus Professor of Anatomy at Cardiff University (the UK). He is a Past President of the International Federation of Associations of Anatomists (IFAA) and was President of the European Federation for Experimental Morphology (EFEM) and of the Anatomical Society of Great Britain and Ireland. He was the founder, and first President, of the Trans-European Pedagogic Anatomical Research Group (TEPARG).

**Odile Plaisant** M.D., Ph.D. was a Senior Lecturer in Anatomy at the University of Paris Cité, Paris (France), and a psychiatrist at the Epilepsy Unit and Pain Centre of the University Hospital of Pitié Salpêtrière (APHP) in Paris. She was a member of the Research Unit on Development, Imaging and Anatomy (URDIA, EA4465) at Paris. She was a founder member and past President of the Trans-European Pedagogic Anatomical Research Group (TEPARG).

**Diogo Pais** M.D., M.Sc., Ph.D., Agg is Full Professor and Chairman of the Department of Anatomy at NOVA Medical School, NOVA University of Lisbon (Portugal), where he is the Donation Officer and the President of the

Research Ethics Committee. He is currently the Secretary-General of the International Federation of Associations of Anatomists (IFAA) and Secretary-General of the International Committee of Symposia on Morphological Sciences (ICSMS) and was President of the European Federation for Experimental Morphology (EFEM) and of the Portuguese Anatomical Society (SAP-AAP). He was a founder member of the Trans-European Pedagogic Anatomical Research Group (TEPARG) and is currently one of its vice-presidents.

# 2 The Human Cadaver as an Educational Tool: Classification and Comparison with Other Educational Tools

Erich Brenner, Elisabeth Pechriggl, and Marko Konschake

**Abstract**

The human cadaver, often referred to as the medical student's 'first patient', is a topic of heated debate. Advocates of its use argue that the deceased body, with its unique characteristics, holds a significance that goes beyond being a mere representation. They assert that the cadaver, with its individuality, transcends the status of a model. However, some oppose this view, pointing out that the cadaver lacks a crucial aspect of a patient, that is, its vitality, and could potentially pose a health risk.

To establish a distinctive role for the human cadaver within the spectrum of educational tools, we identified several properties that allow classification and comparison of educational tools: (1) vitality and activity; (2) morbidity; (3) mortality; (4) variability; (5) dimensions (two- or three-dimensional); (6) quality of haptic experience; (7) level of abstraction (individual vs. standardised model); (8) health hazard to students and staff; (9) hazard to tool; (10) availability; (11) costs per student and, finally, (12) ethical issues.

We will compare (a) the human cadaver with (b) the real patient, (c) the (standardised) simulated patient, (d) the 'virtual human', (e) the (anatomical) model and (f) the book.

In its unique role as an educational tool, the human cadaver is a non-vital, morbid, mortal, variable, three-dimensional individual. It offers a high-quality haptic experience with low health hazards and relatively moderate costs per student. Importantly, it is ethically firm and cannot be physically harmed by the students.

**Keywords**

Education, medical, undergraduate · Anatomy · Dissection

In memory of Herbert Maurer (1941–2023)

E. Brenner (✉) · E. Pechriggl · M. Konschake
Institute for Clinical and Functional Anatomy, Medical University of Innsbruck, Innsbruck, Austria
e-mail: erich.brenner@i-med.ac.at; elisabeth.pechriggl@i-med.ac.at; marko.konschake@i-med.ac.at

## 2.1 Introduction

The human cadaver is often discussed controversially as the medical students' '*first patient*' (Aziz and McKenzie 1999; Aziz et al. 2002; Parker 2002; Souza et al. 2020). An e-mail discussion several years ago in the American Association of Clinical Anatomists (AACA) revealed different points of view. Supporters of the concept of a '*first patient*' argue that the cadaver is *more* than a model, mainly due to its individuality.




I. D. Keenan et al. (eds.), *International Anatomical Education*, Biomedical Visualization 9,
https://doi.org/10.1007/978-3-031-91849-0_2

Opponents, on the other hand, argue that the cadaver lacks a significant property of a patient, the vitality, and can be a health hazard.

Another aspect is to address the donated cadaver as the '*great teacher*', which adds a unique ethical aspect (Bohl et al. 2011; Ghosh 2017b; Lin et al. 2009; Prakash et al. 2007; Souza et al. 2020; Winkelmann and Guldner 2004). In a symposium at the 2017 meeting of the AMEE in Helsinki, the discussion topic revolved around whether cadavers continue to be indispensable for teaching anatomy in undergraduate medical education (McMenamin et al. 2018). In this symposium, a 'no, we do not' house was argued against a 'Yes, we do' house based on the Oxford tradition. Before the debate commenced, the audience predominantly supported the utilisation of cadavers. However, following the discussion, opinions on the matter of employing cadavers in undergraduate medical education were equally divided among the attendees.

This chapter aims to categorise the educational significance of the human cadaver and establish a unique position for the bodies as separate and distinct educational tools among various alternatives. We will compare the human cadaver to other educational tools, as there is the real patient, the simulated patient, standardised or not, and the '*virtual human*'—represented as an electronic medium that is derived from several sources such as the Visible Human Dataset (Ackerman et al. 1995; Robb and Hanson 2006; Ackerman 2022)—the anatomical wax-, rubber- or plaster-cast as well as 3D-printed models, and finally the most common source, the classic textbook. Therefore, we identified several properties which will allow for classification and comparison: (1) vitality and activity, (2) morbidity, (3) mortality, (4) variability, (5) dimensions (two- or three-dimensional), (6) quality of haptic experience, (7) level of abstraction (individual vs. standardised model), (8) health hazard to students and staff, (9) hazard to the tool, (10) availability, (11) costs per student and, finally, (12) ethical issues (Fig. 2.1).

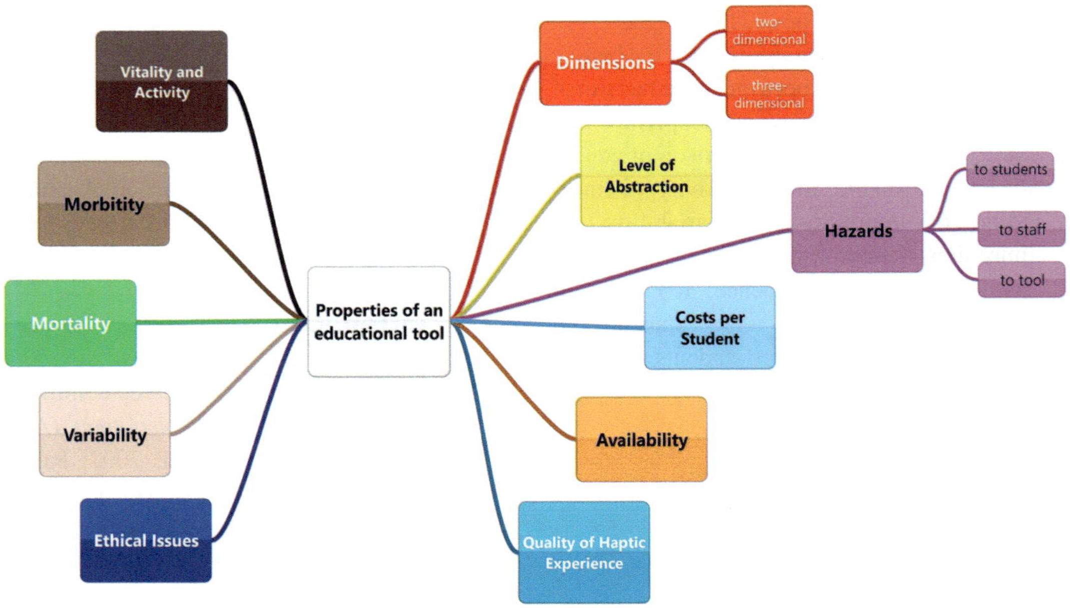


**Fig. 2.1** Properties of an educational tool

## 2.2 Properties of an Educational Tool

### 2.2.1 Vitality and Activity

#### 2.2.1.1 Human Cadaver

A human cadaver, being the deceased body of a person, naturally lacks vital signs such as heartbeat, respiratory rate, reflexes and active motion, which implies that specific clinical examinations relying on these signs cannot be conducted. Moreover, the absence of active muscle motion means that any functional tests that require active participation, such as the evaluation of strength or range of motion, are not feasible. Nevertheless, this lack of dynamic movement restricts the range of assessable parameters, enabling researchers and educators to concentrate on a more theoretical and impartial evaluation of anatomical structures and their correlation. This perspective was also highlighted by Aziz and McKenzie (1999), suggesting that studying cadavers can lead to a dispassionate understanding of human function and anatomy.

Furthermore, a cadaver undergoes various post-mortem changes, particularly if it is preserved for extended periods. These changes arising from decomposition processes and preservation techniques can manifest as tissue consistency, elasticity and colour alterations. For instance, muscles might become stiffer, and skin may adopt a different hue due to the chemicals used for preservation. Although these alterations can introduce variables that researchers and medical students need to account for, studies suggest that specific preservation techniques can mitigate these changes. This ensures that the cadaver retains a state as faithful to its natural condition as feasible (Thiel 1992; Kennel et al. 2018). Implementing such techniques is crucial in elevating cadaveric dissection to a valuable instrument for educational and research purposes within the medical field (Thompson et al. 2022).

#### 2.2.1.2 Real Patient

Every real patient embodies an individual and dynamic entity, a living being capable of displaying an array of active motions (Wass et al. 2001; Patten 2007). Unlike static models or cadavers, these individuals display vital signs, such as a beating heart, respiratory rate, blood pressure and body temperature, which can be monitored, recorded and analysed. These vital signs provide critical information about a patient's current health state and physiological functioning. Furthermore, patients might also display pathological signs, which are abnormal indicators arising from diseases or injuries. Such pathological signs can encompass a spectrum of observable symptoms, such as rashes or swelling, and subtler manifestations discernible through sophisticated diagnostic instruments, like irregular heart rhythms or abnormal laboratory test values. Interacting with a real patient also introduces the intricacies of emotion, communication and personal history, constituting critical aspects of comprehensive medical care. Furthermore, this direct interaction enables healthcare professionals to gather holistic information, ensuring accurate diagnoses and tailored treatment plans considering patient care's physical, emotional and psychological aspects.

#### 2.2.1.3 Simulated Patient

A similar principle applies to the simulated patient, whether standardised or not (Barrows 1968; Wass et al. 2001; Patten 2007). Simulated patients are individuals who are trained to act out symptoms and conditions in a consistent and standardised manner. These simulated patients play a significant role as valuable tools in medical education and evaluation, providing medical students with a controlled setting and an opportunity to practice diagnostic skills, patient interaction and therapeutic communication. While these simulated patients can display many symptoms like real patients, they do not suffer from underlying medical conditions. Instead, they undergo rigorous training to convincingly mimic the signs and symptoms of various diseases and conditions. Progress in makeup and prosthetics can aid in generating lifelike physical representations of specific conditions, such as rashes, wounds or swellings, thereby elevating the realism of the simulation even further. This 'makeup' and their acting skills allow for a near-authentic patient encounter. Another crucial facet of simulated patients is the immediate feedback they can provide post-interaction. This feedback loop can be instrumental in refining a medical professional's approach and techniques. However, while simulated patients provide a realistic representation, they cannot replace the comprehensive learning experience of interacting with real patients and their multifaceted needs and challenges.

#### 2.2.1.4 Virtual Human

The concept of a '*virtual human*' encompasses digital or computer-generated depictions of human beings, commonly utilised in fields like medical simulation, video gaming or virtual reality (Ackerman et al. 1995; McMenamin 2023; Bui et al. 2024; Yang 2023; Kumar et al. 2021; Zhang et al. 2023). Despite the sophisticated technology that goes into creating these models, they inherently lack living individuals' authentic vitality and unpredictability. Contemporary computational algorithms and artificial intelligence can imbue them with specific responsive behaviours or programmed motions; however, they remain constrained by predetermined parameters established by their creators. This means that their reactions, movements and behaviours, no matter how advanced, are confined to predefined rules and lack the spontaneity and variability inherent in real humans. Moreover, the virtual human's inability to provide emotional cues, such as authentic facial expressions, nuanced tones or gestures that arise from genuine feelings, often lacks depth of interaction. The sensory feedback, the warmth of human touch or the intricacies of eye contact cannot be wholly replicated in a digital environment. Moreover, despite significant progress in graphics and animation, capturing the nuanced intricacies of human motion, encompassing micro-expressions on the face and subtle shifts in posture, remains challenging. Hence, even though virtual humans serve a valuable role in specific contexts like training or entertainment, they cannot completely encapsulate the essence and intricacy of a living, breathing individual.

#### 2.2.1.5 Anatomical Model

Models made from materials like wax, rubber, plaster or through 3D-printing technologies are frequently used in medical training, art and various other fields due to their ability to replicate human anatomy or objects with remarkable detail. However, these models are inherently static, devoid of the organic vitality and the myriad spontaneous activities of living beings. Their rigidity and lack of dynamism mean they cannot accurately represent the full scope of human behaviours, reactions or the subtleties of physiology. Technological and material science advancements have paved the way for more sophisticated models, particularly in the medical training sector. Some of these high-end models, often called manikins or dummies, come with a (too) hefty price tag (to be purchased in large numbers). These advanced models can simulate specific functions or activities, such as a beating heart, respiratory motion or even pupil dilation. For instance, reanimation dummies are designed explicitly for cardiopulmonary resuscitation (CPR) training and can mimic chest rise, pulse and other vital signs to offer a quasi-realistic training experience. They can provide feedback on the efficiency of chest compressions, ventilation accuracy and more. Despite this technological sophistication, even the most advanced and expensive models have limitations. While they can mimic certain functions, they cannot replicate the complexities of human emotions, unpredictable reactions or the intricacies of human tissues. Nonetheless, they play a pivotal role in education and training, offering learners a controlled environment to practice, learn from errors and refine their abilities without exposing them to real-world repercussions.

#### 2.2.1.6 Textbook

An anatomical textbook, while being an invaluable resource, is inherently static, offering a fixed representation of the human body. These texts provide detailed descriptions, illustrations and sometimes even photographs of human anatomy, serving as a foundational reference for medical students, educators and practitioners alike. Due to the inherent characteristics of print media, these depictions cannot capture the dynamic movements, functions and processes that occur within a living body. Beyond the lack of motion, textbooks fail to replicate the tactile experience of examining a real or simulated patient. There is a vast difference between observing a printed image of a lung and palpating a real (or even simulated) one, feeling its texture, elasticity and other physical properties. However, an anatomical textbook's strength is its ability to provide a consistent, standardised and detailed overview of human anatomy. Annotations, cross-sectional views, detailed labelling and thorough explanations can offer readers deep insights into the structure and relationships of various anatomical components. Additionally, with the advent of technology, many modern anatomical textbooks now come with supplementary digital resources, including interactive 3D models, videos and animations, aiming to bridge the gap between static 2D-print images and dynamic representation. Still, while these digital enhancements can elevate the learning experience, they cannot fully capture a living human's intricacies, vitality and variability. As with all educational tools, anatomical textbooks are best used in conjunction with other resources to offer a comprehensive understanding of the subject matter.

## 2.2.2 Morbidity

### 2.2.2.1 Human Cadaver

Beyond being an invaluable educational and research tool, the human cadaver is a testament to a person's life, replete with its unique medical history. When a person suffers from various diseases or conditions during their lifetime, these ailments can leave discernible marks, lesions or changes in the anatomy of their body. These alterations provide invaluable insights for medical students and researchers, serving as tangible manifestations of textbook descriptions. For instance, a cadaver that suffered from cardiovascular diseases might display arterial plaques, thickened heart walls or even damaged heart valves. Likewise, an individual who has struggled with chronic respiratory conditions possesses lungs marked by scars, fibrotic changes or other distinct pathological alterations. Organs affected by tumours, infections or other diseases also present unique characteristics that can be palpably and visually different from healthy counterparts. Studying these real-life manifestations can give medical students a deeper understanding of disease progression and its effects on the human body (Geldenhuys et al. 2016). They can see firsthand the effects of diseases like osteoarthritis with its worn-down joint surfaces, or the damage caused by chronic alcohol abuse on the liver. However, while cadavers offer a wealth of information, it is crucial to approach them with respect and reverence. Behind every anatomical finding is a story, a history of a person's life, struggles and experiences. This holistic approach not only enhances the learning experience but also instils a sense of empathy and responsibility in medical professionals, reminding them of the profound connection between disease, life and the human experience (Pirri et al. 2021; Sañudo et al. 2021a, b).

### 2.2.2.2 Real Patient

In contrast to cadavers or models, the real patient presents an active and ongoing narrative of disease and health. When they arrive at a surgery or hospital, it is often a testament to their search for relief, understanding or a cure. Their presentation is not just a static snapshot of a disease's manifestation; it is a dynamic display of symptoms evolving in real time, influenced by various factors like genetics, environment, lifestyle and other coexisting conditions. Beyond the physical manifestations of their illness, real patients carry with them the psychological and emotional weight of their condition. They might be dealing with anxiety about their diagnosis, stress about treatment costs, or even the fear of the unknown. Their symptoms, whether pain, fatigue or discomfort, are authentic and immediate. Each patient's experience with their disease is deeply personal, influenced by their pain thresholds, coping mechanisms and support systems. Interacting with a real patient also highlights the importance of a thorough medical history and the art of patient communication. Effective dialogue can provide crucial clues about their symptoms' onset, progression and triggers. Moreover, the nuances of a patient's description of their symptoms, non-verbal cues and overall demeanour can provide insights that might not be readily apparent in clinical tests or imaging. The real patient embodies health's social, cultural and economic determinants. Their access to healthcare, understanding of their condition, ability to comply with prescribed treatments, and support structures all play a vital role in their overall health and well-being. In essence, while cadavers and models offer a controlled and consistent environment for understanding disease, real patients provide a comprehensive, multifaceted, and profoundly human perspective on health and illness.

#### 2.2.2.3 Simulated Patient

An ideal simulated patient is designed to portray specific symptoms or conditions to offer students and professionals a controlled, consistent, and reproducible learning environment. Unlike real patients, whose medical conditions come with a complex interplay of symptoms, histories and comorbidities, simulated patients present a curated experience based on the training objectives. These individuals undergo rigorous training to emulate specific diseases or conditions convincingly. For example, if a simulated patient is trained to showcase asthma symptoms, they might be taught to exhibit shortness of breath, wheezing and a cough, while omitting unrelated symptoms. This targeted depiction ensures that learners can concentrate on specific diagnostic or therapeutic skills without the distractions or complexities that real-world scenarios might present. The consistent presentation allows standardisation across multiple training sessions or different training locations. This is particularly useful in medical exams or competency assessments, where standardised patients ensure that all examinees face consistent challenges.

#### 2.2.2.4 Virtual Human

The '*virtual human*' typically refers to digital representations of the human body, often crafted using computer graphics, 3D modelling and sometimes even accurate anatomical data. The extent to which these virtual entities can portray morbidity largely depends on the datasets and algorithms they are built upon. For instance, if the source material for a virtual human is derived from the scans of healthy individuals, it may not effectively showcase the nuances of specific diseases or conditions. Conversely, the virtual human can provide a more comprehensive view of morbidity if the base data is rich and varied, with inputs from diverse pathological cases. However, computer-based simulation technology goes beyond just visual representation. It can dynamically emulate physiological and pathological processes, offering learners an interactive experience. For example, a simulator might replicate the hemodynamic changes occurring in the body during heart failure or demonstrate the distribution of a drug throughout the body. Such simulations allow medical students to experiment, intervene and observe outcomes in a risk-free environment. This experiential learning (Kolb et al. 2001; Kolb 2015), where one can make decisions and witness their impacts without real-world consequences, can significantly enhance skills development. As highlighted by Pedowitz et al. (2002), one of the critical advantages of computer-based simulations is that they provide a hands-on learning experience without the potential harm or morbidity associated with real-life clinical scenarios and are a modern, practical and achievable approach that can additively improve cadaver-based teaching, allowing review of previously dissected areas (Pasricha et al. 2023). This ensures that learners can make mistakes, reflect on them, and improve, all without compromising patient safety. Furthermore, as technology advances, the realism and complexity of these simulations continue to grow. Incorporating artificial intelligence and machine learning can lead to more responsive and adaptive scenarios, tailoring the experience to the individual learner's needs. However, while virtual and computer-based training tools offer undeniable benefits, they are ideally used with real-world clinical experiences, ensuring a holistic and comprehensive medical education.

#### 2.2.2.5 Anatomical Model

Models, in the context of anatomical education, are primarily tools designed to provide a clear and focused representation of specific anatomical concepts. Most basic models are designed for clarity and simplicity; thus, they often do not showcase the nuanced signs of morbidity. However, models can be designed and modified to represent these changes when there is a need to understand disease processes, abnormalities or specific pathologies. Among the more advanced and realistic representations are plastinated specimens (Chaturvedi et al. 2014; Fruhstorfer et al. 2011; Ogaili et al. 2018) and 3D-printed models (Garas et al. 2018) derived from real cadavers. These plastinated bodies and organs retain their original anatomy, including the pathological changes they might have had during life, thus providing invaluable insights into various diseases and their manifestations in the human body. 3D-printed models have gained prominence with the advancement of imaging technologies and 3D printing capabilities; by using medical scans from real patients, such as MRIs or CT scans, it is possible to create precise and detailed replicas of organs or body parts that exhibit specific diseases or conditions. For instance, a 3D-printed model of a heart with congenital abnormalities, variation or pathology, can be a valuable tool for surgeons planning a procedure, or for educators explaining the details of this condition to students (Chahal et al. 2024; Brumpt et al. 2024). However, while both plastinated and 3D-printed models derived from real cadavers offer a tangible and realistic representation of morbidity, they still lack the dynamic nature of living tissues and represent only a static snapshot. They do not showcase the functional aspects of organs, such as muscle contractility or blood vessel perfusion. Regardless, in medical education, they serve as bridges between textbook knowledge and real-world clinical encounters, offering learners a deeper understanding of the human body in health and disease.

#### 2.2.2.6 Textbook

Anatomical textbooks primarily focus on the structure and relationships of the typical components of the human body. They provide detailed illustrations, descriptions and sometimes photographs of various anatomical elements, offering insights into how the body is organised in its healthy state. These textbooks are foundational references for medical students, educators and practitioners, emphasizing the standard and unaltered human anatomy. While anatomical textbooks describe structures in idealised forms, pathology textbooks can provide contrasting views by illustrating and explaining the deviations from the norm. This comparison helps medical professionals identify, diagnose and treat various conditions. Other medical discipline textbooks cater to specific areas of morbidity. Radiology textbooks, for instance, display how diseases manifest in imaging studies, while pharmacology texts detail the effects of drugs on various pathological conditions. While anatomical textbooks provide the foundation for understanding the human body, medicine's complexity and multifaceted nature require a diverse array of resources, each offering a unique perspective on health and disease.

### 2.2.3 Mortality

#### 2.2.3.1 Human Cadaver

When utilising a human cadaver for educational purposes, we are essentially learning from a '*real patient*' who once lived, breathed and experienced life with all its intricacies. These cadavers establish a concrete connection between life and death, serving as potent reminders of the impermanence and fragility of human existence. In this context, the cadaver stands as a testament to our mortality. For medical students and professionals, the experience of working with a cadaver is profoundly transformative. It is not just about understanding anatomy but also a lesson in empathy, respect, gratitude and humility towards life and death. Each cadaver was once a person who, in their final act, chose to contribute to advancing medical science and education. Recognising this generosity deepens the sense of responsibility and commitment among learners.

#### 2.2.3.2 Real Patient

The real patient, while alive and breathing, embodies the fragile nature of life and the constant interplay of health, illness and the inexorable march of time. Regardless of their current health status, every patient serves as a reminder of this transient journey. Each encounter with a real patient brings the weight of this reality. Their stories, symptoms, fears and hopes glimpse the ever-present human struggle against mortality. For healthcare professionals, this recognition emphasises the importance of holistic care. Beyond the immediate symptoms and diseases, one must address the emotional and psychological aspects of facing mortality. Patients may grapple with existential questions, anxiety about the future, or concerns about their loved ones, especially when faced with serious illnesses or life-threatening conditions. Awareness of a patient's mortality often shapes the ethical dimensions of medical care. In essence, the real patient stands as a living testament to the human condition, marked by its beauty, complexity, vulnerabilities, and the inevitable reality of mortality.

#### 2.2.3.3 Simulated Patient

The simulated patient plays a crucial role in medical education by mimicking specific symptoms or conditions to facilitate clinical training without putting real patients at risk. While their primary function is to portray certain medical conditions or situations, they do not embody the genuine threat of mortality inherent to a real patient with that condition. This controlled environment is designed to provide learners with consistent, replicable scenarios, free from the unpredictability of real-world clinical encounters. However, behind the role they play, simulated patients are, in their essence, real individuals with emotions, experiences and vulnerabilities. The act of simulating a patient with a life-threatening condition, even in a controlled environment, can inadvertently trigger reflections on their mortality. Such introspection can be particularly pronounced if the simulated condition holds personal resonance. These personal reflections can subtly influence their form of portrayal. While training usually aims to minimise such influences to maintain standardised outcomes for learners, it is nearly impossible to detach personal emotions and experiences from the role-play completely. The nature of medical scenarios, which often revolve around health, illness and mortality, can cause simulated patients and learners to grapple with the more significant existential questions about life, health and the inevitability of death.

#### 2.2.3.4 Virtual Human

The '*virtual human*' provides a technological means of studying and interacting with the human body. However, its representation of mortality is inherently abstract. Unlike real or simulated patients, virtual humans do not breathe, feel pain or exhibit genuine emotions. Its existence is rooted in code, algorithms and graphics, making it an inherently static entity without the complexities and nuances of life. The emotional and existential dimensions of mortality are largely absent in a virtual human. While it can graphically depict pathological changes or injuries, it does not convey the emotional weight, the lived experience of illness or the existential anxieties associated with mortality. It provides a detached, clinical perspective devoid of the personal narratives and emotional connections that come with real-life encounters.

#### 2.2.3.5 Anatomical Model

An anatomical model offers a depersonalised and static view of the human body. These models often lack the idiosyncrasies and intricacies that come with individual human beings. Unlike cadavers, virtual humans or even simulated patients, anatomical models are purely educational tools without any historical or emotional context associated with mortality. Because anatomical models are manufactured objects, there is no direct link to the cycle of life and death. As such, the weighty contemplations of mortality, the fragility of life and the emotional complexities of handling natural specimens are absent when working with anatomical models.

#### 2.2.3.6 Textbook

A textbook is primarily a repository of knowledge, distilling vast amounts of information into accessible, structured content. However, it does not cover this topic except for those dealing with mortality (e.g. Kübler-Ross 1993). These texts are designed to provide technical and theoretical knowledge. Most anatomical textbooks are focused on objectivity and rely on evidence-based information to educate their readers. Therefore, the deeply personal and emotional experiences associated with mortality might be sidelined or presented in a more detached manner.

### 2.2.4 Variability

Willan and Humpherson (1999) highlighted that early exposure to anatomical variations through dissection is crucial in medical education. However, current trends reducing practical experience may impact students' ability to practice medicine safely and competently.

#### 2.2.4.1 Human Cadaver

As human cadavers have been living individuals, they show their very own set of anatomical variations. They might present with a rare variation in their cardiovascular system, a common variation in their digestive system or even a scarce but most impressive topographical variation, an inverse site. Zucconi et al. (2002) stated that '*the frequency of clinically important anatomic variation is high enough to allow the gross anatomy laboratory to serve as an excellent teaching platform. Through anatomy, the student is introduced to the concept of patient individuality and the individualisation of medical and surgical therapies*'. This concept can even be enhanced by rotating among different cadavers (Champney et al. 1995). No two bodies are identical, and through repeated interactions with different cadavers, one can appreciate the unique differences and commonalities present in human anatomy (Konschake 2018; Konschake and Brenner 2014; Sañudo et al. 2021a, b).

#### 2.2.4.2 Real Patient

The same principle applies to the real patient and, to some extent, to simulated patients. Nevertheless, the latter was introduced to medical education—among other factors—to reduce biases related to variations, including anatomical ones, among patients and settings.

#### 2.2.4.3 Virtual Human

As an electronic representation of a single individual, the 'virtual human' shows some variability when built from abstract reconstructions; individuality and variability decrease.

#### 2.2.4.4 Anatomical Model

The model is three-dimensional, but in this form, a non-existing 'prototype' with no individuality or signs of variability (Aziz et al. 2002). However, there is an approach to overcome this limitation with statistical 3D-shaped models (Ambellan et al. 2019). 3D statistical-shaped models allow quantitative analysis of anatomical variation, visual exploration and educational visualisation. Future digital anatomy atlases will show a static (average) anatomy and its normal or pathological variation in three or even four dimensions, illustrating growth and disease progression.

#### 2.2.4.5 Textbook

Most recent textbooks provide only a fixed, idealised 'type' or 'norm'. This '*prototypic*' human leads to simplification and is accused of leading to poor training, resulting in misdiagnoses and even malpractice (Willan and Humpherson 1999; Wise 2000). The major problem is the definition of the '*norm*'. In many cases, the '*norm*' was derived from white males. There is, in fact, a sex bias in anatomical textbooks (Giacomini et al. 1986; Gradl-Dietsch et al. 2016), and also gender and diversity biases (Darici et al. 2023; Parker et al. 2017; Mendelsohn et al. 1994).

### 2.2.5 Dimensions (Two- or Three-Dimensional)

#### 2.2.5.1 Human Cadaver

Dissecting a human cadaver is an integral part of many medical curricula around the world, providing an unmatched hands-on learning opportunity. Beyond the immediate anatomical knowledge gained, it offers students a tangible experience of the complex structure of anatomical layers, textures and spatial relationships within the human body. While two-dimensional images in textbooks or virtual simulations can illustrate the placement and appearance of various structures, they often cannot capture the depth, tactile feel and spatial complexities one experiences during an actual dissection (Ghosh 2017a; Morris and Jacques 2018).

Such hands-on lessons deepen students' understanding of human anatomy and help them develop essential spatial visualisation skills. The ability to mentally visualise the three-dimensional layout of anatomical structures is crucial in various medical procedures, from routine examinations to complex surgeries. A physician's ability to recall and mentally process this three-dimensional knowledge on the fly is one of the most essential attributes in clinical practice. It aids diagnostic processes, guides surgical interventions and informs therapeutic decisions (Moore 1998).

Mutyala and Cahill (1996) stated, '*Not only did I learn the art of dissection, I gained considerable knowledge in anatomy (much of it clinically relevant), acquired better skill with surgical instruments and enhanced my logical and critical thinking ability*'. This comprises assessing the similarity of anatomical structures and reading materials, understanding reference materials, reconstructing atlas and text figures in a three-dimensional format and decreasing dependence on rote recall for answers to clinical problems. The authors suggested finally that '*Most of the time spent in the gross anatomy course should be in developing a three-dimensional mental picture of the anatomy of a living patient*'. For instance, Pabst (1987) stated, '*In our experience students learn the three-dimensional anatomy much better by dissection than by viewing prosected material*'. While technological advancements have introduced various tools and methods for studying anatomy, the hands-on experience of dissecting a cadaver remains unparalleled in fostering a deep, three-dimensional understanding of the human body.

#### 2.2.5.2 Real/Simulated Patient

Interacting with a real patient offers medical professionals a tactile, multisensory experience vital for building clinical skills and judgement. Like the simulated patient, the real patient provides an opportunity to practice and perfect physical examination techniques, from simple observation to more complex manoeuvres like auscultation and percussion. The texture of the skin, the warmth of a feverish body, the rigidity or laxity of the abdomen and the subtle nuances in a patient's breath sounds or heartbeats are just some of the many pieces of information that a physician can obtain during a hands-on examination.

However, there are inherent limitations to what can be discerned from a surface examination alone. The skin, muscles and superficial structures can be directly palpated and assessed, but deeper anatomical and pathological features remain obscured. This is where advanced imaging techniques come into play. Modalities like sonography, CT and MRI offer a non-invasive window into the body's internal structures, enabling clinicians to visualise and understand the three-dimensional nature of organs, tissues and pathological entities in situ. Integrating imaging with clinical examination augments the physician's three-dimensional conceptualisation, allowing for more accurate diagnosis, treatment planning and monitoring (Patra et al. 2023; Pillay-Addinall et al. 2023; McBain et al. 2023). As technology continues to evolve, with advancements like 3D reconstructions and augmented reality overlays, the depth and clarity of this internal visualisation will only improve, further enriching the clinician's understanding of the patient's condition.

#### 2.2.5.3 Virtual Human

The concept of a '*virtual human*' has revolutionised medical education, offering a dynamic, interactive and customizable view of human anatomy. These advanced digital tools provide learners with a multifaceted understanding of structures that are difficult to replicate in traditional learning settings. Using cutting-edge graphics and intricate design, many virtual human simulations can precisely mimic the complexity and depth of actual human anatomy. These programs can zoom, rotate and dissect virtual bodies, providing an exceptional spatial understanding. Nevertheless, they are ultimately displayed on a flat, two-dimensional surface. Even with advancements in 3D visualisation technologies like virtual (VR) and augmented reality (AR), the experience remains mainly visual, lacking tactile feedback from direct contact. For instance, no virtual system can replicate the sensation of a distended bladder, the resistance of a mass in the abdomen or the crepitus felt in a joint with certain conditions. Furthermore, reconstruction and distortion artefacts that arise during the creation of the virtual human being, and therefore do not represent reality exactly, should not be disregarded.

#### 2.2.5.4 Anatomical Model

Models, in the context of anatomical study, provide a tangible and manipulable representation of body structures, serving as essential tools for education and training. These physical replications are meticulously crafted to provide a standard representation, emphasising the anatomy's typical features and general organisation. This approach ensures a consistent and unchanging depiction, which can be particularly useful for novices just beginning to familiarise themselves with human anatomy (Aziz et al. 2002; Chan and Cheung 2020; Narang et al. 2021). When models lack variability, they make up for this in clarity and accessibility. Their design often emphaszses clarity of structure and is often augmented with colour differentiation, labelling or even detachable components. This modularity, where a model can be disassembled and reassembled, is incredibly beneficial. It allows for a deeper understanding of spatial relationships between structures, how they fit together and how they function concerning one another. For instance, a model of the human heart might have removable chambers, allowing learners to explore in detail internal structures, pathways of blood flow and the positioning of valves.

#### 2.2.5.5 Textbook

A textbook, irrespective of its visually appealing and informative illustrations, is inherently limited to the two-dimensional plane of its pages. This confines the representation of complex anatomical structures to flat illustrations. The depth, texture and spatial relationships critical to understanding the human body in its three-dimensional reality cannot be fully conveyed on paper—however, the strengths of a textbook lie in other aspects. First, the didactic organisation and narrative descriptions provide a structured and sequential learning journey, elucidating concepts and promoting a deeper understanding of the subject. While a figure can depict the placement of organs or structures, the accompanying text explains their functions, interactions and relevance in a broader context. Illustrations in textbooks often simplify and emphasise critical details, omitting away extraneous information to highlight core concepts. This selective representation can benefit students trying to understand basic ideas before diving into the three-dimensional complexities of the real world. In the digital age, however, the nature of 'textbooks' is evolving. Many are now augmented with online resources, interactive modules and even augmented reality components that attempt to bridge the gap between two-dimensional illustrations and three-dimensional reality. These digital enhancements offer a more immersive and interactive learning experience that caters to modern students' different learning styles and technological inclinations.

### 2.2.6 Quality of Haptic Experience

#### 2.2.6.1 Human Cadaver

The tactile exploration of a human cadaver provides medical students with a rare and valuable opportunity to interact directly with real human anatomy. This haptic feedback, or sense of touch, goes a long way towards developing a deeper understanding of the human body's physical properties, spatial relationships and intricacies (Shapiro et al. 2020, 2023). For instance, while a textbook or digital model might describe the texture of an organ theoretically, nothing can replicate the feeling of an actual liver or the firmness of the heart in one's hands. The haptic feedback from a cadaver also plays a vital role in training future surgeons. Skin resistance during an incision, the texture of diseased versus healthy tissues or the pliability of vascular structures are tactile experiences that cannot be fully replicated through simulations. These nuances become particularly important during surgical procedures where tactile sensitivity can be as crucial as visual perception.

However, the tactile quality of cadaveric tissue might not always be representative of living tissue. Preservation techniques will alter tissues' natural consistency and pliability, making them harder or softer than in life.

#### 2.2.6.2 Real Patient

Interacting with a real patient offers medical students an unparalleled opportunity to understand the nuances and variations of the living human body. Unlike models or cadavers, real patients offer verbal and non-verbal feedback that can be significant in refining a student's diagnostic and palpation skills. The warmth of the skin, the pulsation of an artery, the elasticity of a muscle, and the rigidity or suppleness of specific pathological changes are all aspects that can only truly be appreciated by a living individual. Beyond the tactile experiences, real patients bring an element of dynamism that is absent in cadavers or models. For example, the palpation of a liver might change subtly with the respiratory cycle, or the feel of a pulsating mass might vary with alterations in blood pressure. This dynamic variability challenges students to sharpen their observational and interpretational skills, making their learning experience more prosperous and multifaceted.

#### 2.2.6.3 Simulated Patient

Simulated patients, trained actors or volunteers portraying specific medical conditions add another layer of complexity. While they might not have real pathologies, their portrayal of symptoms and interaction with trainee healthcare professionals provide an invaluable context to the learning process. For instance, a simulated patient might be trained to respond in a certain way when an examining student presses too hard or locates a 'painful' area, providing real-time feedback on technique and bedside manner.

#### 2.2.6.4 Virtual Human

The 'virtual human' inherent limitation in its current form is the absence of tactile feedback, which is crucial for a holistic understanding of human anatomy. While visual and auditory cues can be impressively simulated through advanced graphics and sound systems, the sensory experience of touch, so intrinsic to medical practice, remains elusive in digital formats. However, virtual reality (VR) and augmented reality (AR) are rapidly evolving with the promise of haptic feedback. The introduction of 'digital' or 'haptic' gloves, as well as other wearable technologies, is bridging this sensory gap (Wiet et al. 2002; Syamlan et al. 2022; Temkin et al. 2002). These devices are designed to simulate touch sensation using forces, vibrations or motions, providing users with feedback corresponding to their interactions with virtual objects. For instance, when palpating a virtual liver, the gloves might replicate the firmness and contour of the organ, offering a tactile sensation that closely mimics the experience. Incorporating haptic suits or vests can simulate broader physical sensations, such as the rhythmic expansion of lungs during breathing, or the pulsation of a virtual heart. These technologies, combined with VR headsets and immersive sound systems, can create a multisensory environment that comes incredibly close to replicating real-life interactions. As AI technology improves, the virtual human's responsiveness will also evolve. For example, in future versions, a virtual patient might react to a medical student's touch, indicating areas of pain or discomfort, just as a real or simulated patient would. This dynamic feedback will make the learning experience more interactive and realistic when combined with haptic technology.

#### 2.2.6.5 Anatomical Model

A model, often crafted from materials like silicone, rubber or plastic, offers tangible, hands-on interactions that can significantly aid in understanding human anatomy. The challenge with many models is accurately replicating human tissues' intricate nuances and variability. Real human tissues exhibit a wide range of consistencies, textures and elasticities, which can play a pivotal role in diagnosis and treatment. For example, the palpable difference between healthy liver tissue and a liver with cirrhosis, or the distinct feel of a benign versus malignant tumour, are subtleties that most standard models cannot capture. Advanced, high-end models that more closely resemble real tissue consistency often use sophisticated materials and manufacturing techniques, driving up costs. Some models even incorporate dynamic elements, like fluid-filled chambers, to simulate blood flow or pulsation, further enhancing their realism. These premium models can be valuable for specialised training, such as surgical simulations, where the tactile feedback and realistic representation of tissues can significantly impact the training experience. As technology and materials science continue to evolve, the gap between the haptic feedback of models and natural human tissues is expected to narrow, making these educational tools even more suitable.

#### 2.2.6.6 Textbook

As a primary source of theoretical knowledge, a textbook is inherently limited or incapable of replicating the multifaceted sensory interactions that occur during patient care or cadaver dissections. Although rich in detail, diagrams and descriptions, it cannot capture the tactile nuances, the warmth of living tissue, the subtleties of a pulse or the variances in tissue consistency that medical professionals encounter daily. Textbooks do not provide direct haptic experiences, but are vital in shaping the cognitive framework that guides future tactile interactions. By understanding the theory behind their feelings (Klatzky and Lederman 2011), medical students can better interpret their hands-on encounters, making those interactions more meaningful and informative.

### 2.2.7 Level of Abstraction (Individual Vs. Standardised Model)

#### 2.2.7.1 Human Cadaver

The human cadaver serves as a direct and tangible link to the intricate complexities of human anatomy. Unlike models or virtual simulations, cadavers do not represent a distilled or idealised version of the human body but display the nuanced realities of individual variation, disease processes and the effects of ageing. Therefore, cadavers can be considered the opposite of abstraction in the realm of medical education.

#### 2.2.7.2 Real Patient

The same is true for the real patient, who also represents the lowest level of abstraction since they are the individuals a physician must deal with.

#### 2.2.7.3 Simulated Patient

In the context of medical education, simulated patients bridge the gap between theoretical knowledge and hands-on clinical practice. They allow students to apply what they have learned in textbooks and lectures in a pseudo-real-world setting, building confidence and competence. That means students learning with the educational tool of simulated patients will have to create an abstract concept, which can afterwards be transferred to other individuals. Instructors must ensure that students understand that the abstractions provided by simulated patients are 'stepping stones', foundational experiences that must be supplemented with diverse clinical exposures to prepare them for the complexities of real-world medical practice.

#### 2.2.7.4 Virtual Human

The level of abstraction the 'virtual human' can provide derives mainly from the sources used to create this virtual representation. When real cadavers are used, the level is relatively low; when '*virtual humans*' are based on schematic or stylised constructions, they tend to represent a more generalised or idealised version of the human body. While these representations might be cleaner and easier to understand, they might not capture the full range of human variability. Sometimes, the simplifications can be intentional, designed to emphasise specific elements or concepts without the 'noise' of real-life variations (Mal et al. 2022). Such abstract representations may be beneficial for beginners or for focusing on basic concepts. However, they might not prepare students for the complexities and variances they encounter in real-life clinical scenarios. Additionally, the interface and interaction mechanisms also play a role in the level of abstraction. A highly interactive and immersive virtual reality experience where students can 'navigate' through different layers and systems of the body offers a more tangible experience than a simple 2D screen-based representation. The benefit of a 'virtual human' lies in its adaptability and scalability. Educators can choose between high-fidelity, realistic models or more abstract, schematic ones, depending on the teaching objective. The challenge remains in ensuring that students recognise the differences and understand the implications of the level of abstraction they are working with to equip them for real-world medical practice.

#### 2.2.7.5 Anatomical Model

A model, by its very nature, simplifies complex realities into more digestible and understandable forms. This inherent characteristic gives rise to a high level of abstraction (Chan and Cheung 2020). Models often strive to capture the essence of what they represent without getting lost in every intricate detail. For instance, details may be generalised or omitted to focus on the primary structures. While a real organ might have a plethora of veins, arteries and minor imperfections, a model might only showcase the main blood vessels or overlook the subtle variations in texture and colour found in real life. This simplification facilitates easier learning for beginners, offering a clear overview without overwhelming them with complexities. Moreover, physical models' material and design constraints further contribute to the abstraction. Models of plastic, rubber or other materials can never fully replicate real tissues and organs' tactile feel, elasticity or intricacy. These discrepancies can sometimes make the experience of studying a model distinctively different from handling a real specimen.

#### 2.2.7.6 Textbook

Modern anatomical textbooks are often concerned with clarity and uniformity in visual representations. The focus is on presenting an 'ideal' or 'standard' version of the human body or any subject matter (Shaver 2022; Naicker 2023). This can be attributed to several reasons: first, the primary goal of anatomical textbooks is to offer a clear and comprehensible introduction to complex subjects. Using standardised images or diagrams can help reduce confusion among students and give them a starting point for understanding. Presenting an 'ideal' version of anatomy or any concept allows students to familiarise themselves with a general understanding before delving into the intricacies and variations. Second, with advances in digital graphics and imaging technologies, it has become easier for publishers to produce high-resolution, detailed and consistent images. While these visuals are often stunning and highly informative, they can sometimes come across as sterile or devoid of the natural variations one would find in real-life specimens. Furthermore, in the era of digital publishing and online resources, textbooks are often supplemented with interactive tools, videos and 3D animations. While these supplementary materials help improve understanding, they can further standardise textbooks because they often rely on digital models based on averaged or idealised data. However, this lack of individuality in textbooks can be a double-edged sword. On the one hand, it facilitates learning by providing clear, unambiguous visuals. On the other hand, it might inadvertently give students the impression that what they see in these books is the only authentic version of reality.

### 2.2.8 Health Hazard to Students and Staff

#### 2.2.8.1 Human Cadaver

One central argument against the use of human cadavers in medical education is that they may be a potential source of health hazards for both students and staff (Aziz et al. 2002; Štrkalj et al. 2020; Gundreddy and Gaurkar 2022). The hazards can come from the embalming fluids (Chia et al. 1992; Pabst 1987; Wei et al. 2007), infectious diseases (Gupta et al. 2013) and the stress provoked (Arráez-Aybar et al. 2009; Bati et al. 2013; Bockers et al. 2012; Boeckers et al. 2010; Finkelstein and Mathers 2005; Leboulanger 2011; Shalev and Nathan 1985; Singh et al. 2007; Stewart et al. 1999; Yanılmaz et al. 2023). Hazards related to embalming fluids can be reduced by low concentrations and proper laboratory equipment and ventilation, eliminating excessive fluids and 'odours'. On the other hand, appropriate preservation of the cadaver and hygienic protocols in the lab do limit a potential health hazard by infectious agents to an almost negligible level in most cases (e.g. Platzer et al. 1978; Putz et al. 1974).

#### 2.2.8.2 Real Patient

The real patient's status as a living, morbid individual leads to several considerations in clinical settings, primarily related to the potential transmission of infectious diseases. This is especially pertinent in our modern era, marked by the emergence of new pathogens and the resurgence of some old ones. Patients with active tuberculosis, influenza or even more contemporary diseases like COVID-19 can transmit the pathogens through respiratory droplets. Similarly, patients with diseases like hepatitis or HIV may be a risk through blood and body fluids. Beyond direct transmission, there is also the risk of fomite transmission. This occurs when infectious agents found on the patient's body or personal items meet surfaces, equipment or objects in the clinical setting. Unsuspecting individuals can get infected if they touch these contaminated surfaces and then touch their face, especially the nose, mouth or eyes.

To mitigate these risks, strict infection control measures are implemented in healthcare facilities. These include but are not limited to using personal protective equipment (PPE) like gloves, masks and gowns; frequent hand hygiene practices; isolation of patients with highly contagious diseases; and regular disinfection of equipment and surfaces. In summary, while the real patient presents invaluable learning opportunities, their potential to be a vector for infectious diseases underscores the paramount importance of adherence to safety and hygiene protocols in clinical settings.

#### 2.2.8.3 Simulated Patient

While primarily serving as an educational tool, the simulated patient is still a human with the potential to carry and transmit infectious agents. Even if they simulate a non-infectious ailment, they are exposed to common pathogens in everyday life, just like any other individual. Their proximity to students and instructors during training sessions poses a risk of transmission, especially in scenarios that require physical examinations or close contact.

#### 2.2.8.4 Virtual Human

The health hazards associated with prolonged screen use, as with the '*virtual human*' medium, are multifaceted. The screen's refresh frequency, or refresh rate, can cause visual strain, mainly if it is too low. A low refresh rate can cause flickering, which, although not always consciously noticed, can lead to eye strain and discomfort over extended periods. Staring at the screen for long periods can result in 'computer vision syndrome' (Abudawood et al. 2020; Iqbal et al. 2021) or 'digital eye strain' (Kaur et al. 2022). Symptoms can include dry eyes, blurred vision and even headaches. The blue light emitted by screens is also a subject of concern (Akman 2021; Randjelović et al. 2023). Prolonged exposure to this short-wavelength light has been linked to disrupted sleep patterns because it interferes with melatonin production, the hormone responsible for sleep regulation. Ergonomic factors also play a crucial role. The static posture often adopted during prolonged computer use can lead to musculoskeletal problems. The lack of dynamic movement can cause discomfort or pain in the neck, back and shoulders. The head and neck angle, the keyboard's position, and the monitor's height and angle can influence how the body responds over time. Proper ergonomic setups and practices, such as taking regular breaks and using adjustable chairs and desks, can help alleviate some of these issues. The hand and wrist posture required to use a keyboard and mouse can result in repetitive strain injuries, like carpal tunnel syndrome. This condition arises from the compression of the median nerve, leading to symptoms such as numbness, tingling or weakness in the hand (Benden et al. 2021; Jamil et al. 2023; Pierce et al. 2023; Rashwan 2024).

#### 2.2.8.5 Anatomical Model

Models generally pose minimal health risks, particularly those used in educational settings. However, the materials and manufacturing processes used in their creation may have implications. For example, some plastics used in models may release volatile organic compounds (VOCs) over time, especially when new. Prolonged exposure to these compounds can lead to various health problems, ranging from minor irritations such as headaches or dizziness to more severe respiratory symptoms in sensitive individuals. The type of plastic or material used can also impact those with allergies. Some individuals may be allergic or sensitive to specific materials or the chemical compounds they release, leading to skin irritations or respiratory issues when handling these models. It is essential to know the materials used in the model, mainly if they will be used frequently or by people with known hypersensitivity (Zhang et al. 2022). Moreover, the physical design of some models can pose risks. Sharp edges or parts that can easily break off and become a choking hazard are concerns, mainly if the model is used in a classroom setting. Over time, some parts of the model may also loosen or break due to wear and tear, further increasing potential hazards. Additionally, if a model is not cleaned or appropriately disinfected after use in medical training, it can become a vector for bacterial or fungal growth (Mudgal et al. 2022). This can pose a risk to subsequent users, primarily if the model is used in a hands-on, tactile learning environment.

#### 2.2.8.6 Textbook

Since a textbook is a static and non-interactive medium, the risk to its users is minimal. However, there are some nuanced considerations even with such seemingly innocuous materials. First and foremost, older textbooks, especially those printed in previous decades, might have been bound using adhesives that, over time, can degrade and release potentially harmful fumes. Some books, particularly old ones, may contain traces of substances no longer considered safe by modern standards, such as certain inks or dyes. The physical weight and size of textbooks should also not be underestimated. Carrying heavy textbooks for extended periods can lead to back or shoulder discomfort, especially in younger students or those with musculoskeletal problems. Furthermore, large textbooks can pose a falling risk when stacked or stored improperly. In terms of health, multiple users' frequent handling of textbooks, as is typical in libraries or shared educational environments, can spread pathogens. Textbooks can accumulate dust without regular cleaning, triggering allergies in sensitive individuals. Mold growth is another problem, especially if books are stored in damp environments. Additionally, the vision strain can also be a problem. Extended reading, particularly in poor lighting or if the print is small, can lead to eye fatigue or exacerbate existing vision problems. Continuous reading in the same posture can lead to ergonomic issues and neck strain, or contribute to a sedentary lifestyle. Although these risks are relatively small compared to other educational tools and methods, they emphasise the need for a holistic approach to safety and well-being in any learning environment, regardless of the medium (Pierce et al. 2023).

### 2.2.9 Hazard to Tool

#### 2.2.9.1 Human Cadaver

Dissection does harm the human cadaver in a way which cannot be neglected. This kind of 'destruction' is one of the causes of stress students are exposed to. Nevertheless, this possibility of 'destruction' or 'deconstruction' is one of the significant advantages of the educational tool 'human cadaver' as it allows for procedures and insights that cannot be provided by any other tool. Dissecting a human cadaver is not merely a physical act but also has profound ethical, emotional and psychological implications. For many students, this is their first direct encounter with death, making it a profoundly introspective experience. It confronts them with the finite nature of life and challenges their perceptions of mortality. Moreover, dissection requires students to navigate the delicate balance between reverence for the once-living and the scientific curiosity that drives the desire to understand the human body in-depth. While the dissection process may indeed be considered destructive, it is essential to frame it within the context of learning and discovery. The cadaver serves as an invaluable educational tool, willingly donated by individuals who understood the significance of their posthumous contribution to the advancement of medical knowledge. This donation is a profound gesture of trust and altruism, allowing future medical professionals to gain hands-on experience. Educators must guide students through this potentially overwhelming process by emphasising the respect and gratitude owed to the donors. Appropriate mentoring can transform what might initially be perceived as a distressing activity into a meaningful rite of passage in a medical student's journey. In summary, while dissection undoubtedly challenges the student's psyche and confronts them with the tangible reality of mortality, it remains an unparalleled tool in medical education, instilling technical knowledge and a profound respect for human morphology.

#### 2.2.9.2 Real Patient

Actions performed by an untrained student may harm the real patient. This learning experience must be carefully managed and monitored by the supervisors to minimise any potential harm or discomfort to the patient. Supervision by experienced clinicians is essential to ensure that examinations and procedures are conducted appropriately and ethically. Additionally, informed consent must be obtained from patients, and they should be made fully aware of the nature of the interaction, the level of experience of the student and their right to withdraw at any stage. Moreover, providing adequate training and preparation to students before interacting with real patients is essential. This includes not only technical skills but also communication and ethical considerations. It is also essential to have a structured feedback system for students and patients. Students should receive constructive feedback on their performance to enhance their learning and development. Similarly, patients should be able to share their experiences, improve the training program, and ensure that it remains respectful, ethical and beneficial for all parties involved. For instance, real patients may be helpful in training residents (e.g. Korenstein et al. 2002) or teaching patient counselling skills to medical students (e.g. Benbassat and Baumal 2002).

#### 2.2.9.3 Simulated Patient

Simulated patients allow students to work through realistic patient problems, make mistakes, learn and be assessed without exposing a real patient to risk (Clifton et al. 2020; Yakura et al. 2022). Again, the possibility of injuring the simulated patient must be minimised as much as possible. The same measures should take place as in real patients.

#### 2.2.9.4 Virtual Human

The '*virtual human*' as a dataset cannot be harmed directly when the software used to display is appropriately programmed. Of course, manipulations on the computer where the dataset is installed may seriously damage the data. This will result in at least the necessity to reinstall the data.

#### 2.2.9.5 Anatomical Model and Textbook

Similarly, improper usage can damage a model; the same applies to any textbook.

### 2.2.10 Availability

#### 2.2.10.1 Human Cadavers

Human cadavers are only available in limited numbers, as is the setting in which dissection can take place. The limitation in number depends primarily on the financial resources and the general acceptance of body donation within the citizenry. If socio-cultural beliefs and customs do not support body donation, dissection with adequate cadavers will be impossible (Akinola 2011; Anyanwu et al. 2011; Kramer et al. 2019; Habicht et al. 2018). Among the essential misconceptions and fears are the belief that removal of parts and organs violates the sanctity of the deceased, personal concerns about being dismembered death, the desire to be buried in whole, a misconception of brain death and the idea that donation violates religious beliefs (Woo 1992; Ullah et al. 2023). The financial aspect, which is often mentioned, will be discussed later. Of course, the dissection of human cadavers will need proper facilities for embalming and storage. The course can take place in these proper labs and specifically skilled staff, as dissection is restricted to medical staff in most countries.

#### 2.2.10.2 Real Patient

The availability of real patients with specific diseases or symptoms is inherently unpredictable and can fluctuate significantly from day to day. This makes it challenging to ensure that medical students can interact with a diverse range of patients and conditions during their training. Additionally, the availability of patients can also be affected by various factors such as the time of year, the location of the training institution and the prevalence of certain diseases in the local population.

Furthermore, the willingness of patients to participate in educational activities can also impact their availability. Some patients may be uncomfortable being examined by students or may decline to participate due to privacy concerns, anxiety or other personal reasons. This further reduces the pool of available patients for educational purposes.

Moreover, patient availability is also restricted by the capacity of the healthcare facility. Hospitals and clinics have limited beds and resources, and patient care always takes priority over enabling educational activities. Students may have limited opportunities to interact with patients, particularly in busy or specialised facilities.

To address these challenges, medical schools and training institutions often partner with multiple healthcare facilities, including community clinics, outpatient centres and specialised hospitals, to increase the opportunities for students to interact with a diverse range of patients. Additionally, simulation-based training, virtual patients and simulated patients supplement the training and ensure that students can practice their skills and develop their clinical reasoning abilities, even when real patients are unavailable.

#### 2.2.10.3 Simulated Patient

Simulated patients may also be insufficiently available. This can be due to the limited number of people willing to serve as simulated patients and limitations in the resources for training such individuals. Further limitations arise from the need for many facilities for the student encounter.

#### 2.2.10.4 Virtual Human

In contrast, the '*virtual human*' can be available throughout the whole campus when proper site licenses have been purchased and adequate workstations are available.

#### 2.2.10.5 Anatomical Model

There are also no restrictions on the availability of models, apart from an appropriate number of acquired and suitable storage facilities.

#### 2.2.10.6 Textbook

Textbooks have the fewest limitations in availability. Everyone can use them everywhere; their only limits are the availability of an appropriate language edition and the purchase of an adequate number.

### 2.2.11 Costs per Student and Hour

#### 2.2.11.1 Human Cadaver

Costs for human cadavers include the transportation of the deceased to the institute, the embalming procedure, storage and finally, the funeral. A short survey on the AACA mailing list in May 2003 revealed average costs per cadaver ranging from US$ 1200.00 to US$ 2500.00. The primary variable is the transportation cost, as all other costs are similar from one cadaver to another. This seems quite expensive, but several additional factors must be considered. The average cost per student decreases as several students dissect one cadaver, for example, eight students. The total cost drops significantly as students work on a cadaver for several hours. At the Medical University of Innsbruck, the average costs per student and hour have been determined at € 2.25 (≈ US$ 3.25).

#### 2.2.11.2 Real Patient

Estimating the cost for a real patient is not simple. Mainly, real patients are recruited from the different wards and practices. Nevertheless, real patients may also be called in from their workplaces or their homes if they suffer from a chronic disease. Time-based remuneration and travel expenses will, or at least should, be covered for all those patients. Based on each €10.00 for travel expenses and time remuneration per hour and assuming four students work with a real patient for 2 h, costs of approximately €3.75 must be calculated.

#### 2.2.11.3 Simulated Patient

Simulated patients can even be more expensive, as additional costs for training, meals, transportation, etc., will have to be calculated.

#### 2.2.11.4 Virtual Human

The average costs for a '*virtual human*' environment cannot be estimated within this chapter, but it would need to be studied. Nevertheless, the software needed is expensive, which, of course, depends on the amount of content, possibilities and sophistication of the program itself. For example, a well-known program for anatomy, in an institutional edition for six concomitant users, was seen for US$ 449.95 on the website. Adding expenses for a reasonable number of licenses, workstations, and regular software and hardware upgrades, and think of the operation system!—will sum up to a noteworthy amount.

#### 2.2.11.5 Anatomical Model

If estimating costs for the '*virtual human*' is already difficult, it is almost impossible for models. Depending on the manufacturer, the quality of detail, material, etc., several models may total very high amounts. Nevertheless, when properly used, models can be reused year after year.

#### 2.2.11.6 Textbook

Books must also be considered here, even though most medical schools require students to buy textbooks (Katz 2019; Jenkins et al. 2020; Nagle and Vitez 2020). When comparing costs, we assume that the school itself also provides books. Since book prices vary greatly, such calculations depend on the school's internal standard. For proper comparison, one must assume that a single book is used by no more than 5–10 students.

### 2.2.12 Ethics

#### 2.2.12.1 Human Cadaver

In most countries, dissection of a human cadaver is only carried out when the individual expresses a wish to donate his or her body to teaching or research (informed consent). In other countries, that request, as with organ donation, can be revoked by close relatives if they do not want the body to be used for that purpose (Brenner and Pais 2014; Brenner et al. 2024). This causes little ethical debate (Baskett and Steen 1999).

#### 2.2.12.2 Real Patient

In contrast, the use of real patients may raise several ethical questions. Can we expect sick people to be presented in front of several hundred students during a lecture or repeatedly examined by groups of students? There are certainly patients willing to do so, but there are also patients who give their consent only because they fear their treatment will suffer (Malpas et al. 2018).

#### 2.2.12.3 Simulated Patient

This is not a problem with simulated patients. They gave informed consent by internal motivation, not subtle external pressure.

#### 2.2.12.4 Virtual Human

The most crucial ethical problem with the '*virtual human*' is the sometimes unknown and potentially suspicious data source (Mackenzie et al. 2006). There is no ethical discussion about the usage itself.

#### 2.2.12.5 Anatomical Model/Textbook

Finally, models and books do not typically raise ethical discussions (Chaturvedi et al. 2014; Cornwall 2016; Gilbert et al. 2018), with some exceptions, where suspicious ethical material might have been used for creation as for instance for Pernkopf's anatomical atlas (Israel and Seidelman 1996; Czech et al. 2021).

## 2.3 Conclusion

The human cadaver must be categorised as a distinctive educational tool as it exists neither as the '*first patient*' nor as a mere biological model. It embodies a non-vital, morbid, mortal, variable, three-dimensional individual with low health hazard and high quality of haptic experience, restricted availability and reasonably moderate costs per student. Notably, the cadaver remains invulnerable to student harm and adheres to ethical principles (Fig. 2.2).

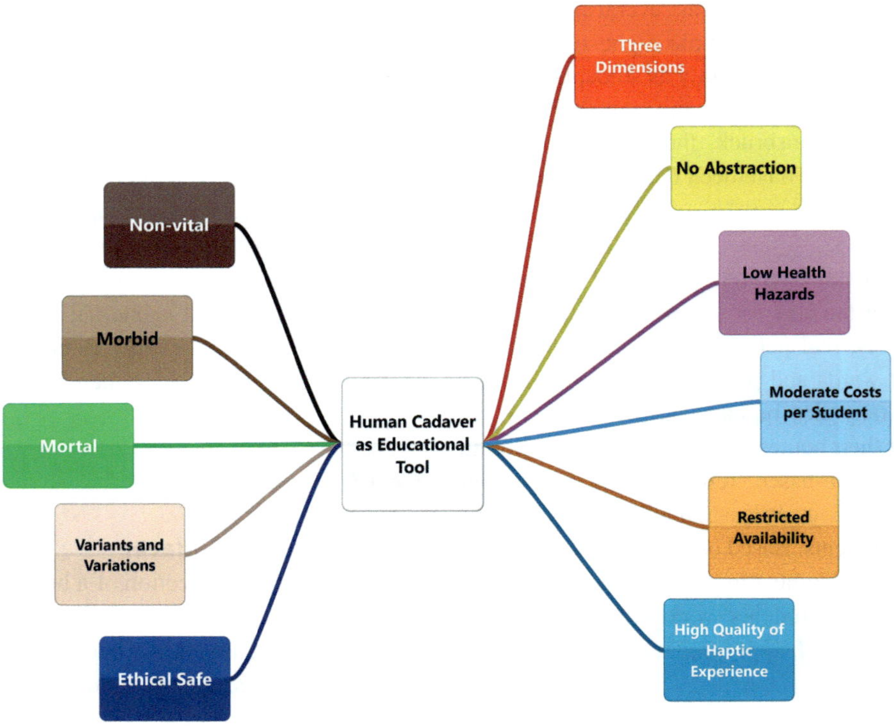


**Fig. 2.2** The human cadaver as educational tool

**Acknowledgements** A preliminary version of this chapter was presented as a poster at the 98. Versammlung der Anatomischen Gesellschaft, Dresden (D), 2003 (Brenner et al. 2003).

## References

Abudawood GA, Ashi HM, Almarzouki NK (2020) Computer vision syndrome among undergraduate medical students in King Abdulaziz University, Jeddah, Saudi Arabia. J Ophthalmol 2020:1–7. https://doi.org/10.1155/2020/2789376

Ackerman MJ (2022) The visible human project. Inf Serv Use 42(1):129–136. https://doi.org/10.3233/ISU-210145

Ackerman MJ, Spitzer VM, Scherzinger AL, Whitlock DG (1995) The visible human data set: an image resource for anatomical visualization. Medinfo 8(Pt 2):1195–1198

Akinola OB (2011) Formal body bequest program in Nigerian medical schools: when do we start? Anat Sci Educ 4(4):239–242. https://doi.org/10.1002/ase.226

Akman M (2021) The influence of virtual learning on the rise in screen time usage and blue light exposure ultimately expanding sleep disturbances amongst students. In: 2021 Academic Festival Sacred Heart University, Fairfield

Ambellan F, Lamecker H, von Tycowicz C, Zachow S (2019) Statistical shape models: understanding and mastering variation in anatomy. In: Rea PM (ed) Biomedical visualisation, vol 3. Springer International Publishing, Cham, pp 67–84. https://doi.org/10.1007/978-3-030-19385-0_5

Anyanwu GE, Udemezue OO, Obikili EN (2011) Dark age of sourcing cadavers in developing countries: a Nigerian survey. Clin Anat 24(7):831–836. https://doi.org/10.1002/ca.21187

Arráez-Aybar L, Castaño-Collado G, Casado-Morales I, Bueno-López JL, Reblet C (2009) Death anxiety levels and emotional reactions to human dissection in anatomy students. Dissection as an educational strategy. In: IFAA (ed) 17th Congress of the International Federation of Associations of Anatomists, Cape Town, South Africa. IFAA, p 386

Aziz MA, McKenzie JC (1999) The dead can still teach the living: the status of cadaver-based anatomy in the age of electronic media. Perspect Biol Med 42(3):402–421. https://doi.org/10.1353/pbm.1999.0014

Aziz MA, McKenzie JC, Wilson JS, Cowie RJ, Ayeni SA, Dunn BK (2002) The human cadaver in the age of biomedical informatics. Anat Rec 269(1):20–32. https://doi.org/10.1002/ar.10046

Barrows HS (1968) Simulated patients in medical teaching. Can Med Assoc J 98(14):674–676

Baskett PJ, Steen PA (1999) Interventions on the recently dead. Resuscitation 43(1):3–5. https://doi.org/10.1016/s0300-9572(99)00108-2

Bati AH, Ozer MA, Govsa F, Pinar Y (2013) Anxiety of first cadaver demonstration in medical, dentistry and pharmacy faculty students. Surg Radiol Anat 35(5):419–426. https://doi.org/10.1007/s00276-013-1075-7

Benbassat J, Baumal R (2002) A step-wise role playing approach for teaching patient counseling skills to medical students. Patient Educ Couns 46(2):147–152

Benden M, Mehta R, Pickens A, Harp B, Smith ML, Towne SD Jr, Peres SC (2021) Health-related consequences of the type and utilization rates of electronic devices by college students. BMC Public Health 21(1):1970. https://doi.org/10.1186/s12889-021-11975-3

Bockers A, Baader C, Fassnacht UK, Ochsner W, Bockers TM (2012) Reduction of mental distress in the dissection course by introducing the body donor experience through anatomical demonstrations of organ systems. Anat Sci Educ 5(6):321–329. https://doi.org/10.1002/ase.1292

Boeckers A, Brinkmann A, Jerg-Bretzke L, Lamp C, Traue HC, Boeckers TM (2010) How can we deal with mental distress in the dissection room?-An evaluation of the need for psychological support. Ann Anat 192(6):366–372. https://doi.org/10.1016/j.aanat.2010.08.002

Bohl M, Bosch P, Hildebrandt S (2011) Medical students' perceptions of the body donor as a "first patient" or "teacher": a pilot study. Anat Sci Educ 4(4):208–213. https://doi.org/10.1002/ase.231

Brenner E, Pais D (2014) The philosophy and ethics of anatomy teaching. Eur J Anat 18(4):353–360

Brenner E, Maurer H, Moriggl B, Pomaroli A (2003) The human cadaver as an educational tool—classification and comparison with other educational tools. Ann Anat 185(Suppl. 98):229–230

Brenner E, Bleys RLWA, de Caro R, Catereniuc I, Chirculescu ARM, Destrieux C, Eppler E, Filgueira L, Kachlik D, Kiss P, Lee C, Matveeva N, Natsis K, Pais D, Paulsen F, Piagkou M, Quondamatteo F, Reglődi D, Şendemir E, Tranum-Jensen J, Tutkuviene J, Vázquez Osorio MT (2024) The legal and ethical framework governing body donation in Europe—2nd update on current practice. Ann Anat 252:152195. https://doi.org/10.1016/j.aanat.2023.152195

Brumpt E, Bertin E, Gabrion X, Coussens C, Tatu L, Louvrier A (2024) Are 3D-printed anatomical models of the ear effective for teaching anatomy? A comparative pilot study versus cadaveric models. Surg Radiol Anat 46(2):103–115. https://doi.org/10.1007/s00276-023-03276-8

Bui T, Ruiz-Cardozo MA, Dave HS, Barot K, Kann MR, Joseph K, Lopez-Alviar S, Trevino G, Brehm S, Yahanda AT, Molina CA (2024) Virtual, augmented, and mixed reality applications for surgical rehearsal, operative execution, and patient education in spine surgery: a scoping review. Medicina (Kaunas) 60(2):332. https://doi.org/10.3390/medicina60020332

Chahal B, Aydin A, Ahmed K (2024) Virtual reality vs. physical models in surgical skills training. An update

of the evidence. Curr Opin Urol 34(1):32–36. https://doi.org/10.1097/mou.0000000000001145
Champney TH, McCord GC, Sampson HW (1995) Rotation of dissecting groups among different cadavers provides students of anatomy with increased opportunity to observe human variations. Clin Anat 8(6):412–413. https://doi.org/10.1002/ca.980080608
Chan LK, Cheung RCC (2020) The use of low-Fidelity models to enhance the learning of anatomy. In: Chan LK, Pawlina W (eds) Teaching anatomy: a practical guide, 2nd edn. Springer International Publishing, Cham, pp 325–335. https://doi.org/10.1007/978-3-030-43283-6_33
Chaturvedi RK, Singh A, Chaturvedi P, Mishra A, Mishra SP (2014) Advantages of plastinated human body in medical education and its legal & ethical aspects. J Evol Med Dent Sci 3(10):2626–2631. https://doi.org/10.14260/jemds/2014/2184
Chia SE, Ong CN, Foo SC, Lee HP (1992) Medical students' exposure to formaldehyde in a gross anatomy dissection laboratory. J Am Coll Health 41(3):115–119. https://doi.org/10.1080/07448481.1992.9936310
Clifton W, Damon A, Nottmeier E, Pichelmann M (2020) The importance of teaching clinical anatomy in surgical skills education: spare the patient, use a sim! Clin Anat 33(1):124–127. https://doi.org/10.1002/ca.23485
Cornwall J (2016) The ethics of 3D printing copies of bodies donated for medical education and research: What is there to worry about? Australas Med J 9:8–11. https://doi.org/10.4066/AMJ.2015.2567
Czech H, Druml C, Müller M, Voegler M, Beilmann A, Fowler N (2021) The Medical University of Vienna and the legacy of Pernkopf's anatomical atlas: Elsevier's donation of the original drawings to the Josephinum. Ann Anat 237:151693. https://doi.org/10.1016/j.aanat.2021.151693
Darici D, Schneider AY, Missler M, Pfleiderer B (2023) Are stereotypes in decline? The portrayal of female anatomy in e-learning. Anat Sci Educ 16(4):720–732. https://doi.org/10.1002/ase.2211
Finkelstein P, Mathers LH (2005) Post-traumatic stress among medical students in the anatomy dissection laboratory. Clin Anat 3(3):219–226. https://doi.org/10.1002/ca.980030308
Fruhstorfer BH, Palmer J, Brydges S, Abrahams PH (2011) The use of plastinated prosections for teaching anatomy—the view of medical students on the value of this learning resource. Clin Anat 24(2):246–252. https://doi.org/10.1002/ca.21107
Garas M, Vaccarezza M, Newland G, McVay-Doornbusch K, Hasani J (2018) 3D-printed specimens as a valuable tool in anatomy education: a pilot study. Ann Anat 219:57–64. https://doi.org/10.1016/j.aanat.2018.05.006
Geldenhuys EM, Burger EH, van Helden PD, Mole CG, Kotze SH (2016) Optimizing the use of cadavers by integrating pathology during anatomy dissection. Anat Sci Educ 9(6):575–582. https://doi.org/10.1002/ase.1617
Ghosh SK (2017a) Cadaveric dissection as an educational tool for anatomical sciences in the 21st century. Anat Sci Educ 10(3):286–299. https://doi.org/10.1002/ase.1649
Ghosh SK (2017b) Paying respect to human cadavers: we owe this to the first teacher in anatomy. Ann Anat 211:129–134. https://doi.org/10.1016/j.aanat.2017.02.004
Giacomini M, Rozée-Koker P, Pepitone-Arreola-Rockwell F (1986) Gender bias in human anatomy textbook illustrations. Psychol. Women Q 10:413–419 https://doi.org/10.1111/j.1471-6402.1986.tb00765.x
Gilbert F, O'Connell CD, Mladenovska T, Dodds S (2018) Print me an organ? ethical and regulatory issues emerging from 3D bioprinting in medicine. Sci Eng Ethics 24:73–91. https://doi.org/10.1007/s11948-017-9874-6
Gradl-Dietsch G et al. (2016) Multidimensional approach to teaching anatomy-Do gender and learning style matter? Ann Anat 208:158–164. https://doi.org/10.1016/j.aanat.2016.03.002
Gundreddy P, Gaurkar SS (2022) Presence of contagious bacterial flora in formalin-fixed cadavers: a potential health Hazard to medical professionals. Cureus 14(10):e30684. https://doi.org/10.7759/cureus.30684
Gupta J, Chaturvedi M, Patil M (2013) Embalmed cadavers—Are they safe to handle, a study to see the microbial flora present in the embalmed cadavers. Int J Pharm Bio Sci 4(1B):383–386
Habicht JL, Kiessling C, Winkelmann A (2018) Bodies for anatomy education in medical schools: an overview of the sources of cadavers worldwide. Acad Med 93(9):1293–1300. https://doi.org/10.1097/ACM.0000000000002227
Iqbal M, Said O, Ibrahim O, Soliman A (2021) Visual sequelae of computer vision syndrome: a cross-sectional case-control study. J Ophthalmol 2021:6630286. https://doi.org/10.1155/2021/6630286
Israel HA, Seidelman WE (1996) Nazi origins of an anatomy text: the Pernkopf atlas. JAMA 276(20):1633; author reply 1634
Jamil K, Baqir SR, Badar S (2023) Workplace evaluation of computer users of the medical university; an ergonomic perspective. Found Univ J Rehab Sci 3(1):3–8. https://doi.org/10.33897/fujrs.v3i1.295
Jenkins JJ, Sánchez LA, Schraedley MAK, Hannans J, Navick N, Young J (2020) Textbook broke: textbook affordability as a social justice issue. J Interact Media Educ 2000(1):3. https://doi.org/10.5334/jime.549
Katz S (2019) Student textbook purchasing: the hidden cost of time. J Persp Appl Acad Pract 7(1):12–18
Kaur K, Gurnani B, Nayak S, Deori N, Kaur S, Jethani J, Singh D, Agarkar S, Hussaindeen JR, Sukhija J, Mishra D (2022) Digital eye strain—A comprehensive review. Ophthalmol Ther 11(5):1655–1680. https://doi.org/10.1007/s40123-022-00540-9
Kennel L, Martin DMA, Shaw H, Wilkinson T (2018) Learning anatomy through Thiel- vs. formalin-embalmed cadavers: student perceptions of embalm-

ing methods and effect on functional anatomy knowledge. Anat Sci Educ 11(2):166–174. https://doi.org/10.1002/ase.1715

Klatzky RL, Lederman SJ (2011) Haptic object perception: spatial dimensionality and relation to vision. Philos Trans R Soc Lond Ser B Biol Sci 366(1581):3097–3105. https://doi.org/10.1098/rstb.2011.0153

Kolb DA (2015) Experiential learning: experience as the source of learning and development, 2nd edn. Pearson Education, Upper Saddle River

Kolb DA, Boyatzis RE, Mainemelis C (2001) Experiential learning theory: previous research and new directions. In: Sternberg RJ, Zhang LF (eds) Perspectives on thinking, learning, and cognitive styles. The educational psychology series. Lawrence Erlbaum Associates, Mahwah, pp 227–247

Konschake M (2018) What do plastic surgeons expect from anatomy in 2018? Clinical anatomy and applied research in plastic surgery. Surg Radiol Anat 40(8):863–864

Konschake M, Brenner E (2014) "Mors auxilium vitae"—causes of death of body donors in an Austrian anatomical department. Ann Anat 196(6):387–393. https://doi.org/10.1016/j.aanat.2014.07.002

Korenstein D, Dunn A, McGinn T (2002) Mixing it up: integrating evidence-based medicine and patient care Acad Med 77:741–742. https://doi.org/10.1097/00001888-200207000-00028

Kramer B, Hutchinson EF, Brits DM, Billings BK (2019) Making the ethical transition in South Africa: acquiring human bodies for training in anatomy. Anat Sci Educ 12(3):264–271. https://doi.org/10.1002/ase.1814

Kübler-Ross E (1993) Sterben lernen, Leben lernen. In: Hermann I (ed) Fragen und Antworten. Silberschnur, Neuwied

Kumar N, Pandey S, Rahman E (2021) A novel three-dimensional interactive virtual face to facilitate facial anatomy teaching using Microsoft HoloLens. Aesthetic Plast Surg 45(3):1005–1011. https://doi.org/10.1007/s00266-020-02110-5

Leboulanger N (2011) First cadaver dissection: stress, preparation, and emotional experience. Eur Ann Otorhinolaryngol Head Neck Dis 128(4):175–183. https://doi.org/10.1016/j.anorl.2011.01.007

Lin SC, Hsu J, Fan VY (2009) "Silent virtuous teachers": anatomical dissection in Taiwan. BMJ 339:b5001. https://doi.org/10.1136/bmj.b5001

Mackenzie JM, Burke MG, Carvalho T, Eades A (2006) Ethics and digital imaging. Microscopy Today 14(1):40–41. https://doi.org/10.1017/s1551929500055188

Mal D, Wolf E, Döllinger N, Botsch M, Wienrich C, Latoschik ME (2022) Virtual human coherence and plausibility—Towards a validated scale. In: 2022 IEEE conference on virtual reality and 3D user interfaces abstracts and workshops (VRW), 12–16 March 2022, pp 788–789. https://doi.org/10.1109/VRW55335.2022.00245

Malpas PJ, Bagg W, Yielder J, Merry AF (2018) Medical students, sensitive examinations and patient consent: a qualitative review. N Z Med J 131(1482):29–37

McBain K, Chen L, Lee A, O'Brien J, Ventura NM, Noel G (2023) Evaluating the integration of body donor imaging into anatomical dissection using augmented reality. Anat Sci Educ 16(1):71–86. https://doi.org/10.1002/ase.2157

McMenamin PG (2023) The third dimension: 3D printed replicas and other alternatives to cadaver-based learning. Adv Exp Med Biol 1421:39–61. https://doi.org/10.1007/978-3-031-30379-1_3

McMenamin PG, McLachlan J, Wilson A, McBride JM, Pickering J, Evans DJR, Winkelmann A (2018) Do we really need cadavers anymore to learn anatomy in undergraduate medicine? Med Teach 40(10):1020–1029. https://doi.org/10.1080/0142159X.2018.1485884

Mendelsohn KD, Nieman LZ, Isaacs K, Lee S, Levison SP (1994) Sex and gender bias in anatomy and physical diagnosis text illustrations. JAMA 272(16):1267–1270. https://doi.org/10.1001/jama.1994.03520160051042

Moore NA (1998) To dissect or not to dissect? Anat Rec 253(1):8–9. https://doi.org/10.1002/(SICI)1097-0185(199802)253:1<8::AID-AR6>3.0.CO;2-Z

Morris H, Jacques S (2018) Dissection should remain a key part of medical education. ANZ J Surg 88(1–2):10–11. https://doi.org/10.1111/ans.14274

Mudgal S, Dabral S, Juyal D, Thakur A (2022) Microbial contamination of the anatomical models used in the museum and dissection hall of the Department of Anatomy. J Pharm Sci Res 14(5):748–752

Mutyala S, Cahill DR (1996) Catching up. Clin Anat 9(1):53–56. https://doi.org/10.1002/(SICI)1098-2353(1996)9:1<53::AID-CA11>3.0.CO;2-9

Nagle C, Vitez K (2020) Fixing the broken textbook market. 2nd edn. U.S. PIRG Education Fund

Naicker J (2023) Medical illustration in anatomy. In: Shapiro L (ed) Graphic medicine, humanizing healthcare and novel approaches in anatomical education. Biomedical visualization, Biomedical visualization, vol 3. Springer Nature, Cham, pp 63–83. https://doi.org/10.1007/978-3-031-39035-7_4

Narang P, Raju B, Jumah F, Konar SK, Nagaraj A, Gupta G, Nanda A (2021) The evolution of 3D anatomical models: a brief historical overview. World Neurosurg 155:135–143. https://doi.org/10.1016/j.wneu.2021.07.133

Ogaili R, Baker SS, Sui H-J (2018) Using of polyester P45 plastinated sheet specimens in teaching anatomy, pathology and radiology courses. Int J ChemTech Res 11(9):393–398

Pabst R (1987) Exposure to formaldehyde in anatomy: an occupational health hazard? Anat Rec 219(2):109–112. https://doi.org/10.1002/ar.1092190202

Parker LM (2002) What's wrong with the dead body? Use of the human cadaver in medical education. Med J Aust 176(2):74–76. https://doi.org/10.5694/j.1326-5377.2002.tb04290.x

Parker R, Larkin T, Cockburn J (2017) A visual analysis of gender bias in contemporary anatomy textbooks. Soc Sci Med 180:106–113. https://doi.org/10.1016/j.socscimed.2017.03.032

Pasricha N, Badyal DK, Goyal PK, Sthapak E (2023) Gross anatomy teaching for medical undergraduates through computer-based simulation: introduction and evaluation of effectiveness. Cureus 15(11):e49517. https://doi.org/10.7759/cureus.49517

Patra A, Pushpa NB, Ravi KS (2023) Visualization in anatomy education. In: Biomedical visualisation, vol 15—Visualisation in teaching of biomedical and clinical subjects: anatomy, advanced microscopy and radiology. Springer, pp 171–186

Patten D (2007) What lies beneath: the use of three-dimensional projection in living anatomy teaching. Clin Teach 4(1):10–14. https://doi.org/10.1111/j.1743-498X.2007.00136.x

Pedowitz RA, Esch J, Snyder S (2002) Evaluation of a virtual reality simulator for arthroscopy skills development. Arthroscopy 18(6):E29. https://doi.org/10.1053/jars.2002.33791

Pierce SM, Heiman AJ, Ricci JA (2023) Evaluating the current state of ergonomics education offered to students in US medical students. Am Surg 89(5):1798–1806. https://doi.org/10.1177/00031348211063555

Pillay-Addinall S, Japhta NL, Govender-Davies S (2023) The use of biomedical imaging in visuospatial teaching of anatomy. In: Microscopy techniques for biomedical education and healthcare practice: principles in light, fluorescence, super-resolution and digital microscopy, and medical imaging. Springer, pp 145–178

Pirri C, Stecco C, Porzionato A, Boscolo-Berto R, Fortelny RH, Macchi V, Konschake M, Merigliano S, De Caro R (2021) Forensic implications of anatomical education and surgical training with cadavers. Front Surg 8:641581. https://doi.org/10.3389/fsurg.2021.641581

Platzer W, Putz R, Poisel S (1978) [New system for the preservation and storage of anatomical matter]. Acta Anat (Basel) 102(1):60–67

Prakash, Prabhu LV, Rai R, D'Costa S, Jiji PJ, Singh G (2007) Cadavers as teachers in medical education: knowledge is the ultimate gift of body donors. Singapore Med J 48(3):186–189; quiz 190

Putz R, Poisel S, Tiefenbrunner F (1974) [Problems with preservation fluids in anatomical preparations. 1]. Acta Anat (Basel) 90(3):394–402

Randjelović P, Stojanović N, Ilić I, Vučković D (2023) The effect of reducing blue light from smartphone screen on subjective quality of sleep among students. Chronobiol Int 40(3):335–342. https://doi.org/10.1080/07420528.2023.2173606

Rashwan ZI (2024) The relationship of computer misuse-related body pain with awareness of workstation ergonomics during digital learning era. J Dow Univ Health Sci 18(1). https://doi.org/10.36570/jduhs.2024.1.2031

Robb RA, Hanson DP (2006) Biomedical image visualization research using the visible human datasets. Clin Anat 19(3):240–253. https://doi.org/10.1002/ca.20332

Sañudo JR, Talarico EFJ, Duparc F, Vázquez T, Valderrama F, Mompeo B, Mirapeix R, Elizondo-Omaña R, Larkin T, McHanwell S, Maranillo E, Moriggl B, Konschake M (2021a) Practical pregraduate teaching in human anatomy: a review. Eur J Anat 25(S1):81–90

Sañudo JR, Vázquez T, Talarico EF Jr, Duparc F, Paulsen F, Macchi V, Feigl G, Elizondo-Omaña R, De Caro R, Aragonés P, McHanwell S, Moriggl B, Konschake M (2021b) Practical postgraduate teaching in human anatomy: a review. Eur J Anat 25(S1):91–99

Shalev A, Nathan H (1985) Medical students' stress reactions to dissections. Isr J Psychiatry Relat Sci 22(1–2):121–133

Shapiro L, Bell K, Dhas K, Branson T, Louw G, Keenan ID (2020) Focused multisensory anatomy observation and drawing for enhancing social learning and three-dimensional spatial understanding. Anat Sci Educ 13(4):488–503. https://doi.org/10.1002/ase.1929

Shapiro L, Hobbs E, Keenan ID (2023) Transforming musculoskeletal anatomy learning with haptic surface painting. Anat Sci Educ 16(4):677–693. https://doi.org/10.1002/ase.2262

Shaver SD (2022) Anatomy, art, and the desire for divine beauty: an exploration into the creation and perpetuation of the standardized body types of early modern anatomical illustrations. Master thesis, University of Windsor (Canada), Windsor

Singh S, Sethi J, Sood S (2007) Cadaver dissection—need to cope the inevitable stressor. J Indian Med Assoc 105(2):99

Souza AD, Kotian SR, Pandey AK, Rao P, Kalthur SG (2020) Cadaver as a first teacher: a module to learn the ethics and values of cadaveric dissection. J Taibah Univ Med Sci 15(2):94–101. https://doi.org/10.1016/j.jtumed.2020.03.002

Stewart SM, Lam TH, Betson CL, Wong CM, Wong AM (1999) A prospective analysis of stress and academic performance in the first two years of medical school. Med Educ 33(4):243–250. https://doi.org/10.1046/j.1365-2923.1999.00294.x

Štrkalj G, El-Haddad J, Hulme A (2020) A global geography of body acquisition for anatomy education: issues, challenges and prospects. In: Teaching anatomy: a practical guide, pp 223–235. https://doi.org/10.1007/978-3-030-43283-6_24

Syamlan A, Fathurachman DK, Vander Poorten E, Pramujati B, Tjahjowidodo T (2022) Haptic/virtual reality orthopedic surgical simulators: a literature review. Virtual Reality 26(4):1795–1825. https://doi.org/10.1007/s10055-022-00666-y

Temkin B, Acosta E, Hatfield P, Onal E, Tong A (2002) Web-based three-dimensional virtual body structures: W3D-VBS. J Am Med Inform Assoc 9(5):425–436. https://doi.org/10.1197/jamia.m1106

Thiel W (1992) [The preservation of the whole corpse with natural color]. Ann Anat 174(3):185–195

Thompson B, Green E, Scotcher K, Keenan ID (2022) A novel cadaveric embalming technique for enhancing visualisation of human anatomy.

Adv Exp Med Biol 1356:299–317. https://doi.org/10.1007/978-3-030-87779-8_13

Ullah K, Dogar AW, Ochani S, Shoaib A, Shah HH, Ur Rehman ME (2023) Obstacles to the deceased donor transplantation in Pakistan. BMJ Open Gastroenterol 10(1):e001101. https://doi.org/10.1136/bmjgast-2022-001101

Wass V, Jones R, Van der Vleuten C (2001) Standardized or real patients to test clinical competence? The long case revisited. Med Educ 35(4):321–325. https://doi.org/10.1046/j.1365-2923.2001.00928.x

Wei CN, Harada K, Ohmori S, Wei QJ, Minamoto K, Ueda A (2007) Subjective symptoms of medical students exposed to formaldehyde during a gross anatomy dissection course. Int J Immunopathol Pharmacol 20(2 Suppl 2):23–25. https://doi.org/10.1177/03946320070200S205

Wiet GJ, Stredney D, Sessanna D, Bryan JA, Welling DB, Schmalbrock P (2002) Virtual temporal bone dissection: an interactive surgical simulator. Otolaryngol Head Neck Surg 127(1):79–83. https://doi.org/10.1067/mhn.2002.126588

Willan PL, Humpherson JR (1999) Concepts of variation and normality in morphology: important issues at risk of neglect in modern undergraduate medical courses. Clin Anat 12(3):186–190. https://doi.org/10.1002/(SICI)1098-2353(1999)12:3<186::AID--CA7>3.0.CO;2-6

Winkelmann A, Guldner FH (2004) Cadavers as teachers: the dissecting room experience in Thailand. BMJ 329(7480):1455–1457. https://doi.org/10.1136/bmj.329.7480.1455

Wise GE (2000) The new anatomy: a forecast of hope. Clin Anat 13(2):148–149. https://doi.org/10.1002/(SICI)1098-2353(2000)13:2<148::AID--CA11>3.0.CO;2-B

Woo KT (1992) Social and cultural aspects of organ donation in Asia. Ann Acad Med Singap 21(3):421–427

Yakura T, Hatayama N, Kawahara C, Ohmichi Y, Ohmichi M, Ban N, Hirai S, Nakano T, Naito M (2022) The effect of simulation-based education before a cadaver dissection course. Anat Sci Educ 15(2):392–402. https://doi.org/10.1002/ase.2124

Yang J (2023) Technology-enhanced preclinical medical education (anatomy, histology and occasionally, biochemistry): a practical guide. Adv Exp Med Biol 1431:65–93. https://doi.org/10.1007/978-3-031-36727-4_4

Yanılmaz O, Guncu MM, Aksu MB, Ozkan M, Sehırlı US (2023) The failure on the effectiveness of formalin on cadaver disinfection and alternative methods. Marmara Med J 36(1):14–17. https://doi.org/10.5472/marumj.1244383

Zhang Q, Davis AY, Black MS (2022) Emissions and chemical exposure potentials from stereolithography vat polymerization 3D printing and post-processing units. ACS Chem Health Saf 29(2):184–191. https://doi.org/10.1021/acs.chas.2c00002

Zhang J, Lu V, Khanduja V (2023) The impact of extended reality on surgery: a scoping review. Int Orthop 47(3):611–621. https://doi.org/10.1007/s00264-022-05663-z

Zucconi WB, Guelfguat M, Solounias N (2002) Approach to the educational opportunities provided by variant anatomy, illustrated by discussion of a duplicated inferior vena cava. Clin Anat 15(2):165–168. https://doi.org/10.1002/ca.1115

**Erich Brenner** is an esteemed Austrian anatomist and medical educator. He is a specialist in anatomy and holds a Master of Medical Education. His primary research interests include the functional anatomy of the musculoskeletal system, vascular structures and advancements in medical education. Dr. Brenner has made significant contributions to anatomical science, authoring over 170 original papers and reviews, along with more than 30 book chapters and complete volumes. Beyond anatomical research, Dr. Brenner is deeply involved in medical education. His dedication to teaching is further exemplified by his role in developing and evaluating medical curricula, ensuring that educational programs align with contemporary medical standards. Dr. Brenner's career reflects a profound dedication to anatomical research, medical education and ethical practices. His contributions continue to influence both the scientific community and the training of future medical professionals.

**Elisabeth Pechriggl** is the Deputy Director and a specialist in anatomy at the Institute of Clinical and Functional Anatomy at the Medical University of Innsbruck. She leads the Human Development research group, focusing on the prenatal development of human organs. Her recent research investigates the relationship between vascular aging and hearing loss, collaborating with the University Clinic for Otorhinolaryngology and the medical technology company MED-EL. Dr. Pechriggl has also contributed to the study of genital malformations in male newborns, aiming to enhance the understanding of regulatory mechanisms during organogenesis.

**Marko Konschake** is a Full Professor, Managing Director of the Department of Anatomy, Histology and Embryology, and Head of the Institute of Clinical and Functional Anatomy at the Medical University of Innsbruck in Austria. With a strong focus on the integration of surgical applications with advanced imaging and monitoring techniques, he brings a wealth of expertise in clinical functional anatomy. His curriculum vitae shows professional experience in various clinical specialties, underlining a translational approach that combines translational anatomical research and teaching. The working group focuses on applied anatomy with specific areas of interest, including peripheral nerves, surgical anatomy, ultrasound and education.

# A Roadmap Towards an Inclusive Anatomy Curriculum

# 3

Asha Venkatesh and Aydan Rahmanova

**Abstract**

Recent historic events have propelled '*Equality, Diversity and Inclusion*' into the limelight across all walks of life, including the higher education sector. There is an acknowledgement now that students and educators are more diverse than ever in terms of their personal characteristics, life experiences and aspirations. This necessitates examining what and how we teach and assess, whom we represent and include, and whose voices are excluded. There is much ground to cover, whether it be the need to decolonise curricula, ensuring we make higher education accessible to those with disabilities and neurodivergence, or improving the experience and attainment of students and faculty from diverse backgrounds. With finite resources and much to do, it can at times be confusing and overwhelming to embark on this journey towards inclusion in learning and teaching within higher education. Furthermore, the process will inevitably disrupt existing teaching and learning practices and cultures. We sometimes need to ask ourselves uncomfortable questions and face up to challenging answers.

To easily break down the concept of what constitutes '*Inclusive Anatomy',* we have used the journalistic approach of '*Five Ws (and one H)*'. This approach of addressing the questions of 'Why', 'What', 'Where' 'When', 'Who' and 'How' (five Ws and one H) we have created a roadmap, which will aid the reader in their journey towards inclusion in anatomy within their own institution's specific context. We start by making a case for why we need to embark on the quest to make anatomy inclusive, and who the stakeholders for this change need to be. We address how, with finite resources, we may be able to achieve meaningful change. Furthermore, we examine what aspects of a curriculum should be addressed, including curricular domains and the personal characteristics of students and educators that impact these domains. We developed a flexible framework that can be used to allow pragmatic decision-making to make anatomy more inclusive. It is for individual organisations to decide where and when to concentrate efforts, and it might be worth using existing momentum within organisations to make impactful change.

Part of this journey involves looking back at the history of the discipline of anatomy. Some historic practices that have resulted in our body of knowledge would in today's world

A. Venkatesh (✉) · A. Rahmanova
Department of Anatomy, School of Medicine, Medical Sciences and Nutrition, University of Aberdeen, Aberdeen, Scotland
e-mail: a.venkatesh@abdn.ac.uk; aydan.rahmanova@nhs.scot


I. D. Keenan et al. (eds.), *International Anatomical Education*, Biomedical Visualization 9,
https://doi.org/10.1007/978-3-031-91849-0_3

be considered unethical. How should educators and students deal with this troublesome fact, and is it important to do so? As some of the conversations to be had may be disconcerting at times, we believe that at the start of our journey stakeholders need to revisit concepts of unconscious bias and privilege.

We hope that our experiences will help those embarking on a similar expedition, ultimately enabling them to make anatomy inclusive within their own institutions.

**Keywords**

Inclusive curricula and assessments · Curricular domains · Diversity and personal characteristics in higher education · Critical reflection in higher education · Anatomy education

## 3.1 Introduction: Diversity and Inclusion in Our Communities

Over the past decade, our society has seen a huge increase in the diversity of its population. Socioeconomic, cultural and/or political factors have led to our communities becoming multicultural like never before (Shankley et al. 2020; Wells 2020).

When diverse groups of people interact with each other, they have the potential to learn from each other. A diverse community has a broader set of skills and experiences than a less diverse one. Individuals within the community gain a better understanding of other groups, enhancing empathy and personal growth. Diversity therefore results in better creativity and productivity (Gomez and Bernet 2019). Furthermore, diverse groups consistently outperform more homogenous ones in critical tasks such as problem-solving (Aminpour et al. 2021; Watson et al. 1993; Hong and Page 2004).

Inclusion is described as practices or policies we adopt that allow people from diverse groups to participate equally in society (Papay 2019). True inclusion allows individuals to achieve their full potential irrespective of their diverse personal characteristics (Mayo 2004). Recognizing that diversity within our communities is a strength, there is much work to make everyone within our communities feel included. The goal is to foster a sense of belonging among individuals with diverse backgrounds (Scorgie and Forlin 2019).

## 3.2 *Why* Embark on a Journey Towards an Inclusive Anatomy Curriculum?

The diversity of our student and staff population in higher education (HE) to some extent reflects that within our communities. Increasing opportunities for diverse communities to interact and share practice has educational benefits and enriches the environment of the institution. Cletus et al. (2018) showed that diversity improved students' critical thinking, problem-solving abilities and cognitive complexity.

The current political climate within HE predicts a looming crisis due to funding cuts compounded by the pandemic (Featherstone 2023). In such an environment, encouraging diversity amongst our students can aid in increasing institutional viability and making inclusion intentional is critical to maintaining diversity in HE (Huo et al. 2014), further increasing the institution's reach (Van Mol and Perez-Encinas 2022). Most students and staff in HE will agree that we need to work towards creating a socially just society where everyone has equal opportunity. Inclusion then becomes more than just a lofty aspiration; it becomes an ethical obligation. That said, the concept may seem abstract at times and lacking focus. Educators may not be clear on where or when to start, what to focus on and how (D'Arcy 2014).

Our students, upon graduation, will enter our community's workforce and therefore need to be trained to deliver within this diverse environment. This means that our curriculum ought to reflect this diversity. While some might believe that '*hard science*' is based on facts, in truth, there is awareness now that some aspects of biomedical science are based upon standard universal models and do not serve diverse groups equally (Perez 2019; Li et al. 2022). Some examples of this are mentioned in the next section. It is therefore important to change what we teach moving away from a universal normative model.

Educators, institutional leadership, policy-makers and students all have a role in supporting and creating inclusive change.

Anatomy as a discipline has, therefore, rightfully joined this discourse about diversity and inclusion. Various groups of staff and students are researching aspects of diversity such as moves to decolonise curricula (Finn et al. 2022), and how we teach sex and gender in anatomy (Venkatesh and Morris 2024). These in turn are leading to pockets of change. But for a novice adopter, and especially when resources are limited and the pace of change is rapid, it can be difficult to comprehend where to start and how to bring this all together in one comprehensive framework.

This is what we hope this chapter will add. Using a staff–student collaborative approach, we developed a flexible framework that educators can use to make anatomy more inclusive. We believe this will provide colleagues in higher education institutions (HEIs) with a flexible tool and a roadmap to help them get their journey towards inclusion underway.

Embarking on this journey will uncover various emotive and sensitive topics such as race and religion. Educators therefore cannot enter such teaching sessions without a carefully constructed plan to ensure that students feel safe to be able to express their views freely without judgement by peers or educators.

The process may at times challenge existing systems and practices but the long-term gain, that is, better outcomes for students, staff and HEIs, makes this journey worthwhile.

## 3.3 The Universal Model and *Why* We Need to Move Away from It

Despite the diversity of our general population, there is a social norm across many aspects of life—the white, able-bodied, cisgender, heterosexual and socio-economically middle-class male is frequently considered as the universal model (Bowden 2021). We can find examples of this in research, healthcare and HE.

In automotive engineering design, a medium-sized male body (1.75 m; 78 kg) has frequently been used as the norm (Innovations 2019), for example, during crash tests. Those individuals who do not fit this very specific target demographic—including women (a special mention here about pregnant women), elderly or obese individuals were considered as ‘out of position’ drivers. Anatomical differences in these individuals were not accounted for in the design of seatbelts, the placement of steering wheels and foot pedals. This meant that the fatality risk was on average 17% higher for females (Kahane 2013). Carter et al. (2014) also showed that with more than one characteristic as an ‘*outlier*’ the risk of injury in a car crash was compounded due to an interaction of these factors, so an elderly, obese female is much more likely to suffer more serious injury than the ‘*standard*’ or universal male.

Within healthcare research, clinical trials are also frequently formulated for this universal model and do not include people with traits common in specific ancestry, those belonging to lower socio-economic backgrounds or women (Beard et al. 2021). An example is the lack of research on minority ethnic groups within clinical trials investigating cardiovascular disease (Prasanna et al. 2021), despite these patients experiencing a greater prevalence of illness. The study showed that treatment algorithms formulated based on such research would not lead to the best outcomes for patients belonging to these marginalised groups.

Health inequalities and disparities also frequently stem from using such a normative universal model. One example is the difference in maternal death. A published report (Knight et al. 2020) revealed that black women have a fourfold greater risk of mortality during pregnancy and childbirth compared to their white counterparts.

Within HE, the existence of an attainment gap in marginalised groups is well-known (Bunce et al. 2021; Linton 2020) and efforts are being made to narrow this differential attainment through better support targeted at those who need it most (Kaye 2021; Fyfe et al. 2022).

All these studies show conclusively that we must move away from the normative '*universal*' model. When particular personal characteristics such as being white, male, able-bodied are seen by society as '*normal*' others then emerge as deviations from the norm (Morris 2016). When society operates through such a normative lens, it marginalises and therefore disadvantages individuals not possessing these characteristics, with deleterious consequences.

## 3.4 *Who* Should Get Involved in this Exercise? The Profile of Our Stakeholders

We embarked on this project to develop a framework to help educators on their journey to make the discipline of anatomy inclusive for themselves and their students.

**Staff:** Like our general population, the profile of staff who teach anatomy is becoming increasingly diverse. A study by Smith et al. (2022) through a survey administered to all medical schools within the UK and Ireland demonstrated that 46% of the 760 people who taught anatomy were university staff, of whom 30% worked part-time. Of all teachers, 54% were employed in a demonstrator capacity, and 71% of these worked part-time. The study only focused on anatomy teaching to medical students, and only within the UK and Ireland. Furthermore, the study did not describe the profile of staff in relation to racial groups, disability and other personal characteristics. Within the UK, the Higher Education Statistics Agency (HESA) publishes data on staff and students from the mandatory returns of HE providers, which gives us some insight into the profile of racial groups, disability and gender of all UK HE staff. Data for 2020/2021 (HESA 2022) shows that of the 191,435 HE staff in the UK, 1835 taught Anatomy and Physiology. Of all HE staff, 4.2% declared some form of disability, 51.7% of those declared their gender as female and 12.2% of those declared their ethnicity belonged to racialised groups (HESA 2022, Advance, H.E., 2023).

We use the term 'racialised' here because there is now an understanding that the concept of race is not a meaningful descriptor of biological ancestry (Bamshad et al. 2004) and is merely a social construct (Green and Smith 1983). However, it is often used to denote a racial meaning to something that does not have a racial basis (Gonzalez-Sobrino and Goss 2019).

The majority of HE staff (~30%) were in the age range of 36–45 years, followed by individuals who were 26–35 years (~26%), and those in the 46–55 years age range (~24%) (HESA 2022). We recognise that this may not be a uniform picture worldwide, but it does show that our educators are diverse in terms of their age ranges, ethnicity and some variable personal characteristics.

**Students**: UK HESA data (2022) shows that 15.2% of all students in UK higher education in 2020/2021 were either studying medicine/dentistry (and therefore anatomy) or subjects allied to medicine such as nursing or midwifery (that would usually include anatomy). The proportions of these students with various recorded diversity characteristics are shown in Table 3.1.

The above-mentioned reports and studies show that our population of staff and students who teach and study anatomy is diverse. As key stakeholders, involving these diverse groups in discussions about creating an '*Inclusive Anatomy Curriculum*' (IAC) is a strength that brings authenticity to the process.

**Table 3.1** HESA (2022) figures. Proportion of students studying medicine/dentistry and subjects allied to medicine (e.g. nursing) possessing diversity characteristics of gender, racialised group and disability

| Personal characteristic | Proportion of all students studying medicine/dentistry | Proportion of all students studying subjects allied to medicine |
|---|---|---|
| Declared disability | 12.9% | 15.6% |
| Racialised groups | 42.5% | 31.8% |
| Female gender | 61.7% | 79.5% |

**Co-Creating Inclusive Curricula Through Staff–Student Partnerships**: The benefits of co-creating curricula with students have long been recognised (Cook-Sather and Matthews 2021; Cecchinato and Foschi 2017), including within anatomy (McDonald et al. 2021). Szelics et al. (2023) also described how this collaborative approach was used successfully when developing an inclusive curriculum in life sciences. When students are involved in such partnerships, it gives authenticity to the resources created and to the curricular changes made. Pedagogical theories on peer learning also suggest that students engage better with peers, perhaps because they are not in a position of authority (Topping 2005; Tullis and Goldstone 2020). Our own experience is similar, with student-co-created and student-led sessions on racial inclusion within our medical curriculum being rated highly by students in course evaluations.

**Higher Education Institutions (HEIs) as Stakeholders**: HEIs make decisions on their strategy, policy and resource allocation, which pave the way for effecting change in the sphere of '*Equality, Diversity and Inclusion*' (EDI) (LERU 2019). To be comprehensively inclusive, a vast number of changes are needed in all spheres of the curriculum, which will take time to fully implement and embed. Consequently, efforts will need to be prolonged and sustained, which is only manageable when institutional policies and strategies are in place (May and Bridger 2010). National legislation such as the Equality Act (2010) in the UK has provided the legal framework that has made it imperative for public bodies to implement changes (Equality Act: guidance 2010). Within HEIs, links to measurements of performance and funding also act as strong drivers for change at the institutional level (Scott 2016).

We also believe that institutions are beginning to realise that implementing change beyond legal obligations is grounded in ethics and valorisation (Baltaru 2022). Consequently, an ever-increasing number of HEIs in the UK, Europe and worldwide now have EDI strategies (Claeys-Kulik et al. 2019; Tamtik and Guenter 2019).

**Other Stakeholders**: With many universities being public sector organisations, the wider public is also a stakeholder. Specific to anatomy, our donors come from the general population. Once they graduate, our students will join the workforce, taking the culture of inclusion with them into the working environment. Therefore, industry and other organisations where our students of anatomy seek work following graduation are also stakeholders. Staff, students and HEIs are however our prime stakeholders who need to be consulted and invested in the process if change is to be made.

## 3.5 The *When*

The scope of change is vast. But HEIs and their staff are also under pressure like never before. Making many changes at once can be challenging, and perhaps an impossible aspiration. So how can educators decide what changes to focus on at any given time?

As mentioned above, HEIs may have strategies that aspire towards inclusive changes. HEIs also work towards EDI-related charter marks. In the UK, such charter marks include the Athena Swan (AS) charter for gender equality (AdvanceHE n.d.-a), the Advance HE Race Equality Charter (AdvanceHE n.d.-b), the Stonewall Charter (Stonewall n.d.) and Disability Confident (GOV UK 2014), to name a few.

Ovseiko et al. (2020) describe the effect of such national and international initiatives and professional organisations in effecting change. They describe how, within the UK, the National Institute for Health and Care Research (NIHR), a prominent funding body, required departments and institutions to hold at least a silver-level gender equality AS award to be eligible for NIHR funding. HEIs allocated resources towards achieving this award, which led to a tenfold increase in AS awards between 2011 and 2016. Implementing AS action plans led to more inclusive policies for women in HE. There was a rise in the number of women in mid-level leadership positions from 8% in 2011 to 24% in 2016, and an increase in women in senior director positions from 10% in 2011 to 15% in 2016 (Prince and Francis 2023). Similar initiatives within the UK have now been introduced for race equality with the Advance HE Race Equality charter awards (AdvanceHE n.d.-b).

Institutional priorities lead to the creation of action plans with resource allocation towards achieving these priorities. When educators, including anatomists, align their efforts with institutional priorities, it paves the way for change, opening up resources for facilitating these changes. Within our own institution, inclusion is one of our strategic priorities (University of Aberdeen 2022). Our institution's current focus is on gender and race, leading to several institutional drives. One such initiative that is currently underway is to decolonise our curricula (University of Aberdeen 2023). Decolonisation movements are based on challenging eurocentrism and post-colonial narratives (Winter et al. 2024). At the University of Aberdeen, a toolkit has been produced with several practical suggestions. The current focus helps us effect changes within the anatomy curriculum and gives us a starting point in our journey towards an IAC.

A counterargument made by Titchkosky (2010) is that such bureaucratisation of university processes will unintentionally lead to the indirect exclusion of some characteristics in the immediate term. There is perhaps no definite answer as to when to approach different aspects of inclusion and it is perhaps up to each anatomist to consider their specific context whilst making this decision.

There will, however, be some degree of commonality in how we address inclusion across various characteristics and therefore measures introduced in one sphere, for example, racial inclusion may also help to support inclusion of other characteristics such as gender.

## 3.6 Within Which Curricular Domains *(Where)* Should We Introduce Change?

A curriculum is a conceptual framework of what is taught. Broadly, it comprises learning outcomes, teaching methods and materials alongside assessments (Sanchez and Kumar 2022; Wehrli 2010). We have called these the domains of a curriculum (Fig. 3.1).

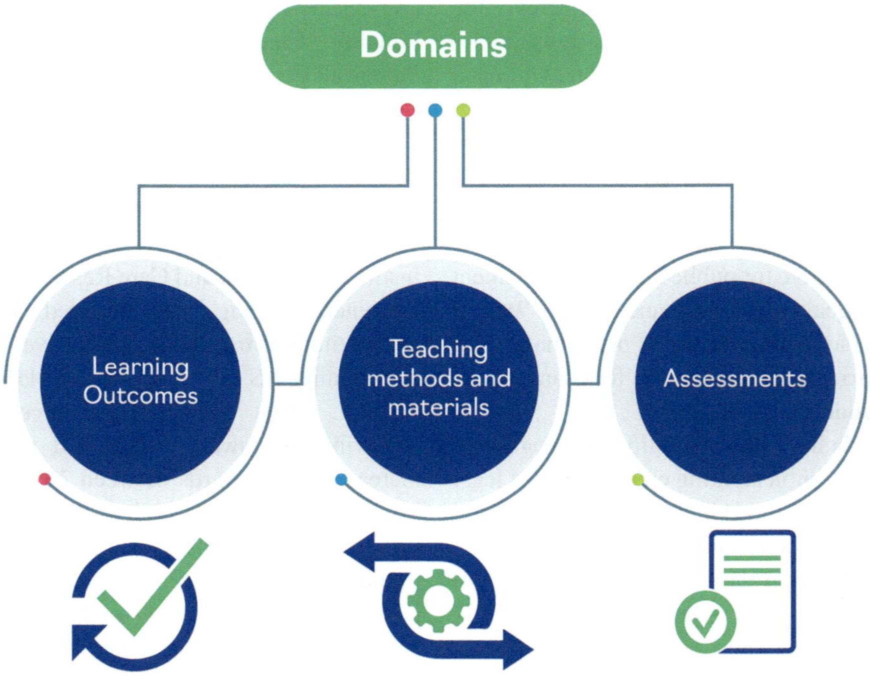


**Fig. 3.1** Three domains of learning that together make up a curriculum. (As described by Sanchez and Kumar 2022)

**Learning Outcomes**: Learning outcomes are navigation tools that help educators plan and help guide students through a course of study (Mahajan and Singh 2017). Conceptual frameworks have been proposed that allow for the integration of inclusion within curricular learning outcomes (Meiman et al. 2019). In anatomy, a series of core anatomy syllabi for various professional courses such as medical, dental and nursing curricula are well known (Smith et al. 2016; Connolly et al. 2018; Matthan et al. 2020). We believe that these core syllabi could be modified to add inclusive outcomes (Venkatesh and Morris 2024). Professional bodies such as the Human Anatomical and Physiological Society have proposed inclusive outcomes to be introduced into curricula (Meyer and Cui 2019), and we believe that doing so will provide the impetus for change in anatomy teaching.

**Teaching Methods and Materials**: Detailing the various teaching and learning pedagogies used in anatomy is outside the scope of this chapter and the reader is directed to expansive literature on this (Sugand et al. 2010; Koh et al. 2023; Pandurangam and Gurajala 2023). Digital pedagogies often became a necessity during the pandemic, leading to further modifications of how anatomy is taught (Xiao and Evans 2022), while also introducing further EDI issues, such as accessibility (Ng et al. 2023).

At every stage during the development and use of teaching methods and materials, educators would need to consider how best to make the material accessible to and inclusive of our diverse range of staff and students. Some examples of making materials accessible include use of alternative text for images and use of Sans Serif fonts in digital material (UK Parliament 2018).

**Assessments**: It is well known that assessment drives student learning (Wormald et al. 2009). Yet there is limited literature examining how to make assessment inclusive than that looking at inclusive teaching practices (Tai et al. 2021). Most recommendations are centred around principles of creating an anti-ableist assessment, expanding the repertoire of provisions for students during assessments, and co-design of assessments with students (especially with those students who we want to include) (Nieminen 2022). Where professional and regulatory bodies influence how assessments are structured, as is the case with many *hard-science* subjects such as anatomy, educators may not have much choice on the assessment instruments used and limited scope to co-create authentic assessments to suit diverse students. We can however involve students in developing assessment literacy (Smith et al. 2013). Smith et al. (2013) defined assessment literacy as 'students' ability to understand the purpose and processes of assessment and accurately judge their own work'. Deeley and Bovill (2017) described how staff–student partnership could achieve improved assessment literacy for students. The authors describe how here, in addition to self-assessment of essays, students partnered with staff to co-create marking criteria for formative and summative essays, leading to the student body's use of assessment *for* learning.

When considering how to ensure the representation of diverse populations within our anatomy assessments, one 'easy win' might be incorporating images of people with diverse characteristics, including gender, ethnicity, body shape and disabilities when assessing students' understanding of surface anatomy (Finn et al. 2022).

With HEIs having a sizeable international student population, another area encompassed in inclusive assessments is tackling presumptions made about international students' understanding of assessment processes (e.g. their understanding of what constitutes plagiarism), so that attainment is not affected (Carroll 2008).

## 3.7 *What* Personal Characteristics Should We Focus On?

Inclusive policies within HEIs usually stem from a drive to achieve social justice (Brennan and Naidoo 2008). Within the UK, the Equality Act (2010) defines nine personal characteristics as 'protected' making it illegal to discriminate against an individual who possesses any of these. These characteristics are age, gender and gender reas-

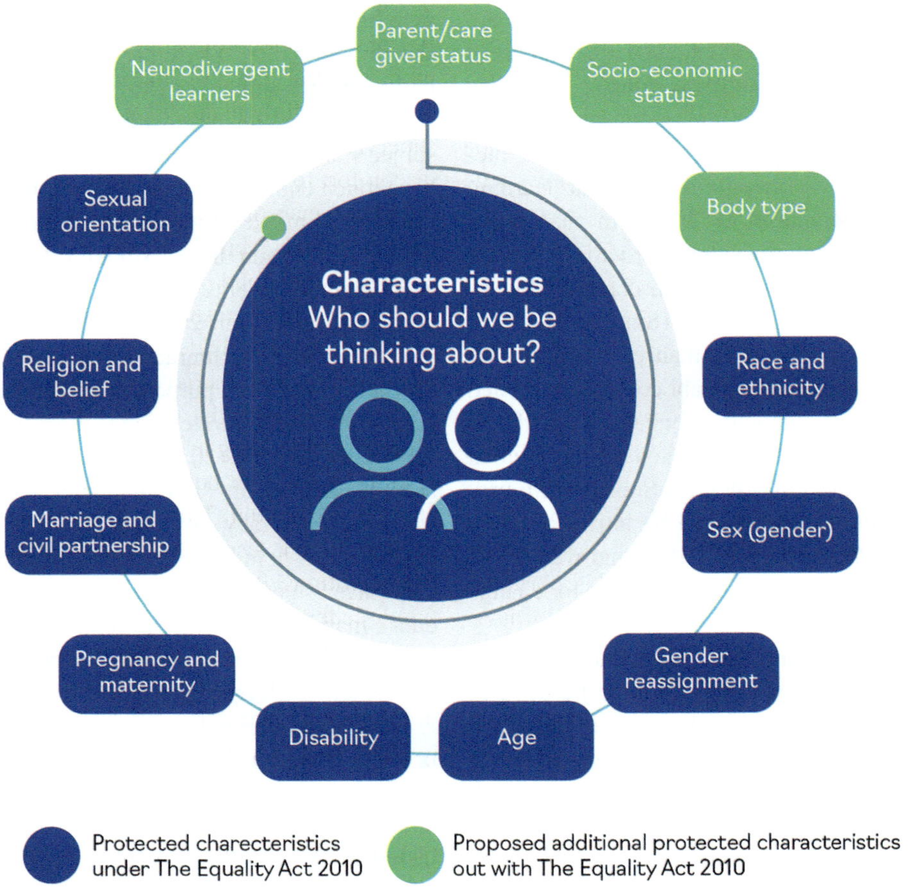


**Fig. 3.2** Proposed characteristics that should be considered when creating an Inclusive Anatomy Curriculum. (Adapted from the Equality Act 2010)

signment, pregnancy and maternity, race, religion or belief, sex, sexual orientation and marriage or civil partnerships. This has paved the way for inclusive practices across all walks of life, including within HEIs. Some authors call for more personal characteristics such as socio-economic status, parent/carer status (Jasim et al. 2021) and neurodiversity (Acevedo and Nusbaum 2020) to be included. Specific to anatomy teaching, another personal characteristic that we feel should be addressed is the representation of differences in body shape (Shaver 2022). Compiling these attributes, we have created the following (Fig. 3.2).

**Intersectionality:** It is worth noting that for many individuals there is often more than one of these abovementioned personal characteristics. The manner in which systems of oppression interact with one another is called intersectionality (Weldon 2008). bell hooks' (1981) groundbreaking book '*Ain't I a Woman: On black women and feminism*' was critical in highlighting intersectionality as an issue. The disadvantage faced by individuals who possess multiple personal attributes is more than an additive effect: intersectionality is known to compound their disadvantages (Cooms et al. 2022).

## 3.8 *How* to Comprehensively Effect Change: The Proposed Framework for Creating an IAC

In earlier sections, we have argued that change can only be brought about in small and incremental steps. Our framework is based on a two-step approach. First, we look at the epistemology of anatomy knowledge with an inclusive lens, and second, we account for the inclusion of the various personal characteristics across the various domains of the curriculum.

### 3.8.1 Step 1: Epistemology of Anatomical Knowledge with an Inclusive Lens

Anatomy departments in many countries across the world, and certainly all within the UK obtain their cadavers through bequeathal programmes that centre around informed consent (Brenner et al. 2024; Smith et al. 2023). Today, in countries where such a system exists, there are laws which govern the functioning of anatomy departments receiving human bequeathals. In Scotland, this is the Anatomy Act 1984 and the Human Tissue (Scotland) Act (2006). However, this was not always the case!

Most anatomists are aware of our gruesome history from 1500 A.D to 1800 A.D. This period was marred by practices such as the (often public) dissection of bodies of non-consenting slaves and executed criminals (King 2017; Bennett 2018). In the 1800s, grave-robbing and even murder were often surreptitiously resorted to, to supply the medical profession with cadavers. Within the UK, these practices ultimately led to the Anatomy Act (n.d.), UK, 1832 (2 & 3 Will. 4. c. 75). The passing of this legislation meant that the only bodies that could be used for dissection were unclaimed bodies or those donated by family members (who were usually too poor to afford burial or cremation for their relative). Such cadavers inevitably came from individuals belonging to marginalised groups, and predominantly those from a poorer socio-economic background (Davidson 2007). We can infer from this that the Anatomy Act, UK, 1832, provided a legal route for the medical profession to obtain bodies for study but indirectly discriminated against these marginalised groups (Davidson 2007).

Across the globe, many countries such as Belarus, Greece, Italy, India, Kenya and Nigeria still rely on using unclaimed bodies for dissection (Bharambe et al. 2019; Brenner et al. 2024), though in some instances such as in South Africa efforts are being made to transition to a system that relies on consented bequeathal (Kramer et al. 2019).

The epistemology of much of the anatomical knowledge that we take for granted is therefore marred by egregious practices that in today's world would be considered unethical. One example is Andreas Vesalius, a revolutionary physician in the Renaissance period who detailed various anatomical structures during his lifetime by carrying out anatomical dissections. He has, among other discoveries, given an accurate description of the human uterus and established through dissection that the heart, rather than the liver, was the '*source of vital spirit and the arterial system*' (Zampieri et al. 2015). In his treatise *De Humani Corporis Fabrica,* Vesalius (1543) is believed to have candidly admitted to resorting to grave robbing to acquire cadavers for dissection (Ghosh 2015).

At this point, we would like to clarify that censorship or erasure of this knowledge is not the intention of raising awareness of the discipline's murky past. Wingfield (2017) recognised this sentiment as a common misconception and misinterpretation of inclusive movements. Rather than deleting this knowledge, our intention in looking at practices is to include the voices of the marginalised and to create a more representative, inclusive and unexpurgated system.

In time, we hope that there will be a critical mass of voices seeking meaningful change. One such example is a growing movement to seek the change of name of a commonly used obstetric vaginal speculum which was originally named after its inventor Marion J Sims. Frequently referred to as the '*Father of Modern Gynaecology*', Sims perfected several gynecological surgical techniques and invented the vaginal speculum by experimentation on the black slaves he held. It is well known that he did so without anesthesia or consent (Zhang 2018). Three of these slaves were Anarcha, Lucy and Betsey (Prunty et al. 2021) who had developed urinary and/or fecal incontinence following traumatic vaginal deliveries. These slaves, being no longer '*useful*' to their slaveholders due to their medical conditions, were sent to Sim for experimental procedures to cure their conditions (Prunty et al. 2021). As a slaveholder himself and a physician, Sims performed experimental surgical techniques on these women, without anesthesia or informed consent. While it is believed that there may have been many more women (Cronin 2020), the fate of these three women was chronicled by Sims in his autobiography *The story of my life* (Sims 1884). Diarised entries describe the pain that Lucy was in because of Sim's surgical experiments. There are calls now to rename the speculum, which is still widely used in modern procedures, as the Lucy speculum (Shalit et al. 2021). Removing the inventor's name from the invention would prevent healthcare professionals unknowingly and indirectly condoning Sims' dark acts for generations to come. Shalit et al. (2021) recognise that this will require significant efforts and a multifaceted approach; however, the authors believe that it is well worth doing.

From an anatomy curricular perspective, we believe that the purpose of introducing such examples into the curriculum is to educate both staff and students about the inequalities that led to the construction of our knowledge. This is a small step of reparation towards these practices. If we did not introduce such examples into the curriculum, how would students ever know about such dark histories; and would the tales and miseries of the many Anarchas, Lucys and Betseys in history forever remain untold? We, therefore, propose that educators try to incorporate such examples into their teaching, so our students can understand the inequalities in our history!

Debate about historical practices in many cases may result in decisions to repatriate artefacts (Curtis and Scholten 2022). They may also result in an understanding that repatriation is not possible, and that there is nothing to be gained from disposal of resources such as bones whose origin is unknown (Jones 2023). It is useful to have these discussions and face up to these uncomfortable challenges.

**Safe Spaces for Discussion**: When engaging in such challenging discussions such as the ethics of historical practices in a classroom environment, staff need to make students feel '*psychologically safe*' (Edmondson et al. 2016), to negotiate learning and participate in the class without disengaging from emotionally charged conversations. Edmondson et al. (2016) described psychological safety as a requirement to allow members in a group to be able to express views without fear of reprisal, allowing members to thrive. Discussions about the role of historical (mal)practices in creating knowledge could raise feelings of discomfort and unease amongst a group of students. Teaching will therefore need to be '*scaffolded*' (Millner 2022) to make students feel psychologically safe and brave enough to enter these discussions without fear of reprisal and prejudice. These topics need to be approached sensitively. Often, many staff and students may shy away from having challenging conversations centred around race (Miller-Kleinhenz et al. 2021). Revisiting concepts of unconscious bias (Marcelin et al. 2019) and privilege (McIntosh 1988) are helpful starting points.

**Privilege, Unconscious Bias, Fragility and Cultural Humility**: Privilege is an inherent advantage a person holds in society as a consequence of possessing certain characteristics (McIntosh 1989). Exploring privilege in terms of race, McIntosh defined it as an '*invisible package of unearned assets*', where the recipient is unaware of their advantage. These privileges grant the person possessing them leverage in society. They also contribute to unfavourable outcomes for those who do not possess them. McIntosh concluded that to dismantle current systemic prejudices, the first step would be for people possessing these privileges to acknowledge the inequality of the playing field. However, discussions around privilege are often uncomfortable and may bring up feelings of fragility (DiAngelo 2015). The term was first used in the context of race (*'white fragility'*) and describes the discomfort a white person elicits when confronted with their role in racial injustice. Fragility is an obstacle to having open and honest discussions about inclusion. Evans-Winters and Hines (2020) described how white privilege acted to resist and impact anti-racist education. They used a conceptual framework to explain this resistance including passive-aggressive behaviours, a hostile stance against racialised staff, and microaggressions such as groupthink (where students conspire with each other to reject the staff member's authority) and lynch mob (where they repeatedly use minor mistakes to justify complaints and attack the staff member's credibility). Fear of such attitudes might also cause students within groups who are from marginalised backgrounds or have views differing from the norm to feel less psychologically safe. They may therefore find it difficult to come forward in open honest discussions. Exercises designed to help students become aware of their own privileges (Gallavan 2005) will be helpful in such situations. It helps *'privilege-holders'* uncover their privileges but without raising a feeling of discomfort. This will prevent the reactions characteristic of *'fragility'* and allow an openness and inquisitiveness that will in turn encourage those without the privilege to share their experience in an open and honest manner.

We also believe that introducing the concept of cultural humility (Lokugamage et al. 2020) can aid educators by raising awareness and insight. Henderson et al. (2018) introduced the theoretical model of achieving cultural humility by being open, aware and sensitive through deep self-reflection and moral reasoning (Fig. 3.3).

For inclusion to be intentional, educators would need to actively create a psychologically safe environment for all individuals to share their views, alongside spaces to provide feedback and evaluate current practices. Therefore, it is important to carefully think through how they can introduce such potentially emotive and sensitive topics for discussion. We must maintain open and respectful dialogues and respond with respect, irrespective of whether we agree or disagree with opinions. Including students as co-creators of the curriculum will also be beneficial. While it is the educator's responsibility to create a safe space for this discussion, work done by fellow students can create a sense of interest, relevance and acceptance (Lerner 2022).

### 3.8.2 Step 2: The Domains and Personal Characteristics

There can be an element of subjectivity when thinking about inclusive practices in every domain of the curriculum and various personal characteristics. Using critical reflection methodology (Fook 2011) offers a framework that allows educators to move away from this subjectivity. It bridges the gap between teaching practices and traditional research.

**The Critical Reflection Methodology**: Reflection is a process widely used amongst practitioners, including educators, to gain a deeper and more complex understanding of their practice, thus facilitating improvements (Fry et al. 2003). Reflective cycles such as Gibb's cycle of reflection (1988) are commonly used by practitioners as models. Fook (2011) described critical reflection as a way of learning by reanalysing this experience. Critical reflection is also used as a research framework to explore assumptions about pedagogy and knowledge (Hickson 2016; Fook and Gardner 2007). Researchers in small groups discuss and bring in their own experiences. Existing educational practices are analysed, and an understanding is drawn about the power connections between the individuals and their social context. This process will expose any fundamental assumptions that we have made in choosing a particular teaching practice. Ultimately, it will lead to an understanding of gaps, blind spots and discrepancies. Thereafter, a new and better system that will continue to align with one's personal belief system while also being *'more responsive to the social environment of practice'* (Fook 2011) can be established. Using this research methodology allows us to transform our educational practice in a manner that is mindful of our unique and complex context.

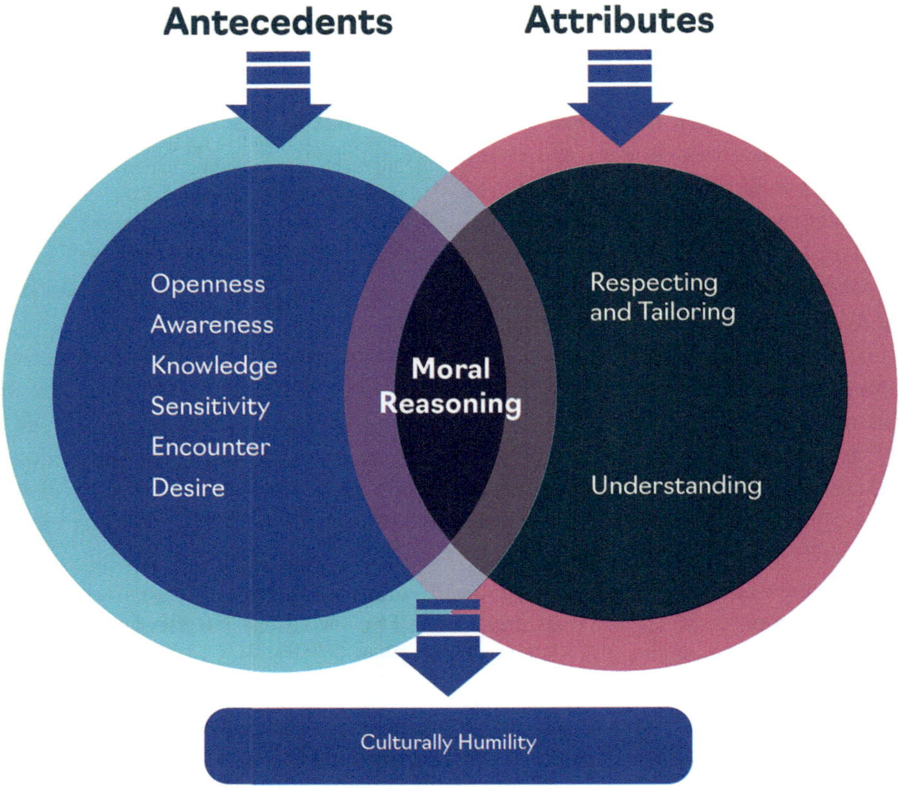


**Fig. 3.3** A theoretical model of cultural humility (adapted from Henderson et al. 2018). This model describes the attributes, antecedents and consequences of cultural competence. The model also proposes that moral reasoning is another valuable antecedent

Utilizing the critical reflection methodology, the team seeking to effect change might list alterations that could be made across each of the personal characteristics in a specific domain, which would, in turn, lead to improving inclusion. A review of the literature will then assist them in understanding whether such a change can or should be implemented as well as any consequences of implementing it. They would then consider whether this is the appropriate change to adopt within their department and institution. If the team decides that the proposed idea for change is indeed something they can implement in their department, the final step should be to ensure that the teaching is scaffolded as described above, so that both educators delivering the content and students participating in the learning feel able to engage in a psychologically safe environment.

Our framework is designed to be flexible, so one could start with a personal characteristic (e.g. race) and apply the framework across domains of the curriculum to identify how to make an anatomy curriculum comprehensively inclusive for race. It is also possible to use the framework by beginning with a particular domain. For instance, if the assessment team in a department has launched a project to make an assessment more inclusive, we can use this framework by applying it across the various personal characteristics to make assessments inclusive for all personal attributes.

Lastly, the framework can also be used for specific topics or regions. For instance, if one wanted to look specifically at neuroanatomy, we would investigate the literature about the epistemology of neuroanatomical knowledge and then how to make neuroanatomy teaching inclusive across various domains and personal characteristics.

We hope that this will give colleagues the flexibility to apply the framework to their own anatomy curriculum, irrespective of their own personal, departmental and institutional context.

## 3.9 The Framework in Action: A Worked Example

Our institution has a current focus on race and decolonisation (University of Aberdeen 2022, 2023). Consequently, in this example, we describe using our proposed IAC framework to make anatomy inclusive for race across the various domains of outcomes, teaching methods and assessment.

**Critical Reflection and Positionality:** We (co-authors AV and AR) undertook this exercise by using the critical reflection methodology. At the outset of this exercise, as with all qualitative research, we acknowledged our positionality because our personal experiences and viewpoints may have biased how we view aspects of the process (Abu Moghli and Kadiwal 2021).

**Step 1: The Epistemology of Knowledge:** When studying a specific topic or region in anatomy, a review of literature can usually lead to some general or specific examples of how the anatomy knowledge came to be. We encourage educators to introduce this into their teaching to contextualise anatomy concerning colonial history. Specific examples relating to the topic under study might be found or general examples can be used as appropriate. For instance, anatomy students often use photographic atlases and video dissections for their study. One such atlas, no longer in use today, is Eduard Pernkopf's *Topographische Anatomie des Menschen*, (Atlas 2001). The atlas was accurate and of very high quality but was created by dissecting at least 1377 victims of the Nazi holocaust. The controversy surrounding the creation and subsequent censorship of this atlas might be mentioned as part of a lecture or as part of a discussion with a small group of students, inviting arguments for and against its use in a safe and non-judgemental manner. As previously mentioned, and in light of the sensitive nature of the discussion, the educator needs to make consideration of how students may view and react to opposing views. Laying ground rules before the session to allow students to freely express controversial views is needed. The specific fear here is that if students freely express views questioning the benefits of deleting or cancelling this resource, they risk being labelled as someone harbouring antisemitic views. One possible method to allow students to freely express their views is to do this anonymously using audience response systems (Kay and LeSage 2009) which are now widely prevalent and easy to access.

Introducing this example and creating an environment that allows opposing views to be discussed encourages critical thinking about the epistemology of anatomy knowledge among students.

**Step 2: Inclusion of Race Across Domains of the Curriculum**: Discussions between co-authors AV and AR resulted in some proposed changes to make anatomy more inclusive for race. However, using the critical reflection methodology resulted in clarifying which changes could be feasibly made for our specific context (Fig. 3.4).

**Learning Outcomes and Race**: We noticed that the observable anatomical differences that we generally associate with people of different races were never mentioned within our anatomy curriculum. We therefore set out to investigate why this was using critical reflection. It is now fairly well known that race is only a social construct (Green and Smith 1983) and that there is no genetic basis for it (Reich 2018). But within healthcare, we often speak of health inequities in the context of race. Some examples are the vast amount of literature describing COVID-19 disproportionately affecting those belonging to racial minority groups more severely (Golestaneh et al. 2020); or of black women having poorer maternity outcomes and an increased prevalence of maternal mortality (Valerio et al. 2023). Yet another example is the increased prevalence of type 2 diabetes in the South Asian population (Gholap et al. 2011). Race is therefore consid-

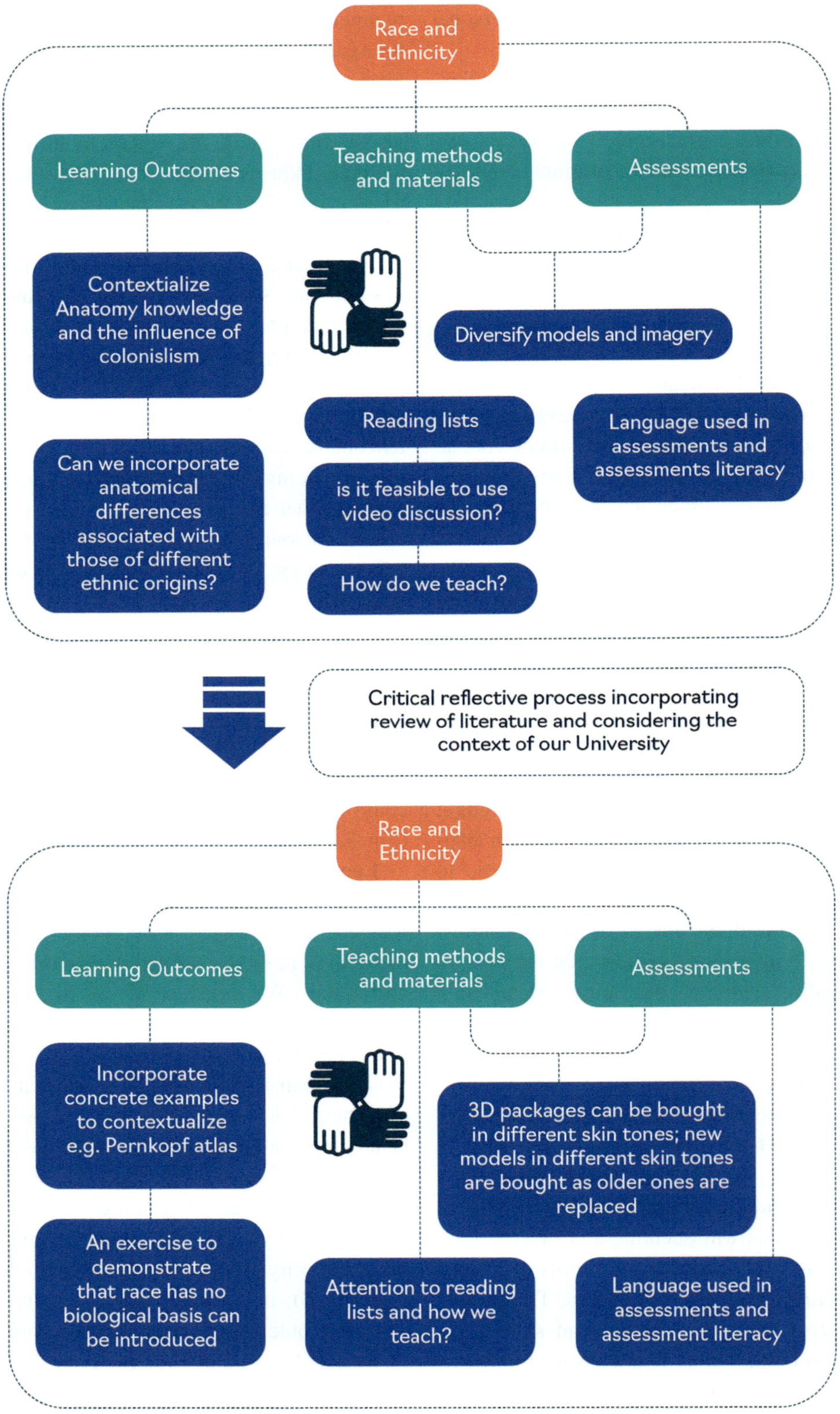


**Fig. 3.4** Making our curriculum inclusive for race and ethnicity. The figure demonstrates how initially the various domains of a curriculum are analysed concerning race inclusion. Next, critical reflection is used to clarify what changes can feasibly be made within the domains of our curriculum

ered a normal part of the decision-making process in healthcare. Witzig (1996) proposed that other factors such as socio-economic, cultural and environmental elements are not taken into consideration when race is used in this manner and called the phenomenon '*medical reductionism*'.

The origins of the concept of race are frequently traced back to Carl Linnaeus and his classification of humans into four '*varieties*' in his seminal work *Systema Naturae* in 1735 (Müller-Wille 2021). Later, an American anthropologist, Hooton (1926) listed traits specific to three '*varieties*' of skull. These misguided attempts may have paved the way to assign a scientific basis for race. But with the advent of the discipline of genetics, we now know that there is no genetic basis for race (Reich 2018). External features commonly associated with race such as skin and hair colour, and the shape of the eyes or nasal apertures are independent traits. Individuals with similar features can be traced back to having a common ancestry from particular geographic regions. It is therefore believed that the inheritance of these traits was a result of evolutionary adaptation to geography (Christensen et al. 2019; Dupré 2008; Jorde and Wooding 2004). For example, having a short and wide nose helped cool the air for those closer to the equator, whereas a narrow nose is more efficient in warming cool air. Similarly, the strength of pigmentation of the skin depended on the exposure to UV rays, and therefore populations living in areas with such high exposure evolutionarily inherited the capacity to produce more melanin. The two traits (skin colour and nasal aperture shape) are unrelated and therefore assuming that black people have a short and broad nose just because they are black is incorrect (Cheng 2008).

An interesting and eye-opening discussion to have with students might be that anatomical differences are a result of ancestry due to evolution and natural selection. This will drive home the point that there is no biological or anatomical evidence for race.

Furthermore, in today's world of travel and migration, people frequently do not share a homogenous ancestry. When someone has parents from different geographical ancestries, morphological traits are expressed differently. This makes a '*trait-list*' approach to determining ancestry problematic (Christensen et al. 2019).

Many anatomy departments house historical bone collections that may include skulls. An exercise can be created to demonstrate how sex might be identified, followed by a discussion about ancestry and traits associated with an individual's '*race*'.

**Teaching Materials: Imagery**: Given how visual anatomy is, one would expect that images used in textbooks would naturally represent the diversity amongst our population, but this is far from the truth. Parker et al. (2017) analysed the images from 17 anatomy textbooks and concluded that 78% of the images were of white bodies, 16.8% of black, 3.6% depicted Asian bodies and 1.2% of South Asian. A similar study by Louie and Wilkes (2018) considering three anatomy texts and one clinical skills book showed that 74.5% of the images depicted a light skin tone, 21% medium skin tone and 4.5% a dark skin tone. An internal audit of our teaching resources (lectures, workbooks and handouts) carried out in 2022 yielded very similar results. Efforts should be made to diversify images in our teaching material, perhaps by building a database of images from consenting individuals (in higher education, we have a vast population of diverse students some of whom might volunteer). At our institution, we have started this process and have thus far gathered some images of individuals with varying skin tones. These images have been incorporated into living or surface anatomy teaching and assessment. It should be mentioned that this is just one of the steps needed to make the curriculum truly inclusive. Merely adding some photos of racialised individuals to '*tick a box*' of inclusion is tokenistic. However, if this is just one of several steps, then it will give staff and students confidence in our true intentions.

**Cadavers and Anatomical Models**: Various constraints such as the geographical location of the anatomy department and therefore the diversity of the local population and religious beliefs influence the diversity of our donor population. As anatomy staff or students, we have little or no control over this. Research is being carried out into belief processes and how they influence the diversity of our donors (Smith et al. 2023). However, given that one of the first steps in dissection is the removal of skin, we wonder if much can be gained from investing resources towards this. There is a critical difference here that emerged from our discussions using the critical reflection methodology, about why it was important in healthcare education for students to use models of diverse skin tones in healthcare courses such as clinical skills (where students learn skills to assess and examine patients and perform basic procedures like venepuncture) as opposed to anatomy. In the former, not being able to recognise a vein in a dark-skinned individual means the student would not have achieved clinical competence. However, the gross and histological anatomy remains the same irrespective of skin colour. Therefore, the availability of models and cadavers with different skin tones is more an issue of representation rather than clinical competence and patient safety. We therefore argue that when resources are limited there is not much to be gained from using these limited resources trying to increase the diversity of bequeathals. Physical anatomical models in different skin tones have long been available (e.g. SOMSO® models have been available since the 1980s) but are still not as common as the white skin-tone model, which is more or less the default. The authors of this paper, however, believe that using white skin-toned models to teach anatomy does not raise the potential of clinical incompetence. Therefore, new or replacement models can be considered in different skin tones, and we suggest that departments not be tempted to replace existing perfectly useable models as a tokenistic gesture towards inclusion. Online 3D models in different skin tones are now available, and this can be an easier step in many instances to increase representation within anatomy teaching resources.

**Reading Lists**: Like imagery in teaching material, our internal audit revealed that our reading lists were predominantly white. Adewumi et al. (2022) describe a staff–student collaborative project that resulted in the diversification of reading lists in social sciences and humanity. We believe that undertaking a similar project could result in the diversification of our reading list. University libraries play a role, and we suggest that they are excellent additions to the team undertaking the project to diversify reading lists (Wilson 2022).

**How We Teach**: In an increasingly diverse classroom, do we spend time thinking about our teaching pedagogy and whether this suits everyone? Students from different parts of the world are likely to have adopted different learning preferences. Reflecting on personal experience, when author AV first began teaching in the UK, not long after completing her medical education in India, the emphasis placed on a problem-solving approach and the lack of emphasis on memorisation of what she now recognises as esoteric facts did take some getting used to. As Felix and Friedberg (2019) said, as educators we have the responsibility to deeply reflect on how we teach, and the pedagogies we use as an essential first step to be truly inclusive of all our students. We suggest study groups to help international students develop important study skills, and where possible an intentional effort to ensure that groups for small group teaching have a mix of students of different ethnicities. **Assessment:** A variety of assessment instruments might be prescribed for different courses and programmes. However, looking at any type of assessment through an inclusive lens can highlight ways in which an educator could make it more inclusive for race. One commonly used format of anatomy assessments is the spotter-type objective assessment. This may include questions on surface or living anatomy. Like with teaching resources, educators could diversify the imagery used here. Assessing applied or clinical anatomy might include an image of a clinical sign asking students to correlate this with the anatomical defect and might involve the use of open-source images. The study by Charani et al. (2023) is worth mentioning here. The authors of this study analysed the use of imagery in global health and found a disproportionally high representation of women and children from low- and middle-income countries suggesting a power imbalance. Images used to examine students on clinical anatomy may frequently be from patients from racialised communities. We urge anatomists to exercise caution, however. It might seem like a '*quick fix*' to use images of patients of colour to test clinical anatomy (e.g. images of embryological defects in Asian children can be sourced online). Increasingly, however, there is realisation of a biased approach to using imagery in healthcare and global health (Charani et al. 2023). We therefore urge the reader to critically appraise the source of the images that demonstrate an aspect of clinical anatomy and ensure that these are ethically sourced, with appropriate consent.

The language of assessment in anatomy is another area that requires attention. A diverse community of students introduces the co-existence of diversity in '*cultural capital*' but often with one majority or dominant culture (MacKinnon and Manathunga 2003). Any assessment that directly or indirectly requires the proficiency of the student in this dominant culture may give an unfair advantage. Some examples include assessment questions that use English words requiring a high level of language proficiency, or that present scenarios that presume a knowledge of local tradition, culture, film or television.

With the written report or essay format of assessment, it is also important to consider students' understanding of citing literature and plagiarism (Song-Turner 2008). While the study discusses this predominantly from the viewpoint of international students, this holds true essentially for any student with a range of study skills, language proficiency and socio-economic status (Fig. 3.4).

As assessment drives learning (Wormald et al. 2009), educators need to pay particular attention to this domain of the curriculum. However, the path may not be straightforward. If we are to avoid the traps of tokenism, we need to attempt to understand the true nature of the problem that results in differential attainment. We need to put the spotlight on our assessment tools and ensure that they do not inadvertently discriminate. These changes will necessarily take longer to implement than others. Deep reflection can be helpful to guide us towards making the right choices for each educator's particular context.

## 3.10 Conclusion

The path to making curricula truly inclusive is challenging (Stein et al. 2021). Within anatomy as a discipline, we have merely begun on this journey. There is much to do, and sustained effort will be required. One of the challenges for educators is to know where to start and how to make comprehensive change. Our framework provides the flexibility for staff and students alike to consider this through a process of critical reflection. However, at times, the process will raise challenging and often uncomfortable questions.

As bell hooks said (2014) '*... there can be, and usually is, some degree of pain involved in giving up old ways of thinking and knowing and learning new approaches*'.

Discussions around race, religion and other characteristics might also raise uncomfortable questions, and at times emotive sentiments, to the surface (DiAngelo 2015). Therefore, introducing such topics cannot be taken lightly and all aspects of how to facilitate such discussions, and how to provide support to students must be well considered. Where there is the likelihood of sensitive discussions or debates arising during teaching, educators need to create ground rules for discussion and a safe space for these challenging discussions to be had (Brigley Thompson 2020).

We also recommend wider discussion of the lesson plan with other staff, including seeking the opinion of any with a role pertaining to inclusion within the HEI. This might be helpful in ensuring the right tone is adopted and contingencies prepared for.

Ultimately, we believe the process of making anatomy truly inclusive is a journey that begins with small steps. We hope that the framework and our reflection of the process will be helpful in starting readers off on their own journeys by providing a roadmap to make the anatomy curriculum inclusive in their setting.

**Acknowledgements** The authors are deeply grateful to the individuals who facilitate anatomy teaching at the University of Aberdeen with their generous body donations.

## References


Abu Moghli M, Kadiwal L (2021) Decolonising the curriculum beyond the surge: conceptualisation, positionality and conduct. Lond Rev Educ 19(1):1–16

Acevedo SM, Nusbaum EA (2020) Autism, neurodiversity, and inclusive education. In Oxford research encyclopedia of education

Adewumi B, Bailey LR, Mires-Richards E, Quinlan KM, Agyeman E, Alabi A, Jeyasingh M, Konadu-Mensah C, Lavinière W, Mighton P, Shortridge T (2022) Cross-disciplinary, collaborative and student-led: developing a change process for diversifying reading lists. Lond Rev Educ 20(1):n1

Advance HE (n.d.-a) Athena Swan charter. https://www.advance-he.ac.uk/equality-charters/athena-swan-charter. Accessed 27 Sept 2023

Advance HE (n.d.-b) The race equality charter. https://www.ecu.ac.uk/equality-charters/race-equality-charter/about-race-equality-charter/. Accessed 27 Sept 2023

Aminpour P, Gray SA, Singer A, Scyphers SB, Jetter AJ, Jordan R, Murphy R Jr, Grabowski JH (2021) The diversity bonus in pooling local knowledge about complex problems. Proc Natl Acad Sci U S A 118(5):e2016887118

Anatomy Act (2 & 3 Will. 4 c. 75: 1832) (n.d.)

Atlas MC (2001) Ethics and access to teaching materials in the medical library: the case of the Pernkopf atlas. Bull Med Libr Assoc 89(1):51

Baltaru RD (2022) The rise of agentic inclusion in the UK universities: maintaining reputation through (formal) diversification. Stud High Educ 47(1):229–242

Bamshad M, Wooding S, Salisbury BA, Stephens JC (2004) Deconstructing the relationship between genetics and race. Nat Rev Genet 5(8):598–609

Beard E, Dixon S, Rai T, Hayward G (2021) Ethnicity and socioeconomic status: missing in research means missing in clinical guidance. BJGP open 5(3)

Bennett RE (2018) Capital punishment and the criminal corpse in Scotland, 1740–1834. Springer International Publishing, Cham, p 159

Bharambe V, Puranam V, Manvikar PR, Bajpayee P (2019) Anatomy acts in India: a review. Eur J Anat 23(6):469–477

Bowden B (2021) Eurocentrism and civilization. In: Routledge handbook of historical international relations. Routledge, pp 162–170

Brennan J, Naidoo R (2008) Higher education and the achievement (and/or prevention) of equity and social justice. High Educ 56:287–302

Brenner E, Bleys RL, de Caro R, Catereniuc I, Chirculescu AR, Destrieux C, Eppler E, Filgueira L, Kachlik D, Kiss P, Lee C (2024) The legal and ethical framework governing body donation in Europe—2nd update on current practice. Ann Anat 252:152195

Brigley Thompson Z (2020) From safe spaces to precarious moments: teaching sexuality and violence in the American higher education classroom. Gend Educ 32(3):395–411

Bunce L, King N, Saran S, Talib N (2021) Experiences of black and minority ethnic (BME) students in higher education: applying self-determination theory to understand the BME attainment gap. Stud High Educ 46(3):534–547

Carroll J (2008) Assessment issues for international students and for teachers of international students. The enhancing series case studies: international learning experience, pp 1–13

Carter PM, Flannagan CA, Reed MP, Cunningham RM, Rupp JD (2014) Comparing the effects of age, BMI and gender on severe injury (AIS 3+) in motor-vehicle crashes. Accid Anal Prev 72:146–160

Cecchinato G, Foschi LC (2017) Flipping the roles: analysis of a university course where students become co-creators of curricula. Teaching and Learning Together in Higher Education 22:1–9

Charani E, Shariq S, Pinto AMC, Farooqi R, Nambatya W, Mbamalu O, Abimbola S, Mendelson M (2023) The use of imagery in global health: an analysis of infectious disease documents and a framework to guide practice. Lancet Glob Health 11(1): e155–e164

Cheng KC (2008) Demystifying skin color and "race". In: Racism in the 21st century: an empirical analysis of skin color, pp 3–23

Christensen AM, Passalacqua NV, Bartelink EJ (2019) Forensic anthropology: current methods and practice. Academic Press

Claeys-Kulik AL, Jørgensen TE, Stöber H (2019) Diversity, equity and inclusion in European higher education institutions. Results from the INVITED Project. European University Association ASIL, Brussel, p 51

Cletus HE, Mahmood NA, Umar A, Ibrahim AD (2018) Prospects and challenges of workplace diversity in modern day organizations: a critical review. Holistica 9(2):35–52

Connolly SA, Gillingwater TH, Chandler C, Grant AW, Greig J, Meskell M, Ross MT, Smith CF, Wood AF, Finn GM (2018) The anatomical society's core anatomy syllabus for undergraduate nursing. J Anat 232(5):721–728

Cook-Sather A, Matthews KE (2021) Pedagogical partnership: engaging with students as co-creators of curriculum, assessment and knowledge. In: University teaching in focus. Routledge, pp 243–259

Cooms S, Muurlink O, Leroy-Dyer S (2022) Intersectional theory and disadvantage: a tool for decolonisation. Disabil Soc 39:453–468

Cronin M (2020) Anarcha, Betsey, Lucy, and the women whose names were not recorded: the legacy of J Marion Sims. Anaesth Intensive Care 48(3_suppl):6–13

Curtis NG, Scholten SC (2022) Repatriation from Scottish museums: a short report. Santander Art Cult Law Rev 2022:427

D'Arcy K (2014) The complexities of teaching 'inclusion' in higher education

Davidson JM (2007) 'Resurrection men' in Dallas: the illegal use of black bodies as medical cadavers (1900—1907). Int J Hist Archaeol 11:193–220

Deeley SJ, Bovill C (2017) Staff student partnership in assessment: enhancing assessment literacy through democratic practices. Assess Eval High Educ 42(3):463–477

DiAngelo R (2015) White fragility: why it's so hard to talk to White people about racism. The Good Men Project 9

Dupré J (2008) What genes are and why there are no genes for race. Revisiting race in a genomic age, pp 39–55

Edmondson AC, Higgins M, Singer S, Weiner J (2016) Understanding psychological safety in health care and education organizations: a comparative perspective. Res Hum Dev 13(1):65–83

Equality Act (2010). https://www.legislation.gov.uk/ukpga/2010/15/contents. Accessed 30 Aug 2023

Equality Act: guidance (2010). https://www.gov.uk/guidance/equality-act-2010-guidance#:~:text=The%20Equality%20Act%202010%20(Specific,and%20to%20set%20equality%20objectives. Accessed 4 May 2024

Equality in higher education: Staff Statistical Report 2022. https://www.advance-he.ac.uk/knowledge-hub/equality-higher-education-statistical-reports-2022. Accessed 25 Sept 2023

Evans-Winters VE, Hines DE (2020) Unmasking white fragility: how whiteness and white student resistance impacts anti-racist education. Whiteness Educ 5(1):1–16

Featherstone D (2023) The crisis in higher education. Soundings 84(84):5–16

Felix M, Friedberg J (2019) To decolonise the curriculum, we have to decolonise ourselves. Wonkhe, London

Finn GM, Danquah A, Matthan J (2022) Colonization, cadavers, and color: considering decolonization of anatomy curricula. Anat Rec 305:938–951

Fook J (2011) Developing critical reflection as a research method. In: Creative spaces for qualitative researching. Brill, Leiden, pp 55–64

Fook J, Gardner F (2007) Ebook: Practising critical reflection: a resource handbook. McGraw-Hill Education (UK)

Fry H, Ketteridge S, Marshall S (2003) Understanding student learning. In: A handbook for teaching and learning in higher education. Routledge, pp 22–36

Fyfe M, Horsburgh J, Blitz J, Chiavaroli N, Kumar S, Cleland J (2022) The do's, don'ts and don't knows of redressing differential attainment related to race/ethnicity in medical schools. Perspect Med Educ 11:1–14

Gallavan NP (2005) Helping teachers unpack their "invisible knapsacks". Multicult Educ 13(1):36

Gholap N, Davies M, Patel K, Sattar N, Khunti K (2011) Type 2 diabetes and cardiovascular disease in south Asians. Prim Care Diabetes 5(1):45–56

Ghosh SK (2015) Human cadaveric dissection: a historical account from ancient Greece to the modern era. Anat Cell Biol 48(3):153–169

Gibbs G (1988) Learning by doing: a guide to teaching and learning methods. Oxford Further Education Unit, London. https://thoughtsmostlyaboutlearning.files.wordpress.com/2015/12/learning-by-doing-graham-gibbs.pdf. Accessed 28 Sept 2023

Golestaneh L, Neugarten J, Fisher M, Billett HH, Gil MR, Johns T, Yunes M, Mokrzycki MH, Coco M, Norris KC, Perez HR (2020) The association of race and COVID-19 mortality. EClinicalMedicine 25:100455

Gomez LE, Bernet P (2019) Diversity improves performance and outcomes. J Natl Med Assoc 111(4):383–392

Gonzalez-Sobrino B, Goss DR (2019) Exploring the mechanisms of racialization beyond the black–white binary. Ethn Racial Stud 42(4):505–510

GOV UK (2014) Disability confident employer scheme [online]. GOV.UK. https://www.gov.uk/government/collections/disability-confident-campaign. Accessed 28 Sept 2023

Green DS, Smith E (1983) W. E. B. DuBois and the concepts of race and class. Phylon (1960-) 44(4):262–272

Henderson S, Horne M, Hills R, Kendall E (2018) Cultural competence in healthcare in the community: a concept analysis. Health Soc Care Commun 26(4):590–603

HESA (2022) Higher education: staff statistics: UK 2020/2021. https://www.hesa.ac.uk/news/01-02-2022/sb261-higher-education-staff-statistics. Accessed 27 Sept 2023

Hickson H (2016) Becoming a critical narrativist: using critical reflection and narrative inquiry as research methodology. Qual Soc Work 15(3):380–391

Hong L, Page SE (2004) Groups of diverse problem solvers can outperform groups of high-ability problem solvers. Proc Natl Acad Sci U S A 101(46):16385–16389

Hooks B (2014) Teaching to transgress. Routledge. pp 42

Hooton EA (1926) Methods of racial analysis. Science 63(1621):75–81

Human Tissue (Scotland) Act (2006) Section 53—Amendment of the Anatomy Act 1984. https://www.legislation.gov.uk/asp/2006/4/notes/division/2/5/1. Accessed 30 Jan 2022

Huo YJ, Binning KR, Begeny CT (2014) Respect and the viability of ethnically diverse institutions. In: Towards inclusive organizations. Psychology Press, London, pp 49–66

Innovations G (2019) Inclusive crash test dummies: rethinking standards and reference models. https://genderedinnovations.stanford.edu/case-studies/crash.html#tabs-2

Jasim S, Salam S, Lewis M, Martins RP (2021) Beyond the protected characteristics: insights from public stories of 'the lost voices' UK national campaign

Jones DG (2023) Anatomists' uses of human skeletons: ethical issues associated with the India bone trade and anonymized archival collections. Anat Sci Educ 16(4):610–617

Jorde LB, Wooding SP (2004) Genetic variation, classification and 'race'. Nat Genet 36(Suppl 11):S28–S33

Kahane CJ (2013) Injury vulnerability and effectiveness of occupant protection technologies for older occupants and women (No. DOT HS 811 766)

Kay RH, LeSage A (2009) A strategic assessment of audience response systems used in higher education. Australas J Educ Technol 25(2)

Kaye N (2021) Evaluating the role of bursaries in widening participation in higher education: a review of the literature and evidence. Educ Rev 73(6):775–797

King P (2017) Punishing the criminal corpse, 1700–1840: aggravated forms of the death penalty in England. Palgrave Macmillan, London, p 1

Knight M, Bunch K, Tuffnell D, Shakespeare J, Kotnis R, Kenyon S, Kurinczuk JJ (eds) on behalf of MBRRACE-UK. Saving lives, improving mothers' care—Lessons learned to inform maternity care from the UK and Ireland confidential enquiries into maternal deaths and morbidity 2016–2018. Oxford: National Perinatal Epidemiology Unit, University of Oxford 2020

Koh ZJ, Yeow M, Srinivasan DK, Ng YK, Ponnamperuma GG, Chong CS (2023) A randomized trial comparing cadaveric dissection and examination of prosections as applied surgical anatomy teaching pedagogies. Anat Sci Educ 16(1):57–70

Kramer B, Hutchinson EF, Brits DM, Billings BK (2019) Making the ethical transition in South Africa: acquiring human bodies for training in anatomy. Anat Sci Educ 12(3):264–271

Lerner JE (2022) "Did you get placed with all white ladies in your group?": reflections on co-creating an anti-racist social work (virtual) classroom. Soc Work Groups 45(1):46–60

LERU (2019) Equality, diversity and inclusion at universities: the power of a systemic approach. https://www.leru.org/publications/equality-diversity-and-inclusion-at-universities

Li G, Zhang J, Van Spall HG, Douglas PS, Wang Y, Sun X, Thabane L (2022) Exploring ethnic representativeness in diabetes clinical trial enrolment from 2000 to 2020: a chronological survey. Diabetologia 65(9):1461–1472

Linton S (2020) Taking the difference out of attainment. BMJ 368:m438

Lokugamage AU, Ahillan T, Pathberiya SDC (2020) Decolonising ideas of healing in medical education. J Med Ethics 46(4):265–272

Louie P, Wilkes R (2018) Representations of race and skin tone in medical textbook imagery. Soc Sci Med 202:38–42

MacKinnon D, Manathunga C (2003) Going global with assessment: what to do when the dominant culture's literacy drives assessment. High Educ Res Dev 22(2):131–144

Mahajan M, Singh MKS (2017) Importance and benefits of learning outcomes. IOSR J Human Soc Sci 22(3):65–67

Marcelin JR, Siraj DS, Victor R, Kotadia S, Maldonado YA (2019) The impact of unconscious bias in healthcare: how to recognize and mitigate it. J Infect Dis 220(Supplement_2):S62–S73

Matthan J, Cobb M, McHanwell S, Moxham BJ, Finn GM (2020) The anatomical society's core anatomy syllabus for dental undergraduates. J Anat 236(4):737–751

May H, Bridger K (2010) Developing and embedding inclusive policy and practice in higher education. Higher Education Academy, York, p 13

Mayo M (2004) Exclusion, inclusion and empowerment: community empowerment? Reflecting on the lessons of strategies to promote empowerment. In: The politics of inclusion and empowerment: gender, class and citizenship. Palgrave Macmillan UK, London, pp 139–158

McDonald A, McGowan H, Dollinger M, Naylor R, Khosravi H (2021) Repositioning students as co-creators of curriculum for online learning resources. Australas J Educ Technol 37(6):102–118

McIntosh P (1988) White privilege: unpacking the invisible knapsack. Gender through the prism of difference, 235(8)

McIntosh P (1989) White privilege: unpacking the invisible backpack. Peace and freedom 49:10–12

Meiman M, Brown NE, Hodges A (2019) Diversity, equity, and inclusion: a conceptual framework for instruction

Meyer ER, Cui D (2019) Diversity and inclusion in anatomy and physiology education, degree programs, and professional societies. HAPS Educator 23(2):396–419

Millner N (2022) Unsettling feelings in the classroom: scaffolding pedagogies of discomfort as part of decolonising human geography in higher education. J Geogr High Educ 47:805–824

Miller-Kleinhenz JM, Collin LJ, Seidel R, Reddy A, Nash R, Switchenko JM, McCullough LE (2021) Racial disparities in diagnostic delay among women with breast cancer. J Am Coll Radiol 18(10):1384–1393

Morris M (2016) Standard white: dismantling white normativity. Calif Law Rev 104:949–978

Müller-Wille S (2021) Corners, tables, lines: towards a diagrammatics of race. Nuncius 36(3):517–531

Ng IY, Lim SS, Pang N (2023) Making universal digital access universal: lessons from COVID-19 in Singapore. Univ Access Inf Soc 22(3):1073–1083

Nieminen JH (2022) Assessment for inclusion: rethinking inclusive assessment in higher education. Teach High Educ 29:841–859

Ovseiko PV, Taylor M, Gilligan RE, Birks J, Elhussein L, Rogers M, Tesanovic S, Hernandez J, Wells G, Greenhalgh T, Buchan AM (2020) Effect of Athena SWAN funding incentives on women's research leadership. BMJ 371:m3975

Pandurangam G, Gurajala S (2023) Pedagogical strategies in medical anatomy: a comprehensive review. Natl J Clin Anat 12(4):216–222

Papay M (2019) Harnessing the power of inclusive voice in your organization. https://www.forbes.com/sites/forbessanfranciscocouncil/2019/08/06/harnessing-the-power-of-inclusive-voice-in-your-organization/
Parker R, Larkin T, Cockburn J (2017) A visual analysis of gender bias in contemporary anatomy textbooks. Soc Sci Med 180:106–113
Perez CC (2019) The deadly truth about a world built for men–from stab vests to car crashes. Sat 23:08–59
Prasanna A, Miller HN, Wu Y, Peeler A, Ogungbe O, Plante TB, Juraschek SP (2021) Recruitment of black adults into cardiovascular disease trials. J Am Heart Assoc 10:1–10
Prince LR, Francis SE (2023) Barriers to equality, diversity and inclusion in research and academia stubbornly persist. So, what are we doing about it? Dis Model Mech 16(7):dmm050048
Prunty M, Bukavina L, Hallman JC (2021) Anarcha, Lucy, and Betsey: the mothers of modern gynecology. Urology 157:1–4
Reich D (2018) How genetics is changing our understanding of 'race'. The New York Times, 23
Sanchez MG, Kumar V (2022) Curriculum design in medical simulation. StatPearls Publishing, Treasure Island
Scorgie K, Forlin C (2019) Social inclusion and belonging: affirming validation, agency and voice. In: Promoting social inclusion, vol 13. Emerald Publishing Limited, Bingley, pp 3–15
Scott CL (ed) (2016) Developing workforce diversity programs, curriculum, and degrees in higher education. IGI Global, Hershey
Shalit A, Downie A, Tiwari M, Arnold K, Malek M, Akbany AF, Walker N, Xu Y, McDuff L (2021) Lucy's legacy: why Sim's speculum needs a different name. O&G Magazine 23(4)
Shankley W, Hannemann T, Simpson L (2020) The demography of ethnic minorities in Britain. In: Ethnicity, race and inequality in the UK. Policy Press, Bristol, pp 15–34
Shaver SD (2022) Anatomy, art, and the desire for divine beauty: an exploration into the creation and perpetuation of the standardized body types of early modern anatomical illustrations (Doctoral dissertation, University of Windsor (Canada))
Sims JM (1884) The story of my life, vol 10. D. Appleton, New York
Smith CD, Worsfold K, Davies L, Fisher R, McPhail R (2013) Assessment literacy and student learning: the case for explicitly developing students 'assessment literacy'. Assess Eval High Educ 38(1):44–60
Smith CF, Finn GM, Stewart J, Atkinson MA, Davies DC, Dyball R, Morris J, Ockleford C, Parkin I, Standring S, Whiten S (2016) The anatomical society core regional anatomy syllabus for undergraduate medicine. J Anat 228(1):15–23
Smith CF, Freeman SK, Heylings D, Finn GM, Davies DC (2022) Anatomy education for medical students in the United Kingdom and Republic of Ireland in 2019: a 20-year follow-up. Anat Sci Educ 15(6):993–1006
Smith CF, Munro R, Davies DC, Wilkinson T, Shaw H, Claridge K, Llewellyn S, Mc Ateer P, Ward S, Farsides T (2023) Understanding beliefs, preferences and actions amongst potential body donors. Anat Sci Educ 16(2):224–236
Song-Turner H (2008) Plagiarism: academic dishonesty or 'blind spot' of multicultural education? Aust Univ Rev 50(2):39–50
Stein S, Andreotti V, Hunt D, Ahenakew C (2021) Complexities and challenges of decolonising higher education: lessons from Canada. In: Decolonisation as democratisation: global insights into the South African experience
Stonewall (n.d.) UK workplace equality index. https://www.stonewall.org.uk/build-workplace-works-lgbtq-people/uk-workplace-equality-index. Accessed 27 Sept 2023
Sugand K, Abrahams P, Khurana A (2010) The anatomy of anatomy: a review for its modernization. Anat Sci Educ 3(2):83–93
Szelics A, Frazier S, Kerr H, Knowles J, Prosser D, Ryan L, Paterson V, Veitch N, White S (2023) Students as co-creators of an inclusive equality and diversity teaching resource: an example from life sciences. In: Collaboration in higher education, p 83
Tai J, Ajjawi R, Umarova A (2021) How do students experience inclusive assessment? A critical review of contemporary literature. Int J Incl Educ 28:1936–1953
Tamtik M, Guenter M (2019) Policy analysis of equity, diversity and inclusion strategies in Canadian universities—How far have we come? Can J High Educ 49(3):41–56
Titchkosky T (2010) The not-yet-time of disability in the bureaucratization of university life. Disabil Stud Q, 30(3/4)
Topping KJ (2005) Trends in peer learning. Educ Psychol 25(6):631–645
Tullis JG, Goldstone RL (2020) Why does peer instruction benefit student learning? Cogn Res 5:1–12
UK Parliament (2018) The public sector bodies (websites and mobile applications) (no. 2) accessibility regulations 2018
University of Aberdeen (2022) Antiracism strategy 2022–2025 [online]. p 4. https://www.abdn.ac.uk/staffnet/documents/Antiracism%20Strategy.pdf. Accessed 27 Sept 2023
University of Aberdeen (2023) Decolonising the curriculum [online]. www.abdn.ac.uk. https://www.abdn.ac.uk/about/strategy-and-governance/decolonising-the-curriculum-2581.php. Accessed 27 Sept 2023
Valerio VC, Downey J, Sgaier SK, Callaghan WM, Hammer B, Smittenaar P (2023) Black-White disparities in maternal vulnerability and adverse pregnancy outcomes: an ecological population study in the United States, 2014–2018. Lancet Reg Health Am 20:100456
Van Mol C, Perez-Encinas A (2022) Inclusive internationalisation: do different (social) groups of students need different internationalisation activities? Stud High Educ 47(12):2523–2538

Venkatesh A, Morris A (2024) The need for creation of new learning outcomes in core curricula publications on differences in sex development and transgender anatomy. J Anat 245:510
Vesalius A (1543) De Humani Corporis Fabrica Libri Septem. Officina Ioannis Oporini, Basilea
Watson WE, Kumar K, Michaelsen LK (1993) Cultural diversity's impact on interaction process and performance: comparing homogeneous and diverse task groups. Acad Manag J 36(3):590–602
Wehrli G (2010) The nuts and bolts of curriculum and assessment. J Clin Apher 26:29–46
Weldon SL (2008) Intersectionality. In: Politics, gender, and concepts: theory and methodology, pp 193–218
Wells AS (2020) Racial, ethnic, and cultural diversity across K–12 and higher education sectors: challenges and opportunities for cross-sector learning. Change: The Magazine of Higher Learning 52(2):56–61
Wilson K (2022) Decolonising library collections: contemporary issues, practical steps and examples from the London School of Economics. In: Narrative expansions: interpreting decolonisation in academic libraries, pp 225–250
Wingfield B (2017) What 'decolonised education' should and shouldn't mean [online]. https://www.fabinet.up.ac.za/publication/pdfs/3089-what__decolonised_education__should_and_shouldnt_mean.pdf. Accessed 6 Mar 2023
Winter J, Webb O, Turner R (2024) Decolonising the curriculum: a survey of current practice in a modern UK university. Innov Educ Teach Int 61(1):181–192
Witzig R (1996) The medicalization of race: scientific legitimization of a flawed social construct. Ann Intern Med 125(8):675–679
Wormald BW, Schoeman S, Somasunderam A, Penn M (2009) Assessment drives learning: an unavoidable truth? Anat Sci Educ 2(5):199–204
Xiao J, Evans DJ (2022) Anatomy education beyond the Covid-19 pandemic: a changing pedagogy. Anat Sci Educ 15(6):1138–1144
Zampieri F, ElMaghawry M, Zanatta A, Thiene G (2015) Andreas Vesalius: celebrating 500 years of dissecting nature. Glob Cardiol Sci Pract 2015(5):66. Equality in higher education: Students Statistical Report 2022 published on 16/11/2022
Zhang S (2018) The surgeon who experimented on slaves. The Atlantic, 18

**Asha Venkatesh** M.B.B.S, P.G.Dip (Ophth), M.Sc (Clinical Education), S.F.H.E.A, F.A.S. Following a brief career in Clinical Ophthalmology in India, Dr. Venkatesh now works as a Senior Lecturer in Anatomy at the University of Aberdeen, UK. In addition to teaching gross anatomy, histology and embryology, she is interested in anatomy assessment, having steered the move to platform-based assessment of Anatomy at the University of Aberdeen. She has won three University-level excellence awards for her teaching and supervision. Dr. Venkatesh is also the Equality Diversity and Inclusion lead of her School and is passionate about inclusive Anatomy education. Externally she is the Membership Officer of the Anatomical Society (the UK and Ireland) participating in the charitable actions of this learned society. Since 2022, she has represented the Anatomical Society at TEPARG.

**Aydan Rahmanova** M.B.Ch.B, B.Sc (Hon) is a Foundation Year Doctor working at Aberdeen Royal Infirmary. The work presented in this chapter is based on work performed by Aydan during her Intercalating BSc degree at the University of Aberdeen.

# Staff and Student Perceptions of Case-Based Anatomy Education

4

Chloe Kelly, Taisha Peplowska, Yasmin Elebessy, Yousr Elebessy, Emily Green, and Kayleigh Scotcher

**Abstract**

Case-based learning (CBL) is an approach used by many medical schools in the UK and around the world. Newcastle University has adopted this curriculum style, utilising fictional patient cases to contextualise all elements of teaching, including anatomy, within the first 2 years of the Bachelor of Medicine, Bachelor of Surgery (MBBS) program. There have been limited studies investigating the effect of this curriculum style on anatomy education. Previous research has found that elements of CBL strengthen connections between scientific knowledge and clinical practice, in addition to enhancing long-term retention of anatomical knowledge. In this project, student and staff perceptions of anatomy education within a case-based MBBS curriculum were explored. Additionally, perceptions of challenging anatomical subjects were collated using a mixed quantitative and qualitative questionnaire. Ninety-three responses from students in stages one to five of their medical degree at Newcastle University across both a UK and Malaysian university campus were collected. Perceptions of case-based learning were generally positive, and thematic analysis of free-text comments provided additional useful feedback. These opinions were further explored in focus groups with both staff and students. Findings on challenging topics supported the trends in existing literature, identifying neuroanatomy as the most challenging topic. This suggests that delivering anatomy education within a case-based curriculum does not affect perceptions of challenging anatomical topics amongst medical students. This project has ethical approval from Newcastle University Faculty of Medical Sciences Ethical Review Committee.

**Keywords**

Anatomy education · Case-based learning · Medical students · Case-based curriculum

Chloe Kelly, Taisha Peplowska, Yasmin Elebessy, Yousr Elebessy contributed equally to the work and are therefore assigned as joint first authors.

C. Kelly · T. Peplowska · E. Green (✉) · K. Scotcher
Faculty of Medical Sciences, School of Medicine, Newcastle University, Newcastle upon Tyne, UK
e-mail: c.m.kelly1@newcastle.ac.uk; t.m.peplowska1@newcastle.ac.uk; emily.green@newcastle.ac.uk

Y. Elebessy · Y. Elebessy
Newcastle University Medicine Malaysia, Iskandar Puteri, Johor, Malaysia
e-mail: y.a.a.i.elebessy1@newcastle.edu.my; y.a.a.i.elebessy2@newcastle.edu.my



I. D. Keenan et al. (eds.), *International Anatomical Education*, Biomedical Visualization 9,
https://doi.org/10.1007/978-3-031-91849-0_4

## 4.1 Introduction

### 4.1.1 Newcastle University Background

Newcastle University has a large medical school with approximately 370 students per year group (2023/2024) at its UK campus. The Newcastle University School of Medicine delivers two versions of the MBBS degree: an undergraduate course of 5 years (A100) and a graduate-entry, accelerated four-year program for health professionals or those with a previous degree (A101). Both programmes utilise an integrated, spiral curriculum where ideas and concepts are revisited and built upon throughout the course. As previously described (Ben Awadh et al. 2022; Keenan and Ben Awadh 2019), the A100 programme consists of two preclinical years (stages one and two), where foundational knowledge in basic medical sciences is taught, including anatomy, physiology, pharmacology, basic clinical reasoning, communication and clinical skills, as well as covering the practical and ethical principles underpinning medicine. Teaching is delivered using a variety of approaches including clinical skills tuition, hospital, and GP visits, anatomy practical sessions, small group seminars, communication skills seminars and lectures. Early clinical and community experience is embedded throughout these 2 years in the form of five clinical or community visits per year, facilitating early exposure to real patients and the clinical environment. The A101 course combines these two preclinical years into a single extended year of teaching, after which the cohort will join the third year of the A100 programme. Stages 3–5 of the programme are clinically based, with students taught from a variety of clinical and community environments located across the regional medical school. Students move around different local education providers (LEPs) within the north-east of England through a series of rotations, clerkships, assistantships and placements, experiencing a variety of clinical settings, geographical locations and medical specialties.

Newcastle University School of Medicine is paired with a sister campus in Malaysia (NUMed), which hosts smaller cohorts of approximately 100 students per year in the five-year undergraduate MBBS degree. An identical curriculum, including shared assessment, is delivered across the two campuses, and the curriculum is adapted to cover common conditions seen in both the UK and Malaysia (Guilding et al. 2021). This identical curriculum delivery provides the opportunity for UK-based students to study in NUMed, and vice versa, on an exchange programme that takes place in year 2 of the programme.

### 4.1.2 Case-Based Curriculum

Newcastle University School of Medicine introduced a new curriculum structure in 2017. Under this new structure, the first 2 years of the A100 curriculum are characterised using a case-led or case-based approach (this structure is used within the first year of the A101 programme). Case-based learning (CBL) at Newcastle University involves using a series of fictional patient cases to contextualise the teaching of basic medical sciences and encourage clinical application of knowledge at an early stage. As previously described (Keenan et al. 2023), a total of 25 cases are used: cases 1–13 in stage 1 of the programme, and cases 14–25 in stage 2. All 25 cases are delivered within the first year of the A101 course. This differs from a modular approach; in CBL, all course subjects such as anatomy, physiology, pharmacology, clinical reasoning, clinical skills and so on are integrated and delivered together within the context of a clinical case, rather than as stand-alone elements. For example, Case 3 discusses two patients with different types of kidney disease, and this is used to contextualise and deliver teaching on kidney anatomy, physiology, treatment and management of kidney disease, as well as facilitating discussion around the doctor–patient relationship and evidence-based medicine. Teaching sessions, which are often small-group in nature, link to the fictional patient case to provide consistent clinical context and relevance, and the cases generally culminate in a 'case round-up', a session usually involving clinical reasoning that encourages students to use and apply what they have learned. CBL is a well-established pedagogic approach and has been used successfully in medical schools across the globe (McLean 2016), although there is no set definition for 'case-based learning', meaning delivery and style of this curriculum structure varies between higher education institutions (HEIs). Thistlethwaite et al. (2012) suggest the following definition: '*The goal of CBL is to prepare students for clinical practice, using authentic clinical cases. It links theory to practice, through the application of knowledge to the cases, using inquiry-based learning methods*'. The CBL curriculum at Newcastle University was originally designed to align with this definition, but in practice encompasses elements of a systems-based curriculum, integrated with case-based elements, as each case deals with a particular body system. Often, CBL involves small group activities that may be self-directed or led by a facilitator, wherein students are provided with a case scenario or scenarios that relate to learning outcomes. Cases may increase in complexity as the course progresses; in anatomy, this may mean early cases focus on normal anatomy, with pathological changes covered in later cases (Rosenstand 2012).

This project focuses on the delivery of anatomy education within the MBBS case-based curriculum at Newcastle University. Anatomy has been considered a crucial element of medical curricula for hundreds of years and underpins safe clinical practice, particularly in surgery (Estai and Bunt 2016). As discussed, basic medical sciences, including anatomy, are integrated across the 25 cases at Newcastle University. Anatomy topics are delivered within the cases as shown in Fig. 4.1. The case-based curriculum structure provides clinical context for the anatomy teaching, helping students apply the anatomy they learn to a clinical scenario. At the UK campus, anatomy teaching is delivered using a variety of approaches (Keenan and Ben Awadh 2019), and for most cases is centred around the use of cadaveric material within dissecting room (DR) practical sessions. Students are given preparatory teaching in advance of DR sessions, which may take the form of flipped classroom methods utilising asynchronous online resources (such as pre-recorded videos, quizzes, workbooks or labelling games/activities), lectures or small or large group seminars. Students therefore attend practical sessions with a basic understanding of anatomical structures, which they then apply and build on using cadaveric specimens. Rather than students participating in cadaveric dissection, Newcastle University utilises prosections. Students, therefore, spend the practical sessions entirely focussed on identifying structures and enhancing their anatomy knowledge. These sessions are facilitated by anatomy teaching staff. There are also opportunities for students to engage with dissection if they wish to in later stages of the programme. At the NUMed campus, students do not have access to cadaveric material, due to logistical, legal and financial barriers in relation to body donation for anatomical examination. Teaching takes a similar structure within the case-based curriculum, wherein students are prepared for practical sessions using a variety of approaches. Anatomy practical sessions in NUMed utilise anatomical models and digital teaching methods in place of cadaveric material. Though this poses a significant variable between the two student groups, it also provides an opportunity to investigate whether student opinions on the curriculum and challenging anatomical topics are influenced by how students are taught.

Despite the use of CBL within the School of Medicine at Newcastle University and across other institutions in the delivery of anatomy teaching, there is limited literature exploring how this curriculum structure is perceived by students and educators. Some work has shown that CBL can be of benefit in encouraging students to integrate anatomical theory with function, creating connections between the anatomy, physiology and pathology they have covered, whilst also facilitating more student-centred learning (Singh and Bhatt 2011). Eseonu et al. (2013) showed that case-based anatomy teaching can greatly increase students' knowledge and confidence in anatomy, and found this approach is perceived positively by students, largely due to clearer links to clinical practice. Sangam et al. (2021) carried out a retention study on case-based anatomy

**Fig. 4.1** The MBBS programme at Newcastle University uses a case-based curriculum structure, where teaching is contextualised within fictional patient cases. Anatomical topics are taught within various individual cases across the first 2 years of the programme. A summary of the anatomical topics taught in all cases delivered in the preclinical years of the MBBS programme at Newcastle University is illustrated here

curricula and found that CBL for anatomy is significantly more effective than didactic lecture-based methods for improvement and retention of knowledge. Johnstone and Vaughan (2020) suggest that a case-based curriculum may not directly assist in learning of anatomy, but is effective in enhancing relevance to future practice and providing clinical context. However, in many of these studies, the case-based curricula being considered were designed specifically and solely for anatomy; that is, the anatomy curriculum itself is case-based. At Newcastle, the structure differs slightly, in that the curriculum uses a mainly case-based structure (albeit with systems-based elements), and anatomy teaching is built into this model. This may present unique challenges with the structure and order of anatomical topics taught as the overall curriculum has not been designed with anatomy in mind. Nevertheless, a case-based MBBS curriculum within which anatomy teaching is integrated may create novel benefits in encouraging links between anatomy and other subjects. The impacts of this type of CBL structure on anatomy learning, and the staff and student perceptions of this approach, are worth exploring further.

### 4.1.3 Challenging Topics in Anatomical Sciences

Regardless of how it is delivered, anatomy teaching is a core part of the medical curriculum. Despite this, many students find anatomy challenging, with certain topics regarded as significantly more difficult than others. Existing research into challenging anatomical topics tends to focus on student and staff perceptions as there is no way to definitively test the objective 'difficulty' of a topic. In previous studies, 'neuroanatomy' and 'head and neck' anatomy have consistently been shown to be two of the most challenging anatomical topics as perceived by students (Javaid et al. 2018). Studies investigating problem topics in anatomy consistently held neuroanatomy in the top two areas that were perceived as most difficult (Cheung et al. 2021; Hall et al. 2018; Kramer and Soley 2002; Qamar et al.

2021). Additionally, Javaid et al. (2018) found that neuroanatomy was perceived as the most difficult anatomical subject in both region- and systems-based teaching, by both medical and other health sciences students. In fact, neuroanatomy has been identified as an area of perceived difficulty at such frequency that Jozefowicz (1994) coined the phrase '*neurophobia*' to describe the common challenges faced by students when studying this topic. Pelvic anatomy has also been identified as an area of perceived difficulty (Hall et al. 2018; Kramer and Soley 2002; Qamar et al. 2021), although this is often ranked as less difficult than neuroanatomy and head and neck (Cheung et al. 2021). Broader skills in anatomical interpretation and understanding have also been identified as perceived problem areas, such as conceptualisation and visualisation of three-dimensional structures (Kramer and Soley 2002) and interpretation of cross-sectional images (Ben Awadh et al. 2022).

Previous literature suggests several reasons as to why students struggle with learning particular anatomical topics. Difficulty in visualizing structures, orientation and understanding anatomical relations have been identified as reasons underlying why certain topics are more difficult (Cheung et al. 2021; Kramer and Soley 2002). Students find visualisation to be a notable obstacle, particularly in areas of the body with multiple structures in a small space, as seen in neuroanatomy and anatomy of the head and neck. Javaid et al. (2018) suggest that certain teaching modalities could present a hindrance to visualizing structures, such as issues with mistakenly removing small vessels or nerves during dissection, or schematic drawings being overly simplified. The 'abstractness' of information can also make visualisation difficult since not all structures can be observed or be visualised in function leaving students with gaps in their understanding (Cheung et al. 2021). In addition, students found imagining structures in three-dimensions difficult but necessary, and 3D anatomy resources were perceived as an important solution to potentially tackle this challenge (Ben Awadh et al. 2022; Javaid et al. 2018; Kramer and Soley 2002; Qamar et al. 2021).

Further issues relate to the detail and complexity of taught content. In Javaid et al. (2018), the complexity of a given subject was deemed a significant contributing factor in finding certain topics more difficult than others. Certain subjects are perceived by students to be taught in too much detail (Qamar et al. 2021), an opinion echoed by some consultants who believe undergraduates should focus on understanding basic concepts and developing a solid foundation of core information in the first instance (Waterston and Stewart 2005). Interestingly, Hall et al. (2018) found that students attending an undergraduate neurology conference did not find neuroanatomy to be significantly harder than any other area of anatomy. This suggests a link between a student's interest in a topic and how difficult they find it, something which should be explored further and potentially be integrated into the way anatomy is taught.

Overload of information and lack of time is also often mentioned as a contributing factor to subject difficulty (Fitzgerald et al. 2008; Kramer and Soley 2002). Students report that even topics perceived as easier had an overwhelming amount of content to learn (Cheung et al. 2021), and lack of time dedicated to anatomy teaching is a common complaint by students (Mitchell and Batty 2009). Qamar et al. (2021) found that the most common reason for finding gross anatomy difficult was not having enough time, an issue often attributed to the 'congested curriculum' (Drake et al. 2009; Turney 2007), as modern medical programmes dedicate less time to the teaching of anatomy in comparison to more traditional medical curricula. Some say these changes can be attributed to the 'revised curriculum framework' formulated by the General Medical Council (GMC) in 1993, which reduced the amount of factual knowledge taught (Gogalniceanu et al. 2009). It is thought that due to the factual nature of anatomical subjects, reducing the amount of taught anatomy was a way to follow said framework (Turney 2007). These reductions have resulted in concern about the lack of adequate knowledge of students and graduates and the associated risks to patient safety (Waterston and Stewart 2005). Medical professionals and students have identified the need for more vertical integration of anatomy, with greater emphasis on clinically relevant anatomical teaching present throughout the whole course and not just in preclinical years (Babacan et al. 2021; Orsbon et al. 2013).

Although there is significant literature identifying the anatomical topics perceived as challenging by students, anatomy is a subject taught differently in each course, country and institution; therefore, current data exploring this is not representative of all students. Most significantly, there are many differences in the structure of MBBS curricula across previous studies. Despite this, there has been little work on how curriculum structure influences perceptions of difficult topics. In particular, there is little work exploring how case-based anatomy education may influence the perceived difficulty of anatomical topics, or how this curriculum structure may influence the areas of anatomy that students perceive as most interesting.

### 4.1.4 Project Rationale

Although there is a clear pedagogic basis for a case-based medical curriculum (McLean 2016), there are fewer studies investigating the impact of this curriculum structure on anatomy education, and the staff perceptions of this. Furthermore, there is a lack of work investigating whether a case-based curriculum structure influences perceived difficulty of anatomical topics, and whether student interest in a topic can influence this. Investigating these areas will help inform future development and design of case-based curricula for anatomy.

In this project, we aimed to evaluate both student and staff perceptions of anatomy within Newcastle's case-based curriculum, across both the UK and NUMed campuses, in order to inform future development and design of case-based curricula for anatomy. The research questions (below) were addressed through a pragmatic epistemological paradigm, as different elements of the project were suited to different methodological approaches. The extent of staff and student perceptions of anatomy teaching within a case-based curriculum were evaluated through quantitative approaches (questionnaire including Likert-scale questions), aligned with a more post-positivist stance, whereas exploration of the views and experiences underpinning these perceptions were explored through interpretivist approaches, analysing qualitative data gathered through focus groups and free-text questions. Qualitative exploration followed collection of quantitative data, following a mixed-methods sequential exploratory design (Ivankova et al. 2006). Adopting a pragmatic epistemology allowed triangulation of our results in order to fully address our research questions. Constructivist theory provided a conceptual framework for the project, wherein learners actively construct new knowledge which is built upon and influenced by previous experiences and understanding (MacLeod et al. 2022). Constructivism is important in evaluating the function and structure of a case-based curriculum; curricula should consider the role of prior knowledge and the process of knowledge construction within its design (Thistlethwaite et al. 2012).

#### 4.1.4.1 Study Goals

This project aimed to explore student and staff perceptions of anatomy education within a case-based curriculum in order to inform future development and design of case-based curricula for anatomy. The following research questions were developed:

- How do staff and students perceive anatomy teaching within a case-based MBBS curriculum?
- How do students perceive challenging anatomical topics within a case-based MBBS curriculum?
- How can anatomy education with an MBBS case-based curriculum be enhanced?

Our hypothesis was that a case-based curriculum positively influences staff and student perceptions of anatomy education. We hypothesised that both students and staff may find the sequencing of cases, and therefore the sequencing of anatomical topics, the most challenging aspect of the curriculum with regards to anatomy education. It was also thought that perceptions of challenging anatomical topics would mostly align with current literature but that there may be specific outliers which occur due to the structure of anatomy teaching within a case-based curriculum.

## 4.2 Methods

### 4.2.1 Population and Sampling

A pilot questionnaire was administered to MBBS A101 (graduate-entry) students (*n* = 22) towards the end of their extended preclinical year at the UK campus of Newcastle University. Students were invited to participate via an email invitation and advertisements during in-person anatomy teaching sessions. The pilot student questionnaire received 13 responses (response rate of 59.1%).

The final questionnaire was distributed to all MBBS A100 (undergraduate) students from Newcastle University at both the UK (*n* = 1743) and NUMed (*n* = 558) campuses via email. Both the pilot and final questionnaires were distributed as an online form. The final student questionnaire received 93 responses. There was an almost even split of respondents from the UK campus (*n* = 49) and the NUMed campus (*n* = 44). This was a response rate of 2.8% and 7.9%, respectively. It is thought that the low response rate is likely due to the questionnaire being advertised through online learning environments rather than within in-person teaching sessions. Additionally, the high workload of the students may have influenced their decision to take part in the research.

Across all student respondents, 45.2% (*n* = 42) were from the preclinical period of the programme, stages 1 and 2, and 54.8% (*n* = 51) were from stage 3 and onwards. With regards to their educational background and career aspirations, 21.5% (*n* = 20) had studied anatomy previously and 41.9% (*n* = 39) were considering a career in surgery.

Student respondents could express their interest in participating in a focus group by completing a separate online form after completing the questionnaire. All core MBBS anatomy staff, who were not involved in this research, across both the UK (*n* = 5) and NUMed campus (*n* = 5), were invited to take part in a staff focus group via email. The four focus groups took place with NUMed anatomy staff (*n* = 5), UK anatomy staff (*n* = 3), NUMed medical students (*n* = 3) and UK medical students (*n* = 4).

### 4.2.2 Questionnaire Instrument

Initially, a pilot student questionnaire was produced to validate Likert-type scale questions and help develop the final questionnaire. Expert validation was carried out on the pilot questionnaire by anatomy educators not directly involved with the project. Cronbach's alpha was also calculated after the pilot questionnaire responses had been collected. The score fell below the recommended internal consistency threshold of 0.7 as described in Tavakol and Dennick (2011). The questionnaire was edited accordingly to develop a final questionnaire which was again validated using Cronbach's alpha and scored above 0.7.

The final questionnaire was split into three sections (Table 4.1): (A) participant demographic information, (B) anatomy within a case-based curriculum and (C) challenging anatomical topics.

For section A, specific demographics were chosen, to explore whether stage, previous knowledge and/or further aspirations influenced participants' perceptions. Further demographic information was not included to avoid a lengthy questionnaire and for data protection purposes.

Section B included a seven-point Likert-type scale to generate rich data. This contained both negative and positive statements to create a balanced scale which helped to avoid directional bias (Weijters et al. 2010). Free-text questions were included to gain deeper insights of student perceptions.

The 'interesting' and 'challenging' questions in section C were chosen to help better understand

**Table 4.1** Overview of the questions within the final student questionnaire. The author(s) comments provide explanations of why that question type and/or options were included in the research

| Section (Item type) | Topic | Questionnaire item(s) |
|---|---|---|
| Part A | Participant demographic information | This section asked the following questions:<br>• Are you a student at the UK or NUMed campus?<br>• What is your stage (year) on the course?<br>• Have you studied anatomy before?<br>– This was accompanied with the following description: 'This question is asking if you have studied anatomy at a degree level before (e.g. BSc, MSc). Please do not include any compulsory modules completed at GCSE/A-level or equivalent levels'<br>• Are you considering a career in surgery? |
| Part B (i) (Likert-type) | Anatomy within a case-based curriculum | This section asked participants to select the extent to which you agree with the following statements (1 = strongly disagree and 7 = strongly agree):<br>• I would like more hours allocated for in-person anatomy teaching within each case<br>• There is enough time allocated for asynchronous/self-directed anatomy teaching within each case<br>• Learning anatomy within cases helps me make connections between the anatomy, physiology and pathology<br>• Learning anatomy within cases helps me make connections between the anatomy and clinical applications<br>• Learning anatomy within cases helps me make connections between different anatomical topics<br>• Learning anatomy within cases prepares students for clinical practice<br>• Learning within cases works better for other subjects (e.g. pathology, pharmacology, communication skills, etc.)<br>• The anatomy is taught in an order that is easy to follow<br>• I find learning anatomy within cases challenging<br>• I enjoy learning anatomy within cases |
| Part B (ii) (free-text) | Anatomy within a case-based curriculum | Participants were then asked two free-text questions:<br>• What do you think are the main benefits of learning anatomy in a case-based curriculum?<br>• What do you think are the main drawbacks of learning anatomy in a case-based curriculum? |

| | | |
|---|---|---|
| Part C (i) (drop down) | Challenging anatomical topics | Participants were asked to select the area of anatomy they find (1) most interesting, (2) least interesting, (3) most challenging, (4) second most challenging and (5) least challenging. Participants were asked to select from the following topics:<br>• Cardiovascular<br>• Urinary tract<br>• Respiratory<br>• Gastrointestinal tract<br>• Pelvis and perineum<br>• Neuroanatomy<br>• Head and face (not including the brain)<br>• Eye<br>• Neck<br>• Back and spine<br>• Upper limb<br>• Lower limb<br>• Hip |
| Part C (ii) (drop down) | Most challenging anatomical topic and subtopics | Participants were directed to a further subsection based on the 'most challenging' area of anatomy they selected in the previous subsection. They were asked to select which specific topics within the area of anatomy they found challenging. |
| Part C (ii) (free-text) | Most challenging anatomical topic and subtopics | Participants were asked one free-text question:<br>• Explain why you selected this area of anatomy as the topic you find most challenging |

whether there is a link between students' interest and their perceived difficulty of a topic. The 'second most challenging topic' question was included to gain rich data beyond what is evidenced in current literature. The list of anatomical topics from which participants could choose was designed to link directly to the anatomy taught in each case within the preclinical years of the MBBS programme. For example, lower limb and hip are taught in separate cases and are therefore listed as separate topics. This decision aimed to align the questionnaire structure with that of the curriculum in order to help better understand the influence of case-based curricula in students' perceptions of anatomical topics.

### 4.2.3 Focus Groups

Focus groups are an established data collection approach within the field of medical education (Stalmeijer et al. 2014). Four focus groups were designed using themes established from our questionnaire responses to further explore perceptions of anatomy education within a case-based curriculum. These were arranged for NUMed anatomy staff, UK anatomy staff, NUMed medical students, and UK medical students. Staff participants were recruited directly via email. Students were recruited via responses to an additional questionnaire item which allowed participants to express their interest in taking part in a focus group. Due to time differences, across-campus focus groups involving both NUMed and UK participants could not logistically be arranged. However, participants were asked questions based upon the same script to ensure parity between the groups. All focus groups took place online using a video-calling software programme.

Institutional ethical approval for this work was obtained from the Newcastle University Faculty of Medical Sciences Ethics Committee (reference number: 2133/12881). All participants were provided with a participant information sheet and were required to give written consent prior to undertaking any aspect of the research project. No identifiable personal data was collected. All primary data was securely stored and shared only among the researchers.

### 4.2.4 Data Analysis

All quantitative questionnaire data were analysed using SPSS (IBM Corp. Armonk, NY, USA). A mean score for each Likert-type scale question was calculated for students. These data were compared between cohorts and participant demographics using non-parametric testing (Mann–Whitney U test). All 93 student responses for the free-text questions in section B were thematically analysed.

Only responses from students in stages 3–5 ($n = 51$), who would have completed the full programme of taught anatomy in the preclinical years, were analysed in Section C using the same methods stated above.

Focus groups responses were recorded and transcribed. The focus groups were held online via video conferencing software, with integrated automatic transcription. Transcripts were independently double-coded and thematically analysed as a 'student group' and 'staff group', following the protocol described by Braun and Clarke (2006).

## 4.3 Results and Discussion

### 4.3.1 Perceptions of Anatomy in a Case-Based Curriculum

#### 4.3.1.1 Quantitative Questionnaire Data

Likert-type question data reflected a generally positive view of anatomy within the case-based curriculum at Newcastle. Likert-type question data for student respondents is summarised in Fig. 4.2.

Students ($n$ = 93) tend to perceive the case-based curriculum at Newcastle as positive for their anatomy learning, with 78.5% ($n$ = 73) of student respondents agreeing that they enjoy learning anatomy within cases. 89.2% ($n$ = 83) of student respondents agreed that learning anatomy

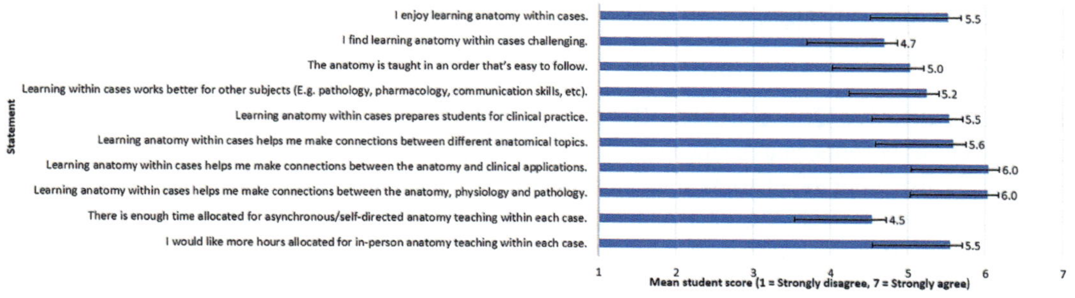


**Fig. 4.2** Bar chart showing the mean score of all student responses to the Likert-type scale questions of the questionnaire. Standard error bars are also present. Students were asked to select the extent to which they agree with a series of statements: (**a**) I enjoy learning anatomy within cases; (**b**) I find learning anatomy within cases challenging; (**c**) The anatomy is taught in an order that is easy to follow; (**d**) Learning within cases works better for other subjects (e.g. pathology, pharmacology, communication skills, etc.); (**e**) Learning anatomy within cases prepares students for clinical practice; (**f**) Learning anatomy within cases helps me make connections between different anatomical topics; (**g**) Learning anatomy within cases helps me make connections between the anatomy and clinical applications; (**h**) Learning anatomy within cases helps me make connections between the anatomy, physiology, and pathology; (**i**) There is enough time allocated for asynchronous/self-directed anatomy teaching within each case; (**j**) I would like more hours allocated for in-person anatomy teaching within each case. All responses were converted into number form (strongly disagree = 1, disagree = 2, somewhat disagree = 3, neither agree nor disagree = 4, somewhat agree = 5, agree = 6, strongly agree = 7)

within cases helps make connections between the anatomy and clinical applications, with 79.6% ($n$ = 74) agreeing or strongly agreeing with this statement. 78.5% ($n$ = 73) of students also agreed that learning anatomy within cases prepares students for clinical practice. This is consistent with previous studies that show that CBL curricula may help strengthen connections between the core topic investigated and their clinical relevance (Garg and Bhanwra 2022; Johnstone and Vaughan 2020). Similarly, the CBL approach at Newcastle University was viewed positively in terms of encouraging students to make links between anatomy and other topics, such as physiology and pathology; 86.0% ($n$ = 75) of student respondents agreed this was the case. However, despite the positive responses regarding the role of case-based anatomy in creating links between clinical practice and other topics, 63.4% ($n$ = 59) of student respondents agreed that they find learning anatomy within cases challenging. This may be due to sequencing challenges; although 75.3% ($n$ = 70) of students agreed the anatomy is taught in an order that is easy to follow, this drops to 51.6% ($n$ = 48) when only considering students who agreed or strongly agreed with this statement, and only 11.8% ($n$ = 11) respondents selected that they strongly agreed with this statement. Challenges to learning anatomy within cases may also be due to more global issues within anatomy and medical education (Turney 2007). The data reflect the observation that students perceive there is a lack of sufficient time for anatomy learning, which is an issue facing many medical schools (Mitchell and Batty 2009). Of the student respondents, 31.2% ($n$ = 29) disagreed that there was enough time allocated for asynchronous or self-directed anatomy teaching, and 81.7% ($n$ = 76) agreed that they would like more hours allocated for in-person anatomy teaching within each case. This is consistent with existing work which describes the reduction in anatomy teaching hours across medical schools (Singh et al. 2015). Lack of anatomy teaching time is a serious issue within medical education, as studies have shown graduates are leaving medicine with potentially inadequate anatomical knowledge, leading to concerns over patient safety (Brunk et al. 2017; O'Keeffe et al. 2019). Despite perceived challenges with the time dedicated to anatomy teaching, it is possible that the case-based approach may have a positive impact on students' clinical anatomy knowledge, as this data shows students appreciate the clinical relevance of their anatomy teaching when delivered in this curriculum style.

#### 4.3.1.2 Qualitative Questionnaire Data

The perceived benefits and challenges of the case-based curriculum for anatomy education are explored further through thematic analysis of responses to free-text questions. Emergent themes from the student qualitative data are explored below.

##### 4.3.1.2.1 Questionnaire Theme 1: Links to Other Topics

Gross anatomy has important links to other topics within a medical degree programme, such as the correlation between structure and function in physiology, the role of abnormal anatomy in pathology, the development of anatomy through embryology and the microscopic anatomy within histology (Louw et al. 2009). It is crucial that students are able to comprehend these links, and it appears the CBL curriculum is encouraging this link-making process.

Students mentioned how the case-based curriculum structure facilitated better link-making between the anatomy and other topics, such as physiology and pharmacology. They felt experiencing these topics in tandem, as opposed to within sequential modules, allowed them to appreciate how anatomy links to function, and how structures are underpinned by their physiology.

*'Learning anatomy alongside the relevant physiology, pathophysiology, pharmacology etc. helps to consolidate learning and gives you a better understanding of all of the subjects.'*

*'Helps with the case at the time to have a basis of the anatomy affected by pathology.'*

*'Learning anatomy in a case-based curriculum helps to link the anatomy to the rest of the topics (i.e. physiology and anatomical pathologies) to give a more complete image of their relevance.'*

*'The main benefit in my opinion is the link between the case and the anatomy which helps relate what we study in other classes to anatomy classes whereas if we studied anatomy just randomly we would not be able to link or relate what we learn in anatomy lab to what we learn in other classes this well.'*

Previous work has discussed improved link-making to other basic sciences (Fortun et al. 2017). Facilitation of broader understanding of how anatomy sits alongside the other medical sciences is a beneficial feature of style of case-based delivery at Newcastle University. By teaching all topics within a case-based structure, students perceive their learning as more holistic, allowing them to consolidate learning of several topics and better understand how they fit together. CBL has also been shown to be effective for enhancing knowledge acquisition and retention when integrating concepts from different disciplines (Fortun et al. 2017).

##### 4.3.1.2.2 Questionnaire Theme 2: Links to Clinical Practice

Students appreciated the role of the cases in encouraging them to make connections between the anatomy and relevant clinical applications. Students noted that learning anatomy in this way enabled them to link anatomy to clinical conditions, apply their knowledge to simulated patients and gain a more thorough understanding of the anatomical basis for certain clinical examinations:

*'It allows us to understand the reasoning behind the examination techniques we learn to assess patients.'*

*'Helps understand why things happen, especially for MSK and Neuro conditions. Otherwise, you would just be rote learning facts with no grounding.'*

*'Being able to link Anatomy to pathology, physiology and Clinical practice and apply that to a simulated patient. Very useful.'*

*'All the teaching is integrated into one case so there is clear understanding of the clinical relevance of the anatomy teaching.'*

This reflects findings from previous studies; it appears well known that a major benefit of a case-based curriculum is that clinical application is integrated throughout the programme of study (McLean 2016; Srinivasan et al. 2007; Thistlethwaite et al. 2012; Zhang et al. 2023). A common challenge of many medical programmes is that anatomy is delivered early in the programme, prior to any prolonged clinical exposure, meaning students may be less aware of the clinical importance of anatomical structures (Babacan et al. 2021; Jurjus et al. 2014; Orsbon et al. 2013). By delivering anatomy in direct relation to clinical cases, students are encouraged to link anatomical structures to clinical impact. This is important as students are more likely to effectively engage with anatomy teaching if they can appreciate its role in clinical practice, for example, by linking anatomical structures to clinical examinations (Boon et al. 2002; Mcnamara and Nolan 2021). It has been shown that students do recognise the importance of anatomy for clinical practice (Priyadharshini et al. 2019), but these results suggest that educators should not assume students will make their own connections between anatomy and clinical applications without clear guidance or support. In an already crowded curriculum (Turney 2007), emphasis should be placed on clinically relevant anatomy (Leveritt et al. 2016) to produce safe and competent graduates (Waterston and Stewart 2005). This should be considered when delivering anatomy teaching in all curriculum types, not only within case-based programmes.

Vertical integration of anatomy is an ongoing challenge in anatomy education more widely (Babacan et al. 2021; Orsbon et al. 2013), as is the level of post-graduate knowledge of anatomy (Waterston and Stewart 2005; Singh et al. 2015), and encouraging students to make connections to clinical practice at an early stage might be one way to tackle this (Baker 2021).

#### 4.3.1.2.3 Questionnaire Theme 3: Chronological Limitations

This theme relates to the challenges in the order and segmentation of the anatomy within the curriculum structure at Newcastle. As many anatomical topics are only taught once, students sometimes felt there were cases that would have benefitted from further anatomical teaching and found it illogical that related anatomical topics were taught elsewhere. Some discussed how they would prefer the anatomy to be spread more evenly across the cases to help workload management:

*'Potentially less linkage to different systems when we are studying one system at a time- e.g., mouth and throat anatomy could come up in a digestive case and in a respiratory case and it can be difficult to link the separate bits of teaching together.'*

*'While at the start of the cases there was anatomy with most cases there was a large gap from the visceral anatomy before we had to learn neuro and limb/ muscular anatomy which were covered in the last few cases. It may have helped to have spread out the limb/ muscle anatomy in other cases as was quite a burden to learn all at once.'*

Although students recognised that case-based learning assisted with link-making between anatomy, other topics and clinical relevance, some noted that it is difficult to make links between the anatomy itself:

*'The anatomy is separated but in fact the anatomy links together, so I'd prefer to learn the anatomy as one entire topic together.'*

*'Hard to correlate anatomy as a whole later on.'*

*'The anatomy is separated but in fact the anatomy links together, so I'd prefer to learn the anatomy as one entire topic together.'*

There was also discussion from students regarding how anatomy teaching is spread across 2 years, with certain topics fragmented throughout the entire timeline. Several studies have discussed the limitations of delivering anatomy education only in the early stages (i.e. first 2 years) of the program and the resulting issues relating to this lack of vertical integration (Babacan et al. 2021; Jurjus et al. 2014; Orsbon et al. 2013). Additionally, there is much variation between institutions in terms of structure and delivery of anatomy (Older 2004; Sugand et al. 2010), which presents challenges when attempting to review and improve anatomy curricula, as existing approaches are so variable. Development of the core regional anatomy syllabus for undergraduate medicine by the Anatomical Society in 2003, reviewed in 2007 and 2015 (Smith et al. 2016) has provided useful guidance for which topics should be taught. However, this framework has not been adopted by all anatomical educators (Smith et al. 2018). When students also struggle to retain anatomical knowledge (as discussed in a later theme), challenging topics being broken up across a long period of time may create a barrier to their global anatomical understanding.

Chronology of anatomical topics may be one of the biggest challenges of teaching anatomy within a case-based curriculum that has not been designed with anatomy in mind. Decisions regarding which anatomical topics are taught, when they are taught, and in what order, are based on the fictional patient cases and their content, rather than anatomical logic or pedagogy. One solution would be to re-work the order and context of clinical cases from an anatomical standpoint, integrating guidance from extant literature (Louw et al. 2009; Smith et al. 2016), but this would likely cause chronology issues with other topics in the curriculum. It should also be noted that selecting which anatomical subjects to teach, and in what level of detail, is a curriculum design dilemma facing many anatomy educators (Stabile 2015), not just those who are teaching within a case-based structure. Furthermore, some have argued that the generic nature of the GMC's Outcomes for Graduates (GMC 2018) makes it unclear which topics in general medical schools should be prioritizing (Sharma et al. 2019), indicating wider challenges with medical curriculum design in the UK. When developing case-based medical curricula, consideration should be given to the acceptable level of trade-off between the order of topics taught across different disciplines within the program, particularly as anatomy is not the only discipline to have concerns regarding underrepresentation within the medical curriculum (Lloyd et al. 2014; Turner and Brewster 2002). Care should be taken when deciding which disciplines to prioritise and how these interlink, in order to achieve optimum delivery across a case-based medical programme.

#### 4.3.1.2.4 Questionnaire Theme 4: Effectively Broken-Down

One of the benefits of the case-based curriculum for anatomy learning identified by students was the 'chunking' of anatomy into manageable subtopics. Students appreciated that this approach reduced the risk of cognitive overload (Sweller 2011) and made their anatomy learning more palatable, encouraging engagement and reducing fatigue that may result from intense study of anatomy in a single module.

*'It is broken down into chunks and spread out through 2 [two] years. That way, we don't study anatomy continually (less overloaded with anatomical topics).*

*'[the case-based curriculum] makes the links easier to understand and breaks it [ anatomy] down into small chunks to work from.'*

*'It breaks it down in a way that seems digestible rather than having it taught in large modules.'*

*'I think the fact that because anatomy is so complicated, it is good to break it down into smaller chunks over a longer period of time (eg the thorax, pelvis etc). If we learnt it all at once it would be quite overwhelming.'*

Interestingly, the way the anatomy is broken down and delivered over a 2-year period is seen as both a benefit and a limitation by students; they appreciate learning the anatomy in smaller doses across a longer period from a cognitive load point of view, but the chronology of topics creates issues with link-making between anatomical topics. This may be an area of potential improvement for this curriculum.

#### 4.3.1.2.5 Questionnaire Theme 5: Lack of Time

Students raised that the lack of time dedicated to anatomy teaching across the curriculum was a possible limiting factor in their learning.

*'Anatomy cannot be learned in enough depth due to lack of time.'*

*'There's not enough time to grasp the concepts, it feels very rushed because there's so many cases to get through.'*

*'Not a lot of time ends up being spent on anatomy overall.*

*'Not as much designated time to learning anatomy, takes the emphasis away from the anatomy, meaning it's not learnt very well.*

As discussed, reduced anatomy teaching hours are a challenge faced by many modern medical curricula (Mitchell and Batty 2009; Gogalniceanu et al. 2009), and this analysis shows that the case-based structure of Newcastle University's MBBS programme is not immune to this. Some students noted that the time allocated to anatomical topics did not always correlate with their perceived difficulty:

*'Some systems are more complex and larger than others, and may not be allocated more time to be taught.*

In further development of the MBBS curriculum, it would be useful to consider previous and current work that analyses the anatomical topics that students find most challenging (data on the topics Newcastle students perceive as the most challenging are discussed further in Sect. 4.3.2) and adapt the timings and number of teaching sessions accordingly. Studies suggest that time on task can result in increased subject knowledge (Verhoeven et al. 2002), so spending longer on anatomical topics perceived as difficult may be beneficial. One option would be to adopt 'spaced education' approaches for more challenging topics (Kerfoot et al. 2007), revisiting these areas more frequently and in smaller doses, which has been shown to improve knowledge retention (Raman et al. 2010). Case-based curricula may lend itself to this approach, as teaching is already fragmented into smaller time periods.

#### 4.3.1.2.6 Questionnaire Theme 6: De-Prioritisation of Anatomy

Studying anatomy in tandem with other topics was perceived as beneficial for link-making and enabling students to take a holistic view of the medical sciences they learn early in the program. This integration, particularly between anatomy and clinical applications, is an important aspect of an effective anatomy curriculum (Smith et al. 2017), not only for the development of medical students as competent future practitioners, but also to enhance their long-term retention of anatomy knowledge (Baker 2021). However, integration of anatomy teaching amongst other curriculum topics meant some students noted that this can result in anatomy feeling 'lower priority' than other areas of the curriculum.

*'Did not focus too much on anatomy as it felt like there were too many other fields to study alongside.'*

*'Usually students neglect anatomy, unless it's very important in the case e.g. MSK [in] case 23/24.'*

This may link to the lack of time dedicated to anatomy teaching as previously discussed, but more likely this relates to the mode of assessment used on the Newcastle MBBS program. Anatomy learning outcomes are assessed in a single best answer (SBA) exam paper, which integrates examination of learning outcomes on all topics across the cases. Anatomy outcomes being assessed alongside other topics naturally means fewer questions are given on anatomical topics than if a solely anatomy-focussed assessment was administered.

*'There is a lot of other content to learn. Anatomy is easy to be deprioritized when revising as it is not high yield in exams.'*

*'when it comes to cramming before exam (me and most students), we just ended up skipping anatomy all together because we cannot remember it anyways and it's too hard.'*

As previously documented, assessment is a common driver for student motivation to learn (Joughin 2009; Wormald et al. 2009) and without anatomy-focussed assessment, students may naturally de-prioritise anatomy learning, particularly in an already intense and busy curriculum. Consideration of how anatomy is assessed may be important in future development of the program.

#### 4.3.1.2.7 Questionnaire Theme 7: Difficulty Retaining Knowledge

Likely because of previously discussed issues, an emergent theme from student responses related to challenges students identified relating to retention of their anatomy knowledge. In particular, students discussed the difficulty in long-term retention of knowledge:

*'Once you have been over the anatomy from the case you don't really go over it again and if you were taught it at the start of year 1, you might not remember it years later when it is likely to be relevant.'*

*'Anatomy often has to be taught early on in the case (to make time for the rest of the topics), but this means that there is little time to commit*

*anatomical knowledge to memory before moving on to other topics.'*

*'Not being able to remember everything [is a drawback of anatomy learning within the case-based curriculum].'*

*'Forgetting a certain part when moving onto the next few cases [is a drawback of anatomy learning within the case-based curriculum].'*

This theme certainly links to 'Lack of Time', 'Chronological Limitations' and 'De-prioritisation of Anatomy' as these all issues likely feed into the challenges with long-term retention of anatomy knowledge. However, this is unlikely to be an issue that is a direct consequence of the case-based curriculum at Newcastle, as lack of anatomy knowledge of graduates from a range of medical schools is seen across the literature (Brunk et al. 2017; O'Keeffe et al. 2019). It seems the benefits of the case-based curriculum in encouraging holistic and contextual understanding of anatomy do not outweigh the challenges to anatomy education more widely (Turney 2007; Singh et al. 2015; Drake et al. 2009), furthering the argument for more time and resources for anatomy education across all medical schools.

### 4.3.1.3 Focus Groups

Ideas and themes established by the questionnaire were further explored in focus groups with students from both the Newcastle and NUMed campuses. The results of thematic analysis of these focus groups are discussed here.

#### 4.3.1.3.1 Student Focus Groups

Students shared many positive comments about their experience of studying anatomy within a case-based curriculum, but also identified pitfalls and areas for improvement. The analysis identified that students are generally satisfied with the way their anatomy teaching is delivered within the case-based curriculum.

##### 4.3.1.3.1.1 Student Focus Group Theme 1: Lack of Time

Students echoed comments made in the questionnaire free-text questions during the focus group. They commented that their schedule does not allow for enough time to study anatomy in their own time, whilst simultaneously saying that there is not enough time dedicated to teaching.

*'Having only one DR [dissection room session] per case sometimes feels a bit limiting.'*

This supports the general trend in literature. In a questionnaire sent to all Australian medical schools, nearly all students who responded (99%) felt they were not taught enough hours of anatomy (Mitchell and Batty 2009). In another study, the most common reason for finding gross anatomy difficult was not having enough time (Qamar et al. 2021). This was also reflected in our questionnaire data. Clearly there is a desire from students to have more time dedicated to their anatomy learning (Mitchell and Batty 2009), and previous work has equated greater teaching time to improved knowledge and appreciation for clinically oriented anatomy (Holla et al. 2009), but this is a difficult challenge to solve within modern medical curricula, where many new and important disciplines jostle for space amongst fundamental sciences and core clinical learning (Bergman et al. 2008).

##### 4.3.1.3.1.2 Student Focus Group Theme 2: Quality of Teaching

Students agreed that the quality of material delivered was of a high standard and appreciated the variety of study resources available to them.

*'I think the anatomy tutorials are one of the best produced in the uni.'*

*'I like how we do like we do the tutorials first, and then we had some like zoom teachings on it as well and then after that we were in the dissection rooms.'*

In addition, dissection room teaching was viewed as very beneficial. Students highlighted the positive impact that demonstrators in dissection room sessions had on their learning.

*'Within 20 min of one of the [demonstrators] just going through it on a model like it, it clicked, and I was like, "oh, that all makes sense".'*

The demonstrators were especially useful in being able to orient students around the cadaver and connect previously taught elements of anatomy to the anatomical area taught in the current session. There has been a significant increase in the number of demonstrators and recent medical graduates involved in anatomy teaching in the UK over the past 20 years (Rhodes et al. 2017; Smith et al. 2021), so it is comforting to see that the demonstrators at Newcastle University have a positive impact on student learning experiences. Student appreciation for teaching quality may also relate to the use of cadaveric material. Cadaveric dissection has long been the mainstay of anatomical teaching (Sugand et al. 2010; Turney 2007), and is known to be effective for anatomical and clinical learning (Elizondo-Omaña and López 2008; Van Wyk and Rennie 2015), but numbers of medical schools moving away from this approach are rising (Ghosh 2016; McLachlan et al. 2004). Students did also note the usefulness of models and other resources, aligning with current literature regarding use of multiple modalities for effective anatomy teaching (Estai and Bunt 2016), particularly important as students on the NUMed programme do not have access to cadaveric material. The positive student view of their teaching experiences may underlie the positive feedback that emerged through the quantitative questionnaire data and raises the question as to whether students appreciate the case-based curriculum structure itself, or whether this reflects the excellent teaching students receive on the program. The likely answer is a combination of the two; regardless of how well-structured a curriculum is, students will always rely on quality teaching delivery by staff to fully engage with their learning (Rhodes et al. 2017). Further work may be required to fully tease apart the role of the curriculum structure itself from the impact of individual teaching interventions used within it.

###### 4.3.1.3.1.3 Student Focus Group Theme 3: Fragmented Learning

Both the '*Chronological Limitations*' and the '*Challenges in Linking Between Anatomical Topics*' themes from the questionnaire were further explored during the focus group, resulting in an emergent theme regarding fragmentation of anatomy learning across the cases. As discussed, one of the perceived challenges of studying anatomy within a case-based curriculum is the dividing up of some anatomical systems across cases.

*'You can't fit like an entire system into one case. So sometimes you're having to break stuff up which can be a little jarring, especially when you have to rely on knowledge that you've learned like a year ago/months ago.'*

One of the reasons why fragmentation of anatomical systems was perceived as challenging, was that systems or topics were not always spread across subsequent cases; instead across cases months away from each other in the course. This resulted in students having to rely on retention of knowledge taught previously, to understand content taught in the current case. As identified through our questionnaire data and previous studies (Jurjus et al. 2014; Brunk et al. 2017), retention of knowledge is an additional challenge faced by students. This can create additional 'hidden' workload for students, requiring them to spend time re-learning previous content to understand the current case. This additional workload may not be taken into consideration by staff when planning and developing preparatory work, although it could be argued that enforced revision of previous topics is a useful exercise for students, helping to encourage retention in the longer term (Kerfoot et al. 2007; Raman et al. 2010). Unfortunately, however, it seems in some cases the fragmentation of topics can actually place limitations on the longer-term retention and link-making between anatomy subjects:

*'A lot of anatomy tutorials was like, "oh, there's these things, but we're going to do them like in case 20, whatever so you can forget about them now".'*

When teaching anatomy in a case-based curriculum, staff should be cautious of how they address topics that are revisited in more detail at later stages of the program, as there may be missed opportunities to encourage longer term retention in students. Additionally, staff should take care not to foster the view that certain topics are less important because they are not being studied at that moment in time, and aim for a longitudinal view of anatomy teaching delivery (McBride and Drake 2014).

Despite the drawbacks of the fragmentation of anatomical topics within a case and across subsequent cases, some students did identify benefits of this structure.

*'I really liked how, in cases 23 and 24 in second year. They broke down a lot of it a lot more than what they [had] done in previous cases.'*

They felt that in some cases this served to reduce the volume of anatomy needed to be learnt at one given time, which is something students find most challenging about studying anatomy (Cheung et al. 2021).

###### 4.3.1.3.1.4 Student Focus Group Theme 4: Links to Other Topics

A benefit of case-based learning further discussed within the focus groups was the integrated nature of the course, and how all the different disciplines are clearly related to each other. Students reported that this makes anatomy easier to study, remember and understand.

*'You know, if you did it [anatomy, physiology, and pathology] separately, it would feel like learning 3 completely different topics.'*

*'Linking the anatomy to the signs and the clinical skills and going beyond just rote learning it (…) such a good way of cementing that knowledge in my head.'*

Interestingly, students not only perceive the case-based structure as beneficial for making links to other topics, but also that these links improve their anatomical understanding. This reflects themes in existing literature, which suggests CBL provides context and encourages unifying links to be made across curriculum topics (Johnstone and Vaughan 2020). By allowing students to see the 'bigger picture' by contextualizing their learning within a case, their understanding of all topics improves (Eseonu et al. 2013).

###### 4.3.1.3.1.5 Student Focus Group Theme 5: Volume of Work

Previous work has shown that even areas of anatomy perceived as easier are still considered by students as involving an overwhelming amount of content to learn (Cheung et al. 2021). In our focus groups, students echoed these views, in particular commenting on the large volume of content they are expected to cover before practical sessions.

*'We just have so much to learn, and so much pre work, and Async [asynchronous material] to do.'*

The focus group data provided an additional insight as to why the large quantities of material are difficult to learn within the case-based curriculum; students expressed that the dissection room sessions require a lot of preparation, which can feel intense and time-consuming. Practical sessions can run for up to 2 h, which students can find difficult to engage with. Large cohort sizes mean students will have variable schedules, some with their practical sessions much earlier in the week than others, which influences their preparation time. Preparation is an important aspect of adult and active learning, encouraging students to build on prior knowledge once entering the dissection room (Hall et al. 2012). Challenges to preparation time and scheduling are the result of the curriculum being delivered within a large medical school with high cohort numbers (Smith et al. 2021) and may not be a direct result of the case-based structure.

###### 4.3.1.3.1.6 Student Focus Group Theme 6: Limited Clinical Application

Although students lauded the benefits of the curriculum in supporting link-making to other topics, some felt the scope for clinical application was limited.

*'You learn a lot of the anatomy that isn't really that applicable to like clinical presentations and pathologies.'*

It should be noted that these focus groups involved students who were early in their medical training, meaning that at this stage they may not appreciate the full scope of the clinical relevance of their anatomy learning (Sbayeh et al. 2016). However, this view was shared by some staff, so this may be something worth considering in a review of the case-based curriculum. Selecting which anatomical topics are the most clinically relevant and most important to medical students' learning is a difficult task, and a challenge faced by many anatomy programmes (Stabile 2015).

#### 4.3.1.3.2 Staff Focus Groups

##### 4.3.1.3.2.1 Staff Focus Group Theme 1: Lack of Time

Many institutions have undergone structural curricular change, often resulting in a reduction in total contact hours. Anatomy is an area of the curriculum often worst affected by this (Turney 2007; Singh et al. 2015). Within the focus groups, staff recalled the changes to the anatomy curriculum at Newcastle University introduced in 2017:

*'In the old curriculum every student had 57 hours in the dissecting room across year one and one, again, that was all before Christmas in second year. And now it's only 28 hours across year one and two. So it has, you know it's almost halved.'*

This is said to be replaced by a better quality of experience as now educators receive less students per teaching session. However, staff admit that ultimately, they would prefer more teaching time.

*'Ideally we would have 57 hours and it would be 8 students in a bay.'*

*'I'll have 6 weeks for head and neck, please, instead of 2.'*

The focus groups took place with anatomy staff specifically, and it would be interesting to gather views from educators involved in other areas of the curriculum. MBBS programmes have the challenge of balancing many crucial elements of medical education (Buja 2019). Consequently, the fluidity of the anatomy course is limited by the interplay it has with all the other elements of the programme, which in a case-based curriculum, need to be taught in the same 'block'. This poses an obstacle for anatomy educators.

##### 4.3.1.3.2.2 Staff Focus Group Theme 2: Quality of Teaching

Both students and staff share an understanding that within a case-based anatomy curriculum, the provided material and approach to teaching is case-dependent, meaning teaching must be variable and suited to each case. This requires a level of adaptability, which creates space for implementing dynamic changes based on feedback and circumstances.

*'I think one thing that we have found really useful, especially during Covid, was the how important the MLE [medical learning environment] tutorials have been and that they are embedded within the cases.'*

During the COVID-19 pandemic, it was important to reduce the amount of in-person material as much as possible, whilst providing the same quality of teaching (Alsoufi et al. 2020; Keenan et al. 2023; Ferrel and Ryan 2020). This comment suggests that teaching anatomy within a case-based curriculum was able to withstand that change, and integration of online asynchronous material works well within this curriculum. Integrating asynchronous resources in combination with in-person sessions reflects student preferences for a combination of approaches seen in previous studies (Totlis et al. 2021; Johnson et al. 2013). Staff generally felt that the teaching materials created within the case-based curriculum were of high quality and able to support learning.

##### 4.3.1.3.2.3 Staff Focus Group Theme 3: Fragmented Learning

Although staff and students believe that the case-based curriculum is successful in integrating anatomy with the rest of the MBBS course, some felt that the content within the anatomy curriculum can be spread thinly and become disjointed. Teaching staff highlight the fine balance of knowing how to compartmentalise different parts of anatomy, so that they are spaced out and presented in manageable chunks, but simultaneously interconnected to maintain the global understanding of how different structures relate to each other. Unfortunately, this is difficult to achieve, and can therefore lead to poor overall anatomical understanding.

*'Some of it can be slightly disjointed.'*

*'Cases spread out and split, and not necessarily joining up with each other.'*

Staff did note that a fragmented approach can be viewed as beneficial within a spiral curriculum (Harden 1999), encouraging revisiting and building upon prior knowledge. As previously discussed, spacing of topics so they can be revisited has been shown to be beneficial for long-term retention (Kerfoot et al. 2007; Raman et al. 2010). However, members of staff suggested a discrepancy between their opinions as educators, and their perceptions of what students think about divisions of anatomical systems across cases.

*'Things permeate through a lot of cases and again, students would call it repetition. Actually, it's more about understanding, learning and putting those pieces together.'*

*'What we're trying to do is support understanding rather than just getting students to remember a specific fact which relates to a learning outcome.'*

This suggests that although content is being revisited multiple times, the quality of student's understanding does not always extend beyond simple recall, and therefore the curriculum structure is limiting development of adequate study skills and approaches to learning. In the cognitive process dimension of Krathwohl's (2002) revised Bloom's Taxonomy, which is a framework that has helped numerous educators structure the students' learning experience, 'recognizing' and 'recalling' fall under the lowest category known as 'Remember'. Whereas higher categories include learning processes such as 'comparing', 'critiquing' and 'producing', falling under the 'Understand', 'Evaluate' and 'Create' categories, respectively. When teaching anatomy in a case-based curriculum, staff may need to take additional steps to 'point out' the links and wider anatomical context to students, encouraging them to take a step back and view the bigger picture. Staff should attempt to promote a meta-cognitive approach to learning, encouraging students to appreciate the value of reflection and repetition for their learning (Cale et al. 2023; Stambuk-Castellano et al. 2024). Currently, it seems that staff can identify the benefits of separating a system across different cases, allowing revisiting and revising of knowledge, but this benefit is possibly being lost as students are not able to utilise the case-based structure as an opportunity for consolidation.

###### 4.3.1.3.2.4 Staff Focus Group Theme 4: Links to Other Topics

Staff members expressed that teaching anatomy within a case-based curriculum integrated well with other disciplines of the MBBS curriculum.

*'it's allowing the framework for us to then develop physiology, clinical skills, everything around that fitting in some anatomy.'*

The role of the case-based curriculum in integrating different subject areas is not only a beneficial approach to learning, but also provides a good '*framework*' for teaching, allowing educators to interconnect different aspects of taught content through curriculum mapping (Mahzari et al. 2023) and scaffolding (Rogan et al. 2020). Moreover, using the cases to contextualise and apply knowledge can give students a better grasp of what is required in clinical years.

###### 4.3.1.3.2.5 Staff Focus Group Theme 5: Links to Clinical Practice

A clear perception identified during the focus group was that case-based teaching allows the content to be clinically relevant by enabling students to link anatomy to patient cases.

*'I think it gives the students a lever on which to learn their anatomy (…) they can see that in their future clinical practice they might need to know this information. I think it just gives some kind of an impetus to learn it.'*

The benefit of a case-based curriculum in supporting link-making between anatomy and clinical practice has been documented in the literature and has been discussed earlier in this chapter. However, there was an additional point raised by staff during the focus group. Staff noted that clarifying the clinical relevance of the anatomy being taught can act as a motivator to drive students to learn, a phenomenon that has been described previously in the literature (Böckers et al. 2014). In an intense curriculum with a large volume of content, students may naturally seek knowledge they perceive to be most useful and relevant to their future career. This has been suggested in the self-determination theory (SDT), which has been used to better understand human motivation. Motivation is thought to relate to an individual's need for 'competence, autonomy, and relatedness' (Deci and Ryan 2000). Burgess and Ramsey-Stewart (2014) explored these themes when investigating medical students' motivations to study anatomy throughout a whole-body dissection elective. The study found that medical students were motivated by 'the pursuit of excellence'; student participants also made a specific mention of tutors providing clinical insights as well as how they felt part of a professional 'surgical society'. Therefore, by contextualizing information within a clinical scenario, the importance of anatomy to their future practice is made clear and, thus, may help motivate students to learn.

### 4.3.2 Challenging Topics Within a Case-Based Anatomy Curriculum

#### 4.3.2.1 Quantitative Questionnaire Data

Overall students found neuroanatomy most challenging ($n = 34$), followed by pelvis and perineum ($n = 8$) and neck anatomy ($n = 4$) (Fig. 4.3).

These results reflect the current literature as several previous studies have identified neuroanatomy as an area of perceived difficulty for students (Jozefowicz 1994; Javaid et al. 2018; Matthias et al. 2013; McCarron et al. 2014; Pakpoor et al. 2014; Sanya et al. 2010; Zinchuk et al. 2010). This is concerning as a recent study found that 85.6% ($n = 202$) of medical student participants selected 'Neuroanatomy' as the reason why they find neurology as a medical specialism challenging (Jukna et al. 2023). Students also ranked neurology as significantly ($p < 0.001$) more challenging than other clinical specialisms and, consequently, did not feel confident in assessing neurological patients.

73.0% ($n = 27$) of UK campus students ($n = 37$) perceived neuroanatomy to be the most challenging, followed by 8.1% ($n = 3$) who found pelvis and perineum anatomy most challenging. Similarly, NUMed students ($n = 14$) also found neuroanatomy to be the most challenging overall at 50.0% of respondents ($n = 7$). However, pelvis and perineum anatomy followed closely, with 35.7% of students choosing it as their most challenging anatomical topic ($n = 5$). This slight discrepancy may be due to the number of UK student responses being more than double that of NUMed students, therefore being more representative. It may be relevant to explore further whether this can be attributed to the difference in teaching approach (i.e. cadaveric vs non-cadaveric anatomy practicals). However, Mitrousias et al. (2020) found that there was no statistical significance ($P = 0.477$) in students' performance on a knowledge test between the cohort utilising cadaveric upper limb prosections compared to plastic anatomical models. Likewise, there was no statistical significance ($P = 0.441$) observed in students' perceptions with each learning approach.

Students in stages 3–5 perceived neuroanatomy as the most challenging. A greater proportion of students in stages 4 and 5 thought

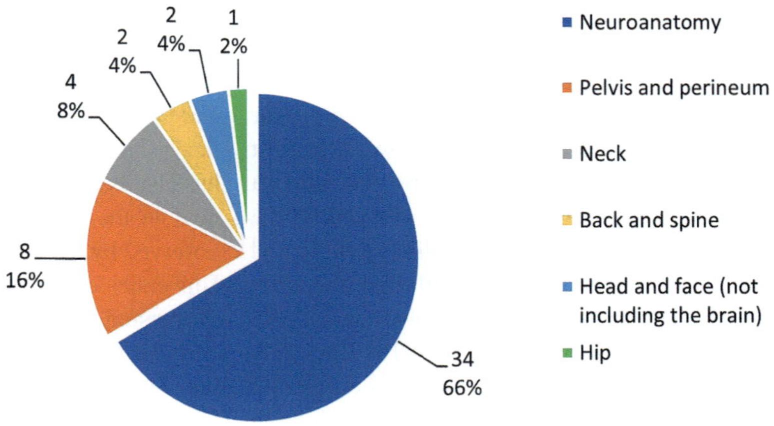


**Fig. 4.3** Pie chart illustrating the number of student responses to each area of anatomy when asked to 'Select the area of anatomy you find the most challenging' in Part C of the student questionnaire

neuroanatomy was the most challenging topic relative to students in stage 3. It may be important to explore whether this is due to retention of the subject or the increasing importance of clinical relevance as students' progress further through the programme and begin to prepare for work in a clinical environment. Previous studies have shown upper-year medical students struggle with clinical application of anatomical knowledge (Brunk et al. 2017), which can be remedied through integration of further gross anatomy teaching at later stages of the course (Lufler et al. 2020).

Previous experience studying anatomy at a higher education level also did not affect which topic students found the most difficult. In fact, a greater proportion of students who had a background of studying anatomy at a higher level found neuroanatomy to be the most challenging (78.8%) compared to students who did not have this prior experience (56.8%). However, the sample of students who did not study anatomy at a higher level was more than double that of students who have. Student views on whether they were considering a career in surgery did not affect their views on neuroanatomy as all three groups (considering, not considering and unsure) found neuroanatomy to be the most challenging, with only a slightly greater proportion of students who do not want to pursue surgery finding it harder relative to other topics.

Further breakdown of responses may suggest that students who find pelvis and perineum anatomy more challenging are less likely to perceive neuroanatomy challenging. This may be due to differences in how the teaching of these two areas is organised in the curriculum. Neurology is introduced in the first teaching block of year 1 and neuroanatomy teaching is split into four cases in the second year of the course, in contrast to the pelvis and perineum anatomy which is first introduced and taught within the first case of the second year. Students who find pelvis and perineum anatomy more difficult may therefore find that learning anatomy of a region in one case is harder than in smaller chunks. However, this suggestion is beyond the scope of this chapter.

Within neuroanatomy, students identified the anatomy of the limbic system, basal ganglia and ascending (sensory) and descending (motor) pathways as the areas of greatest difficulty (Table 4.2).

**Table 4.2** The specific subtopics of neuroanatomy voted as challenging by the students who nominated 'neuroanatomy' as the most challenging area of anatomy within the curriculum. The number and percentage of students who selected each subtopic are listed from highest to lowest number of responses

| Neuroanatomy subtopics | Number of student responses (%) |
|---|---|
| Limbic system | 24 (70.6) |
| Ascending (sensory) and descending (motor) pathways | 22 (64.7) |
| Basal ganglia | 22 (64.7) |
| Brainstem | 19 (55.9) |
| Cranial nerve nuclei | 19 (55.9) |
| Organisation of neurons (e.g. nerve plexuses, spinal nerves, primary, secondary and tertiary neurons) | 16 (47.1) |
| Cerebellum | 12 (35.3) |
| Motor and sensory peripheral nerves | 12 (35.3) |
| Cranial nerves | 11 (32.4) |
| Gross anatomy of the brain and cerebral cortex | 9 (26.5) |
| Visual pathways | 9 (26.5) |
| Divisions of nervous system (e.g. autonomic, peripheral, motor) | 7 (20.6) |
| Structure of a neuron and main cell types of the brain | 6 (17.6) |
| Meninges and venous sinuses | 6 (17.6) |
| Ventricles of the brain | 6 (17.6) |

Javaid et al. (2018) showed that of 127 students who perceived neuroanatomy to be most difficult, responses listed cranial nerves, ventricles and cerebellum as areas students most struggled with. This differs from the areas identified in our work. This variation in results suggests that students struggle with most neuroanatomy topics and there is no one area that is of particular statistical significance. Interestingly, neuroanatomy has also been highlighted to be a point of weakness in the knowledge of medical students, as observed by consultants, in a study conducted at University of Aberdeen teaching hospitals (Waterston and Stewart 2005), showing that perceived difficulty in a topic may also lead to poorer knowledge and retention outcomes.

The questionnaire also allowed exploration into the relationship between students' interest in a topic, and how challenging they perceive it. Neuroanatomy was ranked as the most interesting subject by 11/51 students; 45.5% ($n$ = 5) of these students found neuroanatomy the most challenging compared to 66.7% ($n$ = 34) of all students in stages 3–5. The second subject students found most interesting was the cardiovascular system with 10/51 student responses, and no students chose it as their first or second most challenging subject. This may suggest that the connection between finding a subject interesting and less challenging may not have a strong impact on studying neuroanatomy, but may have an effect on other subjects within anatomy. Previous work has identified neuroanatomy as both an area of difficulty and of interest for students (Hall et al. 2018; Jukna et al. 2023).

An argument could be made that disinterest is linked to finding a topic more challenging. Neuroanatomy was chosen as the least interesting subject by 23.5% ($n$ = 12) of students, followed by the anatomy of the pelvis and perineum by 15.7% ($n$ = 8) of students. As discussed previously, these are the two main topics that students perceived to be the most challenging areas of anatomy. Hence, there is a potential connection between lack of interest and finding the subject challenging. This also is reflected in the comments in free text box answers: '*I didn't find it especially interesting in terms of the clinical applications*' and '*Not a personal area of interest*'.

Lack of interest, although very subjective, may also be influenced by perception of a subject prior to study, a theme prominent throughout staff and student responses. This phenomenon was well described by a staff member in a comment: '*Before the students even start there is the perception that neuroanatomy is difficult which sets them up to struggle*'.

As highlighted by Javaid et al. (2018), the general perception is that neuroanatomy is far more complex relative to other anatomical areas. This results in students having low confidence in their ability to understand it, as captured by some quotes from the student questionnaire: '*The brain is a mystery*', '*I don't have any clue what goes on in there*', '*It's difficult to understand the brain*' and '*It is just confusing*'.

'*Neuroanatomy is complex*' was a staff comment that mirrors the perceptions of the student. It may be that staff perceptions or views about a subject may imprint on students or be the first opinion they hear of a subject. On one hand, it may be important for students to be warned that they may feel out of their comfort zone, and to encourage them to persevere when they face a challenging subject. However, it may also result in a negative attitude, potentially reducing the student's engagement with the content out of belief that it will be beyond their ability. It would be an interesting area of further research, to determine the role and influence an educator has on students' perceptions of challenging topic areas.

Students who selected neuroanatomy as the most challenging area of anatomy found the second most challenging topic to be the anatomy of the head and face (not including the brain) ($n$ = 19), the pelvis and perineum ($n$ = 9) and the eye ($n$ = 4). Neurovasculature was a common topic selected as the most challenging across all anatomical areas. Similar findings have been highlighted in recent literature (Cheung et al. 2021; Rehman et al. 2022).

Overall, as these findings align with current literature, it can be suggested that a case-based curriculum does not significantly influence which anatomical topics students find the most challenging. It also appears there is some relationship between interest in a topic and perceived difficulty, with greater interest in a topic making it less likely to be perceived as difficult, although this correlation was not seen for all anatomical topics; for example, neuroanatomy was perceived as both interesting and challenging. Further work is required to fully establish the extent of influence that student interest in a topic has on its perceived difficulty.

#### 4.3.2.2 Qualitative Questionnaire Data

Thematic analysis was conducted to identify the reasons why neuroanatomy is perceived to be the most challenging anatomical topic. Six themes emerged.

#### 4.3.2.2.1 Questionnaire Theme 1: Difficulty

Student free text responses suggest a generalised opinion of neuroanatomy to be the *'most complex'*, *'confusing'* and *'not very intuitive'*, speaking of the level of difficulty of taught content. This was also relevant to the anatomy of the pelvis and perineum, with one student response stating *' [pelvis and perineum anatomy is] too difficult to make sense'*.

Where anatomical topics were selected as most difficult by fewer students, the reasons listed tended to refer to the amount of content or time constraints. These are factors which can be modified through thoughtful curriculum design (e.g. assigning more time to these topics), which is an easier issue to address pedagogically than a subject's level of complexity. This presents a challenge for teaching neuroanatomy as these results add to a growing body of literature identifying the innate complexity of the topic as an issue for learners (Javaid et al. 2018; McCarron et al. 2014; Matthias et al. 2013; Pakpoor et al. 2014; Sanya et al. 2010; Zinchuk et al. 2010).

During the focus group, a staff member stated, '*the complexity at which we teach is really too much for those students'*. Jozefowicz (1994) also found that complexity of topic was the most common contributing factor for why a subject was more 'challenging' than others. However, although students and staff may agree, the idea of complexity is fluid and widely subjective. It is difficult to determine what specifically makes a subject complex.

#### 4.3.2.2.2 Questionnaire Theme 2: Visualisation

Anatomy learning is heavily reliant on the ability to visualise taught structures within the body. Student comments about neuroanatomy suggest that standard visualisation techniques, such as use of cadaveric specimens, are not as effective when studying neuroanatomy, and the anatomy is harder to visualise than other areas:

*'It's not all visual', 'less picture-like', 'Harder to visualize from diagrams to real life. Specimens aren't great at demonstrating.'*

The 'abstractness' of anatomical topics contributing to the difficulty of understanding concepts has been noted in similar research (Cheung et al. 2021). Approaches to enhance students' understanding and interpretation of two- and three-dimensional images should be integrated into curricula (Keenan and Powell 2020) to support students' visualisation abilities.

#### 4.3.2.2.3 Questionnaire Theme 3: Quantity of Information

'High volume of content' as a reason why an area was chosen to be the most challenging was evident across all areas of anatomy selected. Comments included: *'Too many "stuffs", [I] cannot understand the relationship between stuff's'*, *'very fiddly nerves and vessels and muscles'* and *'too many 'layers'*. This is consistent with previous work that has identified volume of content as a contributing factor to students' perception of anatomy as challenging (Cheung et al. 2021). Anatomy education has also been previously discussed in the context of cognitive load theory (Talip et al. 2021). Cognitive load theory explains that constraints on working memory can hinder the effectiveness of educational interventions when dealing with substantial amounts of content (Sweller 2011, 2019), which may provide a theoretical underpinning to the student views expressed here. Within neuroanatomy, the 'inherent complex nature of the topic' has been previously identified as an underlying reason for students' perception of neuroanatomy as challenging (Javaid et al. 2018), which may contribute to both the perception of 'difficulty' (Theme 1) and 'Quantity of information' as identified here.

#### 4.3.2.2.4 Questionnaire Theme 4: Time Constraints

The amount of time dedicated to teaching anatomy within case-based curricula has been decreasing (Fitzgerald et al. 2008; Turney 2007). This has also been the case at Newcastle University, when in 2017, the systems-based approach was re-developed into the case-based curriculum. This shift resulted in a reduction of in-person contact time by almost half. Students not only commented that there is not enough time to study the area, but that the content is not repeated later, which affects retention of taught material. Regarding neuroanatomy, students commented: '*Content taught once and not repeated*' and '*Not enough time was dedicated to understanding it*'.

This is reflected in extant literature. A study exploring anatomical retention showed that anatomy knowledge decreases by roughly 14.7% within 18 months following an initial anatomy course in a problem-oriented, integrated medical curriculum (Doomernik et al. 2017). Cheung et al. (2021) found that 'repeated exposure' to anatomical topics was essential for effective learning. Additional studies similarly discovered that incorporating longitudinal reinforcement within a curriculum can aid with students' anatomical knowledge retention (Lufler et al. 2020; McBride and Drake 2016).

##### 4.3.2.2.5 Questionnaire Theme 5: Teaching Quality

Teaching quality was mentioned in responses not limited to comments about neuroanatomy. Multiple students expressed the disruption that the COVID-19 pandemic has had on their teaching, consistent with published work on this topic (Alsoufi et al. 2020; Franchi 2020; Ferrel and Ryan 2020; Khasawneh 2021).

'*I missed teaching and exams on it [neuroanatomy] due to COVID interrupting my 2nd year.*'

Teaching quality is associated with other themes discussed earlier, such as time constraints and issues surrounding visualisation. Content delivery is also a key component, as demonstrated by the following comments about the teaching of various areas of anatomy: '*Not taught very thoroughly nor in a stimulating manner*', '*The topic was broad*' and '*It was taught over many cases over two years which made it feel a bit disjointed.*'

This theme highlights that delivery is not only about content of individual sessions, but also the structure of the course and the order in which topics are spread out.

##### 4.3.2.2.6 Questionnaire Theme 6: Anatomical Terminology

Responses about neuroanatomy indicate that the language used to describe it makes it more challenging than other areas. Students feel like they do not have the same level of prior knowledge and experience using these terms; terms may not also follow the same rules as structures in other organ systems, making the learning less intuitive.

'*Lots of new words to learn and [I have] never covered it before.*'

'*The names are more abstract rather than location and description-based like other areas.*'

Mehta et al. (1996) described the background of common anatomical terms and suggested that understanding the origin of these words helps with students' anatomical understanding. It is sometimes presumed that knowledge of Latin or Greek can help students learn anatomy through reducing the time required to understand anatomical terms. However, Pampush and Petto (2011) only found a weak correlation between Latin and Greek knowledge and a student's performance in anatomy.

### 4.3.3 Limitations

Although this project has provided valuable insight into staff and student perceptions of anatomy learning within CBL curricula, this analysis pertained to only a single institution. Interpretation and delivery of CBL vary between institutions, as do methods of anatomy teaching and delivery, so this work may not be transferable to anatomy education in different contexts. Additionally, student participants were all undertaking their first MBBS degree course. Therefore, they do not have any other MBBS curricula to compare their experiences to. This means that there is no 'control' for this study. However, many of the challenges identified by this research, such as the desire for more contact teaching hours, will be of note to any MBBS programme that uses CBL. Fragmented delivery of anatomical topics and limited contact time is a common issue across many MBBS curricula and is not limited to case-based programmes.

Moreover, the low response rate for our student questionnaire data may have impacted how generalisable our results are to the whole MBBS cohort. To address the research questions, it was important that participants had covered most or all the anatomy resources in stages 1 and 2. Many of the students invited to complete the questionnaire and focus group were in their clinical years which meant that they had limited availability to contribute. Students in later years of the programme may also have forgotten much of their preclinical learning experiences, reducing their ability to reflect on their anatomy learning experiences. In addition, there were no incentives provided to students which may have also contributed to the small number of participants recruited. Despite this, the use of focus groups allowed collection of a rich qualitative dataset, allowing us to explore the view of staff and students in more depth, which may give greater insight than a large amount of quantitative data that would have been achieved through a larger questionnaire sample.

## 4.4 Conclusion

Although CBL is a programme approach underpinned by pedagogic research, there is limited information on the impact that this curriculum style has on delivery of anatomy teaching. Anatomy is a crucial element of any medical or health-allied degree, making it important for educators to understand how anatomical topics are best taught in a range of MBBS curricula. This project shows that staff and students perceive anatomy education within a case-based curriculum positively, although it is not without its limitations. Staff and students often identified similar limitations and challenges of CBL for anatomy, consistent with previous literature. Curricular challenges frequently reflect those previously identified as affecting anatomy education within MBBS programmes more widely. A useful follow-up study would be to determine whether a case-based curriculum impacts students' performance in anatomy assessments and their retention of anatomical knowledge.

## References


Alsoufi A, Alsuyihili A, Msherghi A, Elhadi A, Atiyah H, Ashini A, Ashwieb A, Ghula M, Ben Hasan H, Abudabuos S, Alameen H, Abokhdhir T, Anaiba M, Nagib T, Shuwayyah A, Benothman R, Arrefae G, Alkhwayildi A, Alhadi A, Zaid A, Elhadi M (2020) Impact of the COVID-19 pandemic on medical education: medical students' knowledge, attitudes, and practices regarding electronic learning. PLoS One 15(11):e0242905

Babacan S, Çini NT, Kafa IM, Aydin O (2021) Vertical integration of anatomy curriculum in the undergraduate clinical education period: medical students' perspectives. Anatomy 15(3):247–251

Baker KG (2021) Twelve tips for optimising medical student retention of anatomy. Med Teach 44(2):138–143

Ben Awadh A, Clark J, Clowry G, Keenan IDK (2022) Multimodal three-dimensional visualization enhances novice learner interpretation of basic cross-sectional anatomy. Anat Sci Educ 16(1):127–142

Bergman EM, Prince KJAH, Drukker J, Van Der Vleuten CPM, Scherpbier AJJA (2008) How much anatomy is enough? Anat Sci Educ 1(4):184–188

Böckers A, Mayer C, Böckers TM (2014) Does learning in clinical context in anatomical sciences improve examination results, learning motivation, or learning orientation? Anat Sci Educ 7(1):3–11

Boon JM, Meiring JH, Richards PA (2002) Clinical anatomy as the basis for clinical examination: development and evaluation of an introduction to clinical examination in a problem-oriented medical curriculum. Clin Anat 15(1):45–50

Braun V, Clarke E (2006) Using thematic analysis in psychology. Qual Res Psychol 3(2):77–101

Brunk I, Schauber S, Georg W (2017) Do they know too little? An inter-institutional study on the anatomical knowledge of upper-year medical students based on multiple choice questions of a progress test. Ann Anat 209:93–100

Buja LM (2019) Medical education today: all that glitters is not gold. BMC Med Educ 19(1):110

Burgess A, Ramsey-Stewart G (2014) Elective anatomy by whole body dissection course: what motivates students? BMC Med Educ 14:272

Cale AS, Hoffman LA, McNulty MA (2023) Promoting metacognition in an allied health anatomy course. Anat Sci Educ 16(3):473–485

Cheung CC, Bridges SM, Tipoe GL (2021) Why is anatomy difficult to learn? The implications for undergraduate medical curricula. Anat Sci Educ 14(6):752–763

Deci EL, Ryan RM (2000) The "what" and "why" of goal pursuits: human needs and the self-determination of behavior. Psychol Inq 11(4):227–268

Doomernik DE, Van Goor H, Kooloos JGM, Ten Broek RP (2017) Longitudinal retention of anatomical knowledge in second-year medical students. Anat Sci Educ 10:242–248

Drake RL, McBride JM, Lachman N, Pawlina W (2009) Medical education in the anatomical sciences: the winds of change continue to blow. Anat Sci Educ 2(6):253–259
Elizondo-Omaña RE, López SG (2008) The development of clinical reasoning skills: a major objective of the anatomy course. Anat Sci Educ 1(6):267–268
Eseonu O, Carachi R, Brindley N (2013) Case-based anatomy teaching: a viable alternative? Clin Teach 10(4):236–241
Estai M, Bunt S (2016) Best teaching practices in anatomy education: a critical review. Med Educ 208:151–157
Ferrel M, Ryan J (2020) The impact of COVID-19 on medical education. Cureus 12(3):e7492
Fitzgerald JEF, White MJ, Tang SW, Maxwell-Armstrong CA, James DK (2008) Are we teaching sufficient anatomy at medical school? The opinions of newly qualified doctors. Clin Anat 21(7):718–724
Fortun J, Morales AC, Tempest HG (2017) Introduction and evaluation of case-based learning in the first foundational course of an undergraduate medical curriculum. J Biol Educ 51(3):295
Franchi T (2020) The impact of the Covid-19 pandemic on current anatomy education and future careers: a student's perspective. Anat Sci Educ 13(3):312–315
Garg P, Bhanwra S (2022) Case based learning in teaching pharmacology to undergraduate medical students. Cureus 14(9):e29187
Ghosh SK (2016) Cadaveric dissection as an educational tool for anatomical sciences in the 21st century. Anat Sci Educ 10(3):286–299
GMC (2018) Outcomes for graduates. https://www.gmc-uk.org/-/media/documents/dc11326-outcomes-for-graduates-2018_pdf-75040796.pdf. Accessed 22 May 2024
Gogalniceanu P, O'Connor EF, Raftery A (2009) Undergraduate anatomy teaching in the UK. Bull R Coll Surg Engl 91(3):102
Guilding C, Zhi PKL, Krishnan SM, Hubbard PS, McKeegan KS (2021) Insights into delivering cross-cultural medical Education in the UK and Malaysia. Med Sci Educ 31(6):2177–2188
Hall ER, Davis RC, Weller R, Powney S, Williams SB (2012) Doing dissections differently: a structured, peer-assisted learning approach to maximizing learning in dissections. Anat Sci Educ 6(1):56–66
Hall S, Stephens J, Parton W, Myers M, Harrison C, Elmansouri A, Lowry A, Border S (2018) Identifying medical student perceptions on the difficulty of learning different topics of the undergraduate anatomy curriculum. Med Sci Educ 28(2):469–472
Harden RM (1999) What is a spiral curriculum? Med Teach 21(2):141–143
Holla SJ, Ramachandran K, Isaac B, Koshy S (2009) Anatomy education in a changing medical curriculum in India: medical student feedback on duration and emphasis of gross anatomy teaching. Anat Sci Educ 2(4):179–183
Ivankova NV, Creswell JW, Stick S (2006) Using mixed-methods sequential explanatory design: from theory to practice. Field Methods 18(1):3
Javaid MA, Chakraborty S, Cryan JF, Schellekens H, Toulouse A (2018) Understanding neurophobia: reasons behind impaired understanding and learning of neuroanatomy in cross-disciplinary healthcare students. Anat Sci Educ 11(1):81–93
Johnson IP, Palmer E, Brockhouse BM (2013) Online learning resources in anatomy: what do students think? Clin Anat 26(5):556–563
Johnstone S, Vaughan B (2020) We need one more hour solely based on anatomy... Give us anatomy!': Early-year learner perceptions of anatomy within an integrated & case-based learning osteopathy curriculum. Int J Osteop Med 36:49–54
Joughin G (2009) Assessment, learning and judgement in higher Education: a critical review. In: Assessment, learning and judgement in higher education, pp 1–15
Jozefowicz RF (1994) Neurophobia: the fear of neurology among medical students. Arch Neurol 51(4):328–329
Jukna Š, Puteikis K, Mameniškienė R (2023) Perception of neurology among undergraduate medical students—what can be done to counter neurophobia during clinical studies? BMC Med Educ 23:447
Jurjus RA, Lee J, Ahle S, Brown KM, Butera G, Goldman EF, Krapf JM (2014) Anatomical knowledge retention in third-year medical students prior to obstetrics and gynecology and surgery rotations. Anat Sci Educ 7(6):461–468
Keenan ID, Ben Awadh A (2019) Integrating 3D visualisation technologies in undergraduate anatomy education. Adv Exp Med Biol 1120:39–53
Keenan ID, Powell M (2020) Interdimensional travel: visualisation of 3D-2D transitions in anatomy learning. In: Rea P (ed) Biomedical visualisation, Advances in experimental medicine and biology, vol 1235. Springer, Cham
Keenan ID, Green E, Haagensen E, Hancock R, Scotcher KS, Swainson H, Swamy M, Walker S, Woodhouse L (2023) Pandemic-era digital education: insights from an undergraduate medical programme. In: Varsou O, Rea PM, Welsh M (eds) Biomedical visualisation, Advances in experimental medicine and biology, vol 1397. Springer, Cham
Kerfoot BP, Dewolf WC, Masser BA, Church PA, Federman DD (2007) Spaced education improves the retention of clinical knowledge by medical students: a randomised controlled trial. Med Educ 41(1):23–31
Khasawneh RR (2021) Anatomy Education of medical students during the COVID 19 pandemic. Int J Morphol 39(5):1264–1269
Kramer B, Soley JT (2002) Medical student perception of problem topics in anatomy. East Afr Med J 79(8):408–414
Krathwohl DR (2002) A revision of bloom's taxonomy: an overview. Theory Pract 41(4):212–218
Leveritt S, McKnight G, Edwards K, Pratten M, Merrick D (2016) What anatomy is clinically useful and when should we be teaching it? Anat Sci Educ 9(5):468–475
Lloyd S, Tan ZE, Doshi TJ (2014) Development of an ENT undergraduate curriculum using a Delphi survey. Clin Otolarynol 39(5):281–288

Louw G, Eizenberg N, Carmichael SW (2009) The place of anatomy in medical education: AMEE guide no 41. Med Teach 31(5):373

Lufler RS, Lazarus MD, Stefanik JJ (2020) The spectrum of learning and teaching: the impact of a fourth-year anatomy course on medical student knowledge and confidence. Anat Sci Educ 13(1):19–29

MacLeod A, Burm S, Mann K (2022) Constructivism: learning theories and approaches to research. In: Cleland J, Durning SJ (eds) Researching medical education

Mahzari M, Alnahedh T, Ahmed AA, Al Rumyyan A, Shaban S, Magzoub ME (2023) Practical guide to undergraduate medical curriculum alignment and mapping. Adv Med Educ Pract 14:1001–1012

Matthias AT, Nagasingha P, Ranasinghe P, Gunatilake SB (2013) Neurophobia among medical students and non-specialist doctors in Sri Lanka. BMC Med Educ 13:164

McBride JM, Drake RL (2014) Anatomy education 34 in an innovative medical school curriculum. In: Chan L, Pawlina W (eds) Teaching anatomy. Springer, Cham

McBride JM, Drake RL (2016) Longitudinal cohort study on medical student retention of anatomical knowledge in an integrated problem-based learning curriculum. Med Teach 38:1209–1213

McCarron MO, Stevenson M, Loftus AM, McKeown P (2014) Neurophobia among general practice trainees: the evidence, perceived causes and solutions. Clin Neurol Neurosurg 122:124–128

McLachlan JC, Bligh J, Bradley P, Searle J (2004) Teaching anatomy without cadavers. Med Educ 38:418–424

McLean S (2016) Case-based learning and its application in medical and health-care fields: a review of worldwide literature. J Med Educat Curri Develop 3:39–49

Mcnamara JP, Nolan MF (2021) Anatomy of the physical examination: a small group learning approach for increasing engagement and learning in a medical gross anatomy course. Clin Anat 35(2):256–262

Mehta LA, Natrajan M, Kothari ML (1996) Understanding anatomical terms. Clin Anat 9:330–336

Mitchell R, Batty L (2009) Undergraduate perspectives on the teaching and learning of anatomy. ANZ J Surg 79(3):118–121

Mitrousias V, Karachalios TS, Varitimidis SE, Natsis K, Arvanitis DL, Zibis AH (2020) Anatomy learning from prosected cadaveric specimens versus plastic models: a comparative study of upper limb anatomy. Anat Sci Educ 13(4):436–444

O'Keeffe GW, Davy S, Barry DS (2019) Radiologist's views on anatomical knowledge amongst junior doctors and the teaching of anatomy in medical curricula. Ann Anat 223:70–76

Older J (2004) Anatomy: a must for teaching the next generation. Surgeon 2(2):79–90

Orsbon CP, Kaiser RS, Ross CF (2013) Physician opinions about an anatomy core curriculum: a case for medical imaging and vertical integration. Anat Sci Educ 7(4):251–261

Pakpoor J, Handel AE, Disanto G, Davenport RJ, Giovannoni G, Ramagopalan SV (2014) National survey of UK medical students on the perception of neurology. BMC Med Educ 14:225

Pampush JD, Petto AJ (2011) Familiarity with Latin and Greek anatomical terms and course performance in undergraduates. Anat Sci Educ 4(1):9–15

Priyadharshini NA, Dinesh KV, Rajprasath R, Rema D (2019) Relevance of learning anatomy to clinical practice—Perceptive of medical students, interns, and clinicians. Natl J Clin Anat 8(1):32–37

Qamar K, Bashir S, Khalid R, Abaid M, Khadim R (2021) Reasons for difficult topics in anatomy and their solutions as per undergraduate medical students. Khyber Med Univ J 13(3):161–165

Raman M, McLaughlin K, Violato C, Rostom A, Allard JP, Coderre S (2010) Teaching in small portions dispersed over time enhances long-term knowledge retention. Med Teach 32(3):250

Rehman U, Perwaiz I, Sarwar SM, Brennan AP (2022) Are clinical medical students confident about their head and neck anatomy knowledge? Br J Oral Maxillofac Surg 60(7):922–926

Rhodes D, Fogg QA, Lazarus MD (2017) Dissecting the role of sessional anatomy teachers: a systematic literature review. Anat Sci Educ 11(4):410–426

Rogan P, Boohan M, Hamilton P (2020) Introducing case based learning into an undergraduate medical curriculum. In: Irish Network of HealthCare Educators. 1st annual scientific meeting: preparing students for the complexities of practice and transformative experiences

Rosenstand CAF (2012) Case-based learning. In: Seel NM (ed) Encyclopedia of the sciences of learning. Springer, Boston

Sangam MR, Praveen K, Vinay G, Bokan RR, Deka R, Kaur A (2021) Efficacy of case-based learning in anatomy. Cureus 13(12):e20472

Sanya EO, Ayodele OE, Olanrewaju TO (2010) Interest in neurology during medical clerkship in three Nigerian medical schools. BMC Med Educ 10:36

Sbayeh A, Qaedi Choo MA, Quane KA, Finucane P, McGrath D, O'Flynn S, O'Mahony SM, O'Tuathaigh CMP (2016) Relevance of anatomy to medical education and clinical practice: perspectives of medical students, clinicians, and educators. Perspect Med Educ 5:338–346

Sharma M, Murphy R, Doody GA (2019) Do we need a core curriculum for medical students? A scoping review. BMJ Open 9(8):e027369

Singh P, Bhatt R (2011) Introduction of case based learning for teaching anatomy in a conventional medical school. J Anat Soc India 60(2):232–235

Singh R, Shane TR, Gupta K, Singh M, Jones DG, Kumar R (2015) Is the decline of human anatomy hazardous to medical education/profession? A review. Surg Radiol Anat 37(10):1257–1265
Smith CF, Finn GM, Stewart J, Atkinson MA, Davies DC, Dyball R, Morris J, Ockleford C, Parkin I, Standring S, Whiten S, Wilton J, McHanwell S (2016) The anatomical society core regional anatomy syllabus for undergraduate medicine. J Anat 228(1):15–23
Smith CF, Finn GM, Border S (2017) Learning clinical anatomy. Eur J Anat 21(4):269–278
Smith CF, Finn GM, Hennessy C, Luscombe C, Stewart J, McHanwell S (2018) The initial impact of the anatomical society gross anatomy core syllabus for medicine in the United Kingdom: student and teacher perspectives. Anat Sci Educ 12(5):494–506
Smith CF, Freeman SK, Heylings D, Finn GM, Davies DC (2021) Anatomy education for medical students in the United Kingdom and Republic of Ireland in 2019: a 20-year follow-up. Anat Sci Educ 15(6):993–1006
Srinivasan M, Wilkes M, Stevenson F, Nguyen T, Slavin S (2007) Comparing problem-based learning with case-based learning: effects of a major curricular shift at two institutions. Acad Med 82(1):74–82
Stabile I (2015) Designing courses in anatomy. Eur J Anat 19(1):87–104
Stalmeijer RE, McNaughton N, Van Mook WNKA (2014) Using focus groups in medical education research: AMEE guide no. 91. Med Teach 36(11):923–939
Stambuk-Castellano M, Carrera A, Tubbs RS, Alario-Hoyos C, Verdú E, Iwanaga J, Reina F (2024) Personalized strategies for academic success in learning anatomy: exploring metacognitive and technological adaptation in medical students. Clin Anat 37(4):472–483
Sugand K, Abrahams P, Khurana A (2010) The anatomy of anatomy: a review for its modernization. Anat Sci Educ 3(2):83–93
Sweller J (2011) Cognitive load theory. In: Psychology of learning and motivation, vol 55, pp 37–76
Sweller J (2019) Cognitive load theory and educational technology. Educ Tech Res Dev 68:1–16
Talip SB, Ismail ZIM, Hadie SNH (2021) Investigating the benefits of integrated anatomy instruction: a cognitive load theory perspective. Educ Med J 13(3):1–14
Tavakol M, Dennick R (2011) Making sense of Cronbach's alpha. Int J Med Educ 2:53–55
Thistlethwaite JE, Davies D, Ekeocha S, Kidd JM, MacDougall C, Matthews P, Purkis J, Clay D (2012) The effectiveness of case-based learning in health professional education. A BEME systematic review: BEME guide no. 23. Med Teach 34(6):421–444
Totlis T, Tishukov M, Piagkou M, Kostares M, Natsis K (2021) Online educational methods vs. traditional teaching of anatomy during the COVID-19 pandemic. Anat Cell Biol 54(3):332–339
Turner K, Brewster S (2002) Undergraduate urology: a survey of current provisions and guidelines for a core curriculum. BJU Int 90(4):475
Turney BW (2007) Anatomy in a modern medical curriculum. Ann R Coll Surg Engl 89(2):104–107
Van Wyk J, Rennie CO (2015) Learning anatomy through dissection: perceptions of a diverse medical student cohort. Int J Morphol 33(1):89–95
Verhoeven BH, Verwijnen GM, Scherpbier AJJA, Van Der Vleuten CPM (2002) Growth of medical knowledge. Med Educ 36(8):711
Waterston SW, Stewart IJ (2005) Survey of clinicians' attitudes to the anatomical teaching and knowledge of medical students. Clin Anat 18(5):380–384
Weijters B, Cabooter E, Schillewaert N (2010) The effect of rating scale format on response styles: the number of response categories and response category labels. Int J Res Mark 27(3):236–247
Wormald BW, Schoeman S, Somasunderam A, Penn M (2009) Assessment drives learning: an unavoidable truth? Anat Sci Educ 2:199–203
Zhang Y, Ji Z, Zhou P, Dong L, Chen Y (2023) Clinical anatomy teaching: a promising strategy for anatomic education. Heliyon 9(3):e13891
Zinchuk AV, Flanagan EP, Tubridy NJ, Miller WA, McCullough LD (2010) Attitudes of US medical trainees towards neurology education: "neurophobia"—A global issue. BMC Med Educ 10:49

**Chloe Kelly** is an undergraduate student of Medicine at Newcastle University, Newcastle upon Tyne, UK. Chloe has been the treasurer of the Newcastle University Paediatric society, and her research interests include medical education and clinical paediatrics. She contributed to data collection and analysis of the study.

**Taisha Peplowska** is an undergraduate student of Medicine at Newcastle University, Newcastle upon Tyne, UK. Taisha has been the president of the Newcastle University Neurological society and vice-president of its Medical Education society. She contributed to data collection and analysis of the study.

**Yasmin Elebessy** is an undergraduate student of Medicine at Newcastle University Medicine Malaysia (NUMed). She has held key roles in the Medical Education Society as a Research Executive and as a member of the Specialty Interest Groups Committee. Presently, she holds the positions of MedX Series Lead, Education Officer for the Surgical Society, and Operations Manager for the NUMEd Medical Research Society. Yasmin contributed to the literature review, data collection and analysis of the study.

**Yousr Elebessy** is an undergraduate student of Medicine at Newcastle University Medicine Malaysia (NUMed). She served as the Event Executive on the founding committee of the Specialty Interest Group Project under the Medical Education Society, where she also took on the role of Research Executive. Presently, she holds the roles of External Liaison for the Paediatric Society and Education Officer for the Surgical Society. She is committed to fostering research endeavours for fellow medical students through her role as HR Manager with the NUMEd Medical Research Society. For this study, she has contributed to project planning and execution as well as data collection and analysis.

**Emily Green** B.Sc., (Hons), M.Sc., F.H.E.A., is a Lecturer in Anatomy within the School of Medicine at Newcastle University in Newcastle upon Tyne, UK. Her research interests relate to student support and inclusivity within anatomy education, innovative approaches to anatomy education and use of art-based learning and 3D software. She is currently undertaking a part-time Ph.D. at Newcastle University in Medical Education, with a focus on inclusive anatomy education for neurodivergent students and students with specific learning difficulties.

**Kayleigh Scotcher** B.Sc., (Hons), M.Sc., S.F.H.E.A., was a Lecturer in Anatomy within the School of Medicine at Newcastle University in Newcastle upon Tyne, UK. She is currently undertaking a Ph.D. at the University of Melbourne, Melbourne, Australia. Her research interests include multi-scale imaging of the autonomic innervation to the pelvic viscera and the development of innovative teaching approaches in anatomy education. She also has a keen interest in anatomical outreach and public engagement.

# Part II

# Innovating Anatomical Education

# 5 Evaluation of the Effect of Educational Innovation on Learning Outcomes in Anatomy: A Role for Prospective Items in Assessments

Ian P. Johnson

**Abstract**

Educational innovations are usually associated with positive student evaluations. However, their effects on learning outcomes, as determined by assessments, are often mixed or ignored. To better understand the link between educational innovation, student satisfaction and learning outcomes, the results of several of the author's collaborative innovations in anatomy-focused teaching have been compared. All innovations received positive comments from students, but their effects on learning outcomes were varied. An association was found between improved learning outcomes and innovations associated with good teaching practice. This suggests that how an educational innovation is delivered is more important than the new technology associated with it. Using the original assessment format before— and after—the educational innovation may also result in new learning being ignored. To address this, an assessment model that incorporates new learning driven by educational innovations is proposed.

I. P. Johnson (✉)
Adelaide, South Australia
e-mail: ian.johnson@egmpartners.com.au

**Keywords**

Anatomy · Education · Innovation · Evaluation · Learning outcomes · Constructive alignment

## 5.1 Introduction

Educational innovation is characterised by the application of new technology and approaches to facilitate teaching and learning. It ensures education keeps pace with changes in knowledge and changes in the expectations of students and society (Crosscombe 2018). Innovation in medical education has happened at many levels: from large-scale curriculum innovations in the early 1900s (Flexner 1910), to the introduction of problem-based learning in the 1990s (Moust et al. 2005), the widespread use of computers from the turn of this century (Greenhalgh 2001), the adoption of flipped classroom approaches (Chen et al. 2017a) and recent online initiatives driven by worldwide social distancing imperatives (Gewin 2020; Pather et al. 2020). Most commonly, educational innovation is driven by small, incremental changes to learning activities in response to student feedback, such that educators constantly make minor changes to pitch, delivery, and style during a learning activity.


I. D. Keenan et al. (eds.), *International Anatomical Education*, Biomedical Visualization 9,
https://doi.org/10.1007/978-3-031-91849-0_5

Innovation, therefore, can be said to be a tool that is used by most educators most of the time to improve the students' experience.

Despite its widespread adoption, educational innovation is often done in the spirit of action research without ethics approval and so remains unpublished. Moreover, that which is published tends to focus on the effect of the innovation on student satisfaction, with little information on its effect on learning outcomes as measured by assessments (Abdullah et al. 2021; Choi-Lundberg et al. 2016; Tayyem et al. 2019). In some cases, technological advances alone may be confused with educational innovation, and many are simply bolted onto teaching activities without serious consideration of their pedagogical usefulness (Rohlfsen et al. 2020). In a survey-based study of 50 students at a technical college, learning effectiveness (assessed by school grades, number of professional certificates and external examination success) is correlated with the introduction of educational innovation (Lee 2011). In contrast, the introduction of virtual microscopy as an alternative to optical microscopy for groups of 95–104 Biomedical and Audiological science students did not alter learning outcomes (Mione et al. 2013). Mixed results have been reported elsewhere. For example, a study of workshop, lecture-based or self-directed learning for electrocardiogram interpretation for groups of 62–79 medical students found that all three styles resulted in improved post-test and retention scores, but self-directed learning scores were significantly lower than the other two methods (Mahler et al. 2011), a finding that was at odds with previous studies (Rubinstein et al. 2009; Nilsson et al. 2008). Course content, type of assessment and cohort differences may also play a part in evaluating innovation. The introduction of a structured drawing regime improved learning outcomes in asynchronous learning of gross anatomy in a study of 55 medical students but did not improve learning outcomes in practical surface anatomy teaching in a separate study of 154 medical students (Backhouse et al. 2017). A similar mixed picture emerges for the effect of digital game-based learning innovations on learning outcomes (Chen et al. 2017b). Subject matter may also be important, with attention increasingly falling on neuroanatomy as a subject perceived as difficult by students (Hall et al. 2018). In a review of 18 selected studies involving 2165 students using eight different teaching methods to deliver neuroanatomy, 14 (77.7%) resulted in students being able to recognise anatomical structures but only three resulted in long-term (>5 months) retention of knowledge (Sotgiu et al. 2020). Similarly, cranial nerves teaching to groups of 88–102 Year 2 medical students using two flipped classroom approaches compared to traditional lectures did not result in any significant difference in knowledge retention between any of the groups (Anbu et al. 2021). Overall, a mixed picture emerges of the effect of educational innovation on learning outcomes.

Many aspects associated with the delivery of the innovations mentioned above will have been omitted from some of the published reports for reasons of relevance and clarity. It is possible, however, that these features may hold the key to gaining a more comprehensive understanding of the link between educational innovation and learning outcomes. Here, I reflect on the reasons behind an observed mismatch between student satisfaction and learning outcomes for several educational innovations in anatomy and pathophysiology teaching undertaken with colleagues over the past decade. A pattern emerges that leads me to propose that educational innovations improve learning outcomes primarily because they promote good pedagogy. I also suggest that educational innovation can disrupt the constructive alignment of educational activities such that new learnings driven by the innovation are not assessed and propose a strategy to assess these new learnings. A preliminary account has been published (Johnson 2019).

## 5.2 Review of Educational Innovations

Between 2010 and 2022, student satisfaction and learning outcomes were measured after different educational approaches, new technology or both, were introduced into undergraduate courses. The innovations fell into five broad categories: (1) online gross anatomy, (2) online neuroanatomy, (3) reorganised anatomy practicals, (4) ultrasound simulators and (5) flipped classroom and assessment for learning approaches (Table 5.1).

One study has been published in full (Johnson et al. 2013), while the remaining studies have been published as conference abstracts only with no raw data or figures.

Each study is considered below in Sects. 5.2.1, 5.2.2, 5.2.3, 5.2.4, 5.2.5 and 5.2.6, with a description of the rationale, methods and the effect on both student satisfaction and learning outcomes. Statistical significance was set at $p < 0.05$. Where continuous variables approached a normal distribution, a t-test was used to determine statistical significance. For data not conforming to a normal distribution, the non-parametric Mann-Whitney U-test replaced the t-test; where differences between observed and expected values were compared, the Chi Squared test was used, and the Spearman Rank Coefficient was calculated to determine the strength of the association between two sets of ordinal data.

### 5.2.1 Online Anatomy

**Rationale** In response to student feedback requesting an alternative to commercially available online resources, in-house anatomy online resources were created that were tailored to the course (Johnson et al. 2013).

**Methods** Online interactive resources using the same specimens as those in the laboratories, and assessments were created based on clinical scenarios with feedback (Fig. 5.1a). The resources were introduced into semester 2 of Year 3 of the MBBS (Bachelor of Medicine and Surgery) program (159 students). Using the questions shown in Table 5.2, student opinion was obtained (40% response rate). Assessment results were also obtained for semesters 1 and 2 and compared with assessment data for semesters 1 and 2 in the previous year (143 students). Semester 2 was a revision of semester 1 anatomy.

**Results** The survey response rate was 40%. Free comments and focus group discussions were positive, but there was no significant difference ($p > 0.05$) before—and after—the introduction of online anatomy in broad agreement (percentage of responses in the range 5–7 in a seven-point Likert scale) with the statement 'Online materials

**Table 5.1** Summary of educational interventions showing the type of innovation, the title of the degree and the year of study of the cohort involved. The line at the bottom of the table shows the Human Research Ethics approval number from either the University of Adelaide (H prefix) or Macquarie University. Some studies described separately here were covered by the same ethics approval as indicated by the superscript letters a to e

| Innovation | Cohort |
|---|---|
| Online anatomy | Medicine, Year 3 (a) |
| Cranial nerves tool (bolt-on) | Health science, Year 3 (b) |
| Cranial nerves tool (integrated) | Health science, Year 3 (b) |
| Reorganised practicals | Medicine, Year 1 (c) |
| Reorganised practicals | Medicine, Year 3 (c) |
| Ultrasound simulators | Medicine, Year 1 (c) |
| Flipped classroom | Nursing, Year 2(d) |
| Assessment for learning | Medical and clinical science, Year 3 (e) |

Human research ethics approval numbers: (a) H-264-2011; (b) H-2014-195; (c) H-2015-181; (d) HS-2013-037; (e) 520231555850132

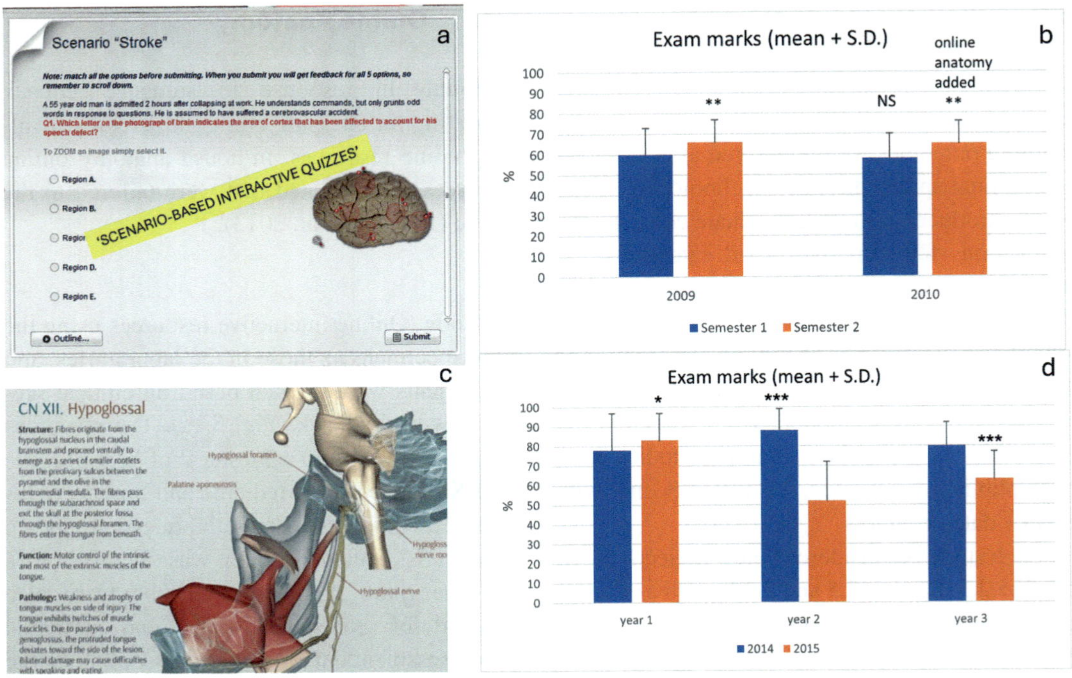


**Fig. 5.1** (**a**) Example of one of the resources introduced to facilitate learning of clinically relevant anatomy. Students are required to click on the brain region whose damage would account for the patient's symptoms given in the brief clinical scenario. To help understanding, more feedback is provided for incorrect responses than for the correct responses. Reproduced with permission from (Johnson et al. 2013). (**b**) Online anatomy: a significant improvement ($p < 0.001$, unpaired t-test) in marks is seen between semesters 1 and 2 for both years (**). The inclusion of online anatomy in semester 2 of 2010 made no significant difference (NS, $p > 0.05$) compared to semester 2 of 2009. (**c**) Screenshots of part of the Prezi canvas showing in the left panel the course, distribution, physiology and pathology of the hypoglossal nerve, and in the right panel, a low-magnification overview of the course and distribution of the optic nerve in relation to the skull. All images were high resolution, so students could zoom in on areas of interest. (**d**) Reorganised anatomy practicals. New style practicals were introduced in 2015 for Years 1 and 3. The new style practicals were associated with a significant increase in mean exam marks for Year 1 but a significant decrease in exam marks for Year 3. Despite the new style anatomy not being introduced in Year 2, there was a significant decrease in exam marks in 2015 compared to 2014

provided by this university help me learn anatomy' (60% vs. 63%). A 10–12% improvement ($p < 0.001$, unpaired t-test) was seen in anatomy assessment scores between semesters 1 and 2 in 2009 (no innovation) and 2010 (innovation). This was unsurprising as semester 2 was simply a revision of semester 1 anatomy. However, when semester 2 was compared between 2009 and 2010, no significant effect of the learning innovation on anatomy scores was detected (Fig. 5.1b).

**Conclusion** Students liked the introduction of online anatomy, but it did not affect learning outcomes. Questionnaire and focus group findings (Johnson et al. 2013), indicated that students liked the resource as it was focused and tailored to their course. Reasons why the educational innovation failed to affect learning outcomes are explored in the context of other educational innovations with similar results in Sect. 5.3 (Discussion).

### 5.2.2 Cranial Nerves Tool

**Rationale** In response to student feedback that neuroanatomy was difficult to understand, and evidence that cranial nerves are considered the most difficult neuroanatomy topic (Richardson-Hatcher et al. 2014), we created an in-house

**Table 5.2** Online anatomy student survey. The attitudes of students to online anatomy were surveyed in an anonymous sample of students in the 2010 cohort using a 26-item questionnaire answered using a seven-part Likert-type scale where 7 = 'strongly agree', 4 = 'undecided' and 1 = 'strongly disagree'

| | |
|---|---|
| 1 | I know what anatomy to learn and what not to learn |
| 2 | The clinical relevance of anatomy is clear to me |
| 3 | The three-dimensional nature of anatomical structures is easy to learn |
| 4 | I have sufficient access to anatomical specimens |
| 5 | I have sufficient access to anatomical textbooks |
| 6 | I have sufficient access to online anatomy learning resources |
| 7 | I have sufficient access to anatomy teachers |
| 8 | Wet specimens (prosections) help me learn anatomy |
| 9 | Potted specimens (museum pots) help me learn anatomy |
| 10 | Lectures help me learn anatomy |
| 11 | Self-directed study during resource sessions help me learn anatomy |
| 12 | Textbooks help me learn anatomy |
| 13 | Commercially available electronic learning aids (e.g. anatomy.Tv, an@tomdeia, ADAM) help me learn anatomy |
| 14 | Past exam questions help me learn anatomy |
| 15 | Discussions with anatomy tutors and demonstrators help me learn anatomy |
| 16 | Discussions with other students help me learn anatomy |
| 17 | Online materials provided by this university help me learn anatomy |
| 18 | Online materials provided by others (for example, other university web pages or via web search engines such as Google) help me learn anatomy |
| 19 | I would like more lectures to help me learn anatomy |
| 20 | I would like more tutorials to help me learn anatomy |
| 21 | I would like more self-directed study time to help me learn anatomy |
| 22 | I would like more access to wet specimens (prosections) to help me learn anatomy |
| 23 | I would like more access to potted specimens (museum pots) to help me learn anatomy |
| 24 | I would like more access to online resources to help me learn anatomy |
| 25 | I would like more access to worked examples of anatomy questions to help me learn anatomy |
| 26 | I would like more access to online tests to help me learn anatomy |

learning tool for cranial nerves (Collins-Praino et al. 2015, 2017).

**Methods** An interactive tool combining anatomy, physiology and pathology of the cranial nerves was created using rendered MR (magnetic resonance) images. We took advantage of the Prezi (www.prezi.com) platform to deploy this, so that students could view and study the course and distribution of all the cranial nerves at a range of magnifications, thereby providing a 'start to finish' overview of cranial nerve anatomy, physiology and pathology (Fig. 5.1c). The link to the tool was made available to 47–49 Year 3 Bachelor of Health Sciences (B. Health Sci.) students taking a functional neuroanatomy course. In the first iteration of the study, the resource was simply bolted onto the existing curriculum. In the second iteration, it was integrated into the curriculum with directed learnings, comprising multiple online-focused tasks associated with specific resources that had clear outcomes and were linked to specific learning goals (Twyman 2021). In both studies, end-of-semester assessment results were compared with the previous year, which was taught without the innovation. As an internal control, approximately 10% of the multiple choice questions and short answer questions and 50% of the neuroanatomy practical questions were repeated for the cohorts with—and without—the innovation. For the written assessments, past questions were available to students, but there was no record available to students of the neuroanatomy practical questions. Students' opinions of the cranial nerves tool were sought in

**Table 5.3** Cranial nerves tool questionnaire. Students scored each item using a seven-point Likert scale from strongly agree (7) to strongly disagree (1)

| | |
|---|---|
| 1 | I find the cranial nerves to be a difficult and challenging topic. |
| 2 | The cranial nerves resource was easy to use. |
| 3 | The cranial nerves resource improved my learning of the material. |
| 4 | The integration of structure, function and pathology on a single slide of the cranial nerves resource significantly contributed to the efficacy of the resource. |
| 5 | The ability to zoom in on various aspects of the slides significantly contributed to the efficacy of the cranial nerves resource. |
| 6 | The three-dimensional nature of the slides significantly contributed to the efficacy of the cranial nerves resource. |
| 7 | The visual aspect of the cranial nerves resource significantly contributed to the efficacy of the resource. |
| 8 | I believe that I understand the cranial nerves better after using the online resource. |
| 9 | I think that the cranial nerves resource should be utilised in the course in future years. |
| 10 | I think that it would significantly improve my learning if similar online resources were developed for other topics in the course. |

**Table 5.4** Cranial nerves tool. Selected free responses from the student questionnaire to the prompt. 'The best aspects (of the cranial nerve tool) were'

| | |
|---|---|
| Response 1 | *'The 3D pictures'* |
| Response 2 | *'Helps to give a more deep understanding of the location and structures of nerves'* |
| Response 3 | *'Explanations of the cranial nerves and locations'* |
| Response 4 | *'Helps with both practical and theory'* |

a ten-item questionnaire delivered at the end of the module (Table 5.3).

**Results** The response rates to the bolt-on and integrated cranial nerves tool survey were 72% and 23%, respectively. There was 86–93% broad agreement (i.e. percentage of responses in the range 5–7) to question 8 'I believe that I understand the cranial nerves better after using the online resource' and this was affirmed for both groups in over half the free responses. Those aspects of the cranial nerves tools valued by the students are shown in Table 5.4.

The cranial nerves online tool had no significant effect on end-of-semester assessment marks, either when bolted onto the existing curriculum or integrated (Table 5.5). While there was a trend for higher marks where resources were integrated, this was not statistically significant ($p > 0.05$). There was no significant difference in mean marks for questions that appeared for cohorts with—and without—the innovation, indicating that the assessments and cohorts were similar.

**Conclusion** Students liked the cranial nerves tool, but it did not affect learning outcomes. Reasons for this discrepancy are discussed in Sect. 5.3 (Discussion).

### 5.2.3 Reorganised Anatomy Practicals

**Rationale** As part of a general adult-learning approach (Knowles 1970), attendance at anatomy practicals was optional and learning was primarily self-directed via the practical notes. This approach, however, was associated with poor attendance and engagement in anatomy practicals. To address this, a system was introduced where learning was heavily directed. This was achieved by requiring all students to answer all workbook questions prior to the practical, breaking the class into small groups (n ~ 12) who would complete specific tasks related to speci-

**Table 5.5** Cranial nerves tool. Breakdown of end-of-semester marks (mean % marks ± S.D.). Compared to the course when no cranial nerves tool was available (2013), there is no significant effect ($p > 0.05$, unpaired t-test) on marks regardless of whether the tool was bolted onto the existing curriculum (2014) or integrated into the curriculum (2015)

| | Cranial nerve questions theory | All neuroanatomy theory | All neuroanatomy practical | Total | *n* |
|---|---|---|---|---|---|
| No resource (2013) | 58 ± 12 | 55 ± 10 | 57.5 ± 14 | 63 ± 10 | 51 |
| Bolted on (2014) | 56 ± 24 | 49.31 ± 23 | 53.28 ± 15 | 56.4 ± 23 | 49 |
| Integrated (2015) | 62 ± 62 | 61.52 ± 19 | 59.51 ± 18 | 67.47 ± 20 | 47 |

mens or images under the close supervision of a tutor, then repeat 4 times with a new tutor and new specimens. Groups would compete in an end-of-session quiz based on laboratory specimen images using an electronic voting system, so that the correct answer was immediately apparent and there was the opportunity for the group to consider the wisdom of their collective response to that question. At the end of the teaching semester, the winning group received a certificate and a bag of chocolate frogs. In addition to all this, there was mandatory attendance that was confirmed by a register (Johnson 2016).

**Methods** Directed small group learning and mandatory attendance in gross anatomy practicals were introduced to Years 1 and 3 of the MBBS course. The laboratory manuals for all years contained clinical anatomy questions for students to answer in their own time. Students in Year 1 ($n = 148$), Year 2 ($n = 136$) and Year 3 ($n = 146$) then completed a survey on their attitudes to anatomy. Learning outcomes for anatomy were compared with those of the previous year.

**Results** The response rates to the surveys were 53%, 34% and 47% for Years 1, 2 and 3, respectively. For Year 3, where directed learning and mandatory attendance were introduced, most students reported that the course was 'improved' or 'much improved' (67% broad agreement). Year 1 students could not be asked to compare with the previous year, although one repeating student was positive about the changes. Free responses in the student questionnaire for Years 1–3 are shown in Table 5.6.

The Year 2 anatomy coordinator declined to change the format in 2015, with 68% of respondents (82/136) reporting that anatomy sessions were 'about the same' as the previous year. The introduction of directed learning and mandatory attendance was associated with a slight (5%) but significant ($p < 0.05$, unpaired t-test) improvement in the Year 1 anatomy assessment marks but a 27% ($p < 0.001$, unpaired t-test) decrease for the Year 3 marks (Fig. 5.1d, Table 5.7). The Year 2 marks for 2015 fell by the largest proportion (36%) compared to the previous year ($p < 0.001$, unpaired t-test) despite there being no change to the format of the anatomy practicals.

**Conclusion** The students liked the reorganised practicals, but there was no consistent association with learning outcomes. Reasons for this discrepancy are discussed in Sect. 5.3 (Discussion).

### 5.2.4 Ultrasound Simulators

**Rationale** In response to student comments that it was difficult to transition from three-dimensional anatomy to two-dimensional anatomical slice images and medical images, and published accounts confirming this view (Ben Awadh et al. 2022; Keenan and Powell 2020), ultrasound simulators were introduced into an anatomy practical for Year 1 medical students (Massey-Westropp et al. 2015). This followed reports that the use of ultrasound simulators improved cardiac gross- and ultrasound- anatomy (Canty et al. 2015).

**Table 5.6** Reorganised anatomy practicals. Free responses from the student questionnaire

| | |
|---|---|
| **Year 1** | |
| Response 1 | *'I have really enjoyed the structure of our anatomy sessions'.* |
| Response 2 | *'As I am repeating, I would like to say how much more I got out of these anatomy sessions than last year'.* |
| Response 3 | *'Love the resource sessions. From what I have heard about the way it has been run in previous years, it seems like it has improved a lot and I'm glad we have this system in place instead of the old one'.* |
| **Year 2** | |
| Response 1 | *'As there is no guidance or "restrictions" on time spent at each station, some students remain at the same station for the entire session making it very difficult for other students to observe the specimens at that station'.* |
| Response 2 | *'We should have the same setup that the 1st and 3rd [Year 1 and Year 3 students] have for anatomy'.* |
| Response 3 | *'In small, allocated groups of possibly 10 students, we should have allotted time to go to a station where there will be a delegated tutor. They could direct the station for about 10 min and answer our questions. Then we would all rotate around the room so we have time at each station, with a tutor'.* |
| **Year 3** | |
| Response 1 | *'I feel that these sessions have improved immensely from last year, because there is much more structure around what is to be covered in each station, and each station has a tutor who is able to answer the relevant questions to that station'.* |
| Response 2 | *'This year, I've really appreciated having tutor-led stations. They are much more useful than self-directed stations because the tutors know a lot more about 3D anatomy than we do'.* |
| Response 3 | *'I would like to commend the tutors for improving resource sessions as I have benefited greatly from it this year'.* |

**Table 5.7** Reorganised anatomy practicals. End of year % marks (mean ± S.D.). Significant increases (Year 1) and decreases (Year 3) in mean marks were seen after reorganizing anatomy practicals. Year 2 anatomy practicals were not reorganised in 2015, yet a significant decrease in marks was seen. The academic standard and format of the assessments remained unchanged for all years

| | 2014 | 2015 |
|---|---|---|
| Year 1 | 78 ± 19 ($n$ = 143) | 83 ± 14 ($n$ = 148)* |
| Year 2 | 88 ± 11 ($n$ = 139) | 52 ± 20 ($n$ = 136)*** |
| Year 3 | 80 ± 12 ($n$ = 141) | 63 ± 14 ($n$ = 146)*** |

*$p < 0.05$, **$p < 0.01$, *** $p$ 0.001, unpaired t-test

**Methods** Three Vimidex ultrasound simulators (CAE Healthcare, Florida, USA) with paired displays showing an interactive ultrasound image linked to an equivalent pseudocoloured rendered anatomical image or medical image were introduced to a single anatomy practical for Year 1 medical students ($n$ = 58). Small groups worked with a clinical tutor, and activities switched between normal and pathological anatomy to emphasise contextual relevance.

Student opinions of the simulators were sought via a survey and students undertook a quiz on the cross-sectional anatomy and imaging before—and after—instruction using the simulators (Fig. 5.2a). This was an anonymous test, but students retained the same radiofrequency voting device (Turning Technologies 2012). In this way, individual responses could be analysed. Students were encouraged to perform the ultrasound examinations for the benefit of their peers, and this was linked directly to the quiz that occurred just before—and after—the simulators were used.

**Results** 53/58 students (91%) completed the questionnaire at the end of the session. There was more than 90% broad agreement (BA) with the statements regarding the usefulness of the innovation (Table 5.8). BA was calculated as the percentage of student responses agreeing with the statement, i.e. in the Likert-type range 5–7.

Mean pre- and post-test scores ($n = 58$) were not significantly different (Fig. 5.2b). However, a comparison of individual scores (Fig. 5.2c) showed that most (55%) students improved their score post-test ($p = 0.05$, Mann-Whitney U test).

**Conclusion** Students liked the ultrasound simulators. Class mean marks were unaffected, but individual scores generally improved. Reasons for this discrepancy are considered in Sect. 5.3 (Discussion).

### 5.2.5 Flipped Classroom

**Rationale** In response to student feedback that a Year 2 nursing unit delivered primarily via lectures and tutorials was unengaging and lacked clinical relevance, it was redesigned (Gladman and Johnson 2017).

**Methods** The redesigned unit in 2017 had a flipped classroom approach (Akçayir and Akçayir 2018) that emphasised clinical relevance. Here, students were asked to view online lectures and complete readings and viewing of specific videos in their own time before coming to class. The subsequent in-class activities required students to use their prior knowledge either alone or in small groups to solve clinical problems. Care was also taken to ensure the assessment aligned with learning outcomes and teaching activities. As part of a team-based learning approach (Fatmi et al. 2013), students also took part in weekly individual readiness assurance tests (IRATs) and group/team readiness assurance tests (TRATs) that contributed 4% and 1% of their final mark, respectively. Students received their IRAT scores but no feedback on which questions they had answered correctly. They then took the same test in small groups (teams) where they had the opportunity to discuss the questions as a group before submitting their responses. Once TRAT responses had been submitted, students were given the correct response for further discussion. Online supporting materials were available in advance.

**Results** An end-of-unit evaluation administered by the university was completed by over 90% of students in both years shortly before the final examinations. In response to the question '*Overall, I am satisfied with this course*', there was 44% broad agreement for the cohort in the traditional style course in 2016 and significantly more (81%) broad agreement ($p < 0.05$, Chi squared test) for the cohort in the flipped course in 2017. The percentage broad agreement was calculated as the percentage of student responses in the Likert range 4–7 where 1 = strongly disagree and 7 = strongly agree.

The mean end of year mark (Fig. 5.2d) increased significantly from 72% for the traditional-style course ($n = 152$) to 77% for the flipped course ($n = 166$) ($p < 0.01$. unpaired t-test). This was reflected in an increase in the proportion of distinctions and high distinctions and a reduction in the proportion of students failing (Fig. 5.2e).

**Conclusion** The students liked the flipped classroom approach and learning outcomes improved. This association between students expressing satisfaction with the educational innovation and improved learning outcomes contrasted with most of the educational innovations described earlier in this article. Reasons for this are considered in Sect. 5.3 (Discussion).

### 5.2.6 Assessment for Learning Approach

**Rationale** In response to pandemic-driven changes to the delivery of anatomy (Pather et al. 2020), an undergraduate unit was delivered online using an assessment for learning approach with a conscious effort to maintain constructive alignment, situated cognition, and contextual rel-

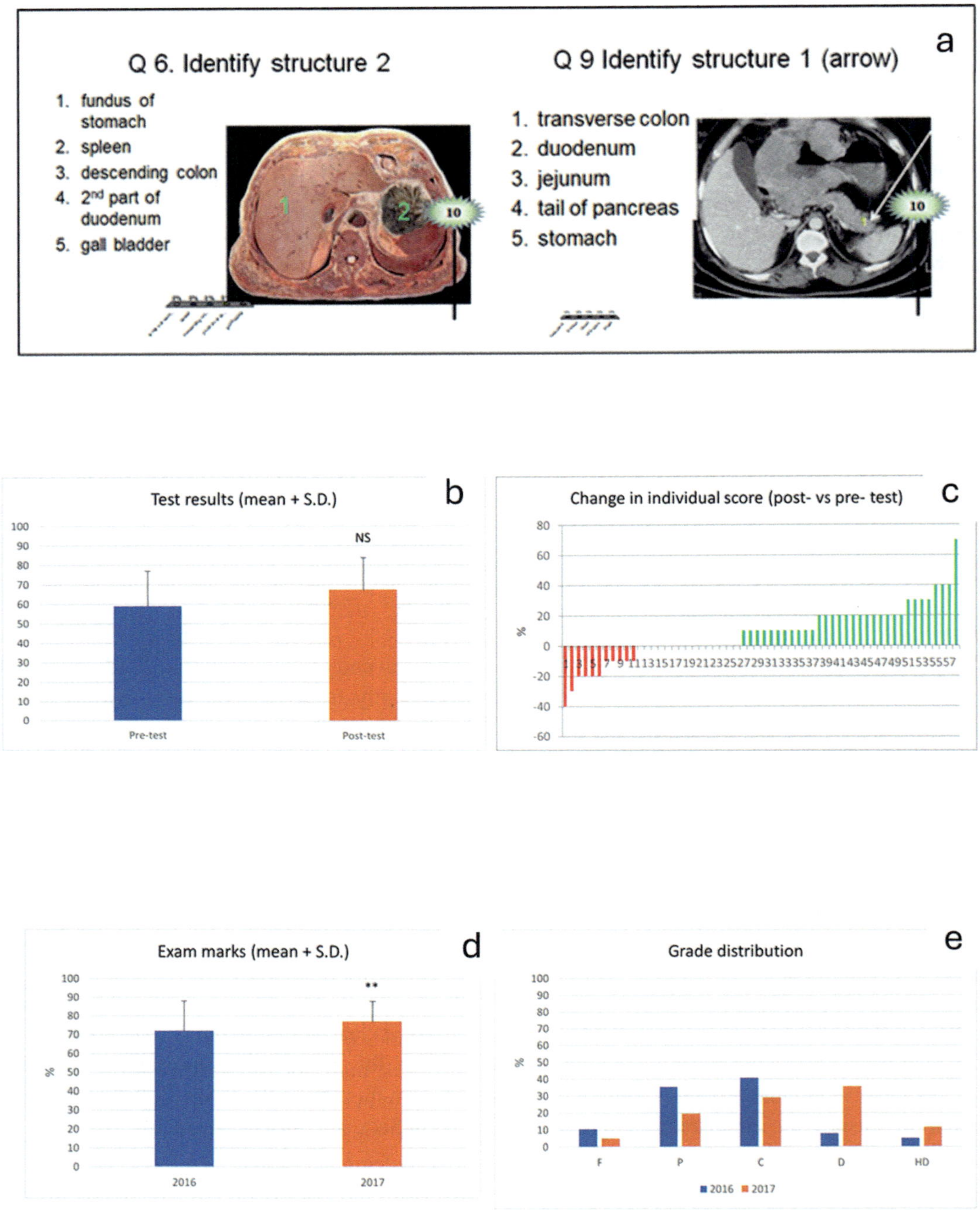


**Fig. 5.2** (**a**) Examples of cross-sectional anatomy quiz questions using the voting system. The timer to the right of each image gives each student 10 s to enter their response without discussion. After 10 s, for both the pre- and post-test, no further responses for that question can be entered. For the post-test only, the small bar chart seen at the bottom of the image with zero entries populates after 10 s with the percentage of the class who have selected options 1, 2, 3, 4 or 5. Where there is a wide spread of responses, this is used to promote in-class discussion immediately following the post-test question that hopefully guides the students towards the correct response in the future, even though their current selection cannot be altered. The correct response is revealed before the next question is displayed. This approach was found to hold students' attention better than leaving the answers and the discussion to the end of the quiz. (**b**) Ultrasound simulators in anatomy. No significant change is seen in the class

**Table 5.8** Ultrasound simulators in anatomy. Free responses from the student survey. BA (broad agreement) corresponding to the percentage of student responses agreeing with the statement, i.e. in the Likert-type range 5–7

| | |
|---|---|
| Response 1 | *'They are easy to use (96% BA)'* |
| Response 2 | *'They help me understand other medical images (91% BA)'* |
| Response 3 | *'They help me understand cross-sectional anatomy (95% BA)'* |
| Response 4 | *'They should be included in all abdomen and thorax anatomy laboratory sessions (96% BA)'* |
| Response 5 | *'Displaying the 3-D image next to the ultrasound image is helpful (91%BA)'* |

evance in the pedagogy (Brown et al. 1989; Biggs 2014; Broadfoot and Black 2004) while emphasising online synchronous student interactions.

**Methods** Using the institutional learning management system, all online recorded lectures included study-focusing questions where content was applied to clinically focused problems and the answers were subsequently explained, emphasizing why wrong answers were incorrect. Asynchronous learning with self-testing and feedback was delivered using the commercial HTML5 package (H5P) (n.d.) embedded in the institutional learning management system. All online practical notes also contained similar questions at the end of every paragraph. These questions were used verbatim or slightly modified in weekly online assessments, and the class performance on individual assessment questions formed the basis of weekly online, synchronous tutorials. The mid-session and end-of-session online assessments had a very similar format to the weekly tests. All assessment questions related closely to the unit learning objectives to ensure students did not doubt what was required of them. Anatomy was live-streamed from the dissection room and teaching centred around clinically relevant questions. Anatomy tutorials were delivered synchronously to small ($n$ = 10) groups in break-out rooms using the Zoom platform https://zoom.us/, assisted by the use of commercial three-dimensional anatomy packages, such as Complete Anatomy (Elsevier, Chatswood, New South Wales, Australia) https://3d4medical.com/), where students would be handed control of the application to rotate the image to illustrate features to their peers. Much use was made of videos in teaching, including narrated medical image stacks (CT, MRI) and examples of physical examinations. Short sections of these videos were selected, then looped and delivered in silent mode for online assessments (Fig. 5.3). Most questions were image-based and clinically relevant. Examples are given in Table 5.9. Many functional anatomy questions used generic videos such as a person walking upstairs, performing a pincer grip to pick up a small item or adopting an unusual posture at the end of a marathon. In such cases, the answer to the question could not be obtained simply through reverse image searching.

During the pandemic, the online assessments were taken offsite and invigilated online. After pandemic restrictions were lifted, the online assessments were taken on-site by students using their own devices and they were invigilated in person. The logistics of the onsite, online invigilated assessments were guided by a colleague's previous experience of these forms of assessment (Hillier and Wijenayake 2020).

**Results** Student feedback (60/171 students, 35% response rate) showed near maximal scores (range 4.8 to 4.9 out of 5) for 6 of the 8 aspects

Fig. 5.2 (continued) mean percentage pre- and post-test marks (unpaired t-test) are seen. (**c**) Ultrasound simulators in anatomy. Most students (green columns) show an increase in their post-test scores. (**d**) Flipped classroom. Exam marks show a small but significant ($p < 0.01$ unpaired t-test) improvement after the introduction of a flipped classroom approach in 2017. (**e**) Flipped classroom. Introduction of a flipped classroom approach in 2017 is associated with a significant shift towards distinction (D) and high distinction (HD) marks with fewer fails (F) ($p < 0.05$, Mann-Whitney U test). Grading criteria: Fail (F), <50%; Pass (P), 50–64%; Credit (C), 65–74%; Distinction (D), 75–84%; High Distinction (HD), 85–100%

**Which shoulder muscle is being tested?**

**Damage to which nerve is associated with this type of gait?**

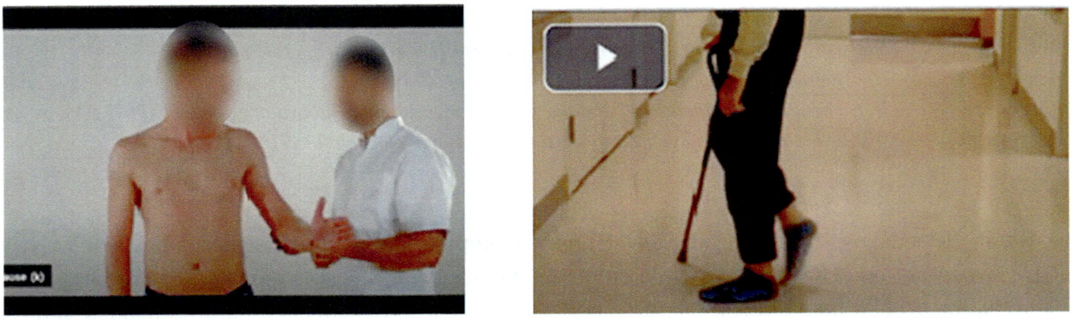

- o Infraspinatus
- o Subscapularis
- o Anterior deltoid
- o Supraspinatus

- o Saphenous
- o Common fibular
- o Perineal
- o Pudendal
- o Tibial

**Fig. 5.3** Use of videos in clinical anatomy MCQs. Short (10–15 s) sections of silent videos associated with a text question would repeat multiple times during the 1 min allocated to the MCQ. Note screenshots with their inherent lower resolution compared to prepared images are shown here to illustrate what students would see during assessments

**Table 5.9** Assessment for learning. Examples of the style of questions used in multiple choice questions so that applied anatomy and reasoning were tested rather than recall

| | |
|---|---|
| MCQ 1 | If this structure was damaged (arrow), the most likely functional defect would be: (*gross anatomy image*) |
| MCQ 2 | Stenosis of the valve best auscultated at this site would be associated with oedema of the…….(*surface anatomy image*) |
| MCQ 3 | Damage to which cranial nerve (A–E) results in this gaze palsy? (*clinical image*) |
| MCQ 4 | A stroke affecting this region of the brain (hatched) will result in the following functional deficit: (*MRI/CT*) |
| MCQ 5 | Which [muscle/nerve/abdominal organ, etc.] is being assessed? (*looped silent video*) |

surveyed by the University (Fig. 5.4a). Only 70% of the students agreed that the online unit allowed them to work with others, which was unsurprising during the pandemic. Fewer (50%) agreed that it had improved their understanding of First Nations perspectives, an aspect that has subsequently been addressed by emphasizing medical conditions that disproportionately affect Australian First Nations people compared to those who have emigrated to Australia since colonisation by Europeans with the landing of the first fleet in 1788. The top five medical conditions disproportionately affecting First Nations people have been identified as mental and substance use disorders, injuries including suicide, cardiovascular diseases, cancer and musculoskeletal conditions (AIHW 2022).

Free responses from students to the broad question 'Any comments you would like to add?' also indicated that deliberate aspects of the design of the unit were apparent to students (Table 5.10). The mean score for the end-of-session assessment when the online invigilated exam was taken offsite on the students' own devices during the pandemic was 86.6%, with a skew of the grade

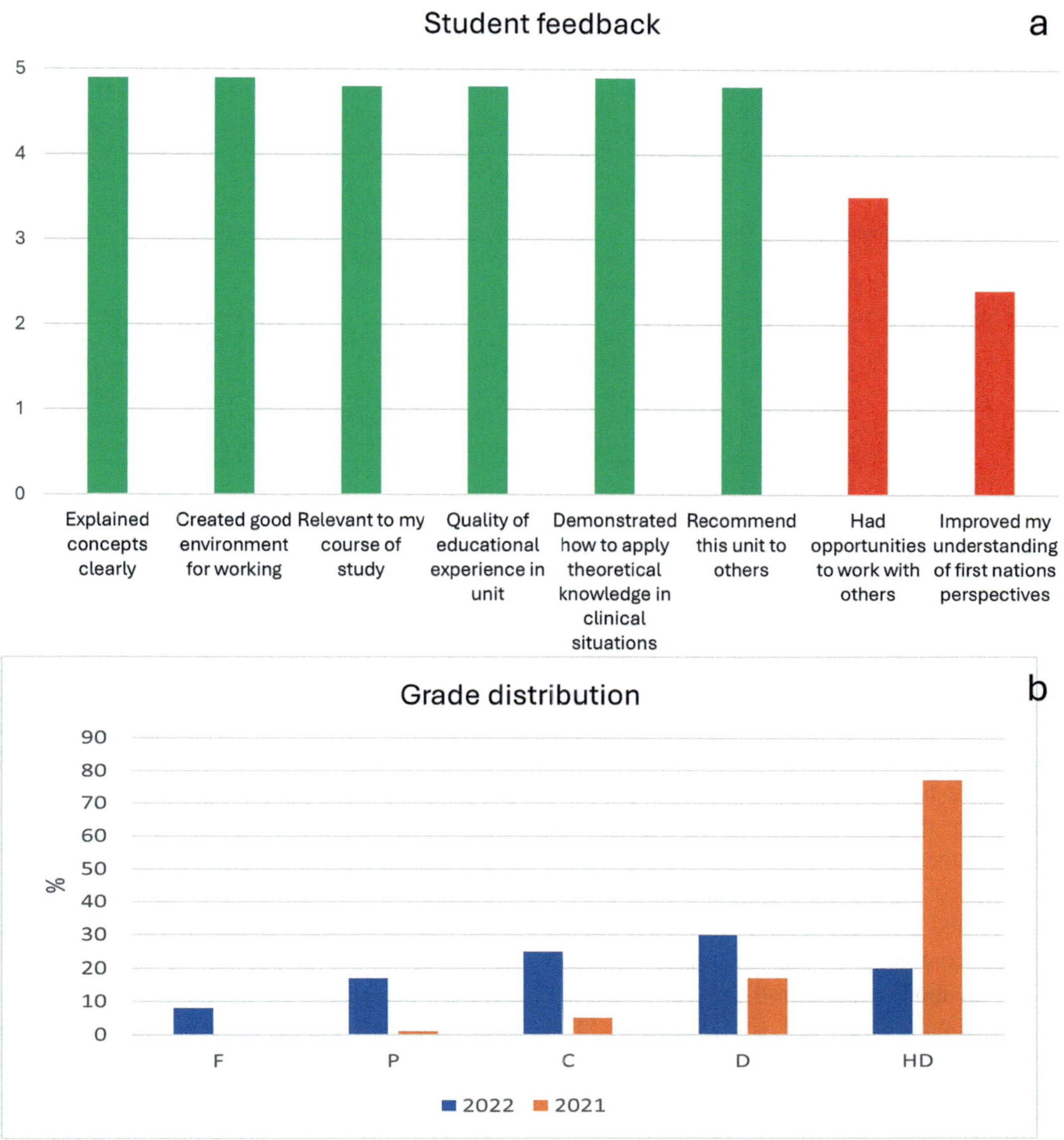


**Fig. 5.4** (**a**) Positive student feedback on the clinical anatomy and medical imaging unit. Mean score out of 5 is shown for each category. (**b**) End-of-session grade distribution is shifted significantly towards the highest grade (HD) ($p < 0.05$, Mann-Whitney U test) when the online invigilated assessment is taken offsite in 2021 (pandemic restrictions in place) compared to it being taken onsite in 2022 (pandemic restrictions eased)

distribution in favour of high distinctions (HD). When pandemic restrictions were eased and the online invigilated assessment was taken on campus ($n = 202$) but still allowing students to use their own devices, the mean mark at 71% was significantly lower ($p < 0.05$, unpaired t-test) and the grade distribution skewed back towards distinction (D) grades (Fig. 5.4b). No breaches of exam integrity were found, but this remains a possible explanation for the difference in grades when the invigilated assessments were conducted offsite versus onsite.

**Table 5.10** Assessment for learning approach. Free responses selected from the student questionnaire indicate that students were aware of the educational approaches underpinning the unit

| Assessment for learning. Free responses from the student questionnaire recognising the pedagogical approaches used in the module | |
|---|---|
| Assessment for learning | *'He gave us study focusing questions at the end of each lecture, ran sessions about the weekly quizzes and explained all the answers (including why answers were wrong)'.* |
| Constructive alignment | *'...hence assessments related to the learning outcomes and it was clear to see how and why. No other unit on campus is able to do this well'.* |
| Situated cognition and contextual relevance | *'Overall, this unit gave me a newfound appreciation for anatomy. Initially, I was hesitant as I previously have found anatomy dare I say slightly boring because I perceived it as memorisation. This unit takes it beyond this and teaches us why it's important to know this and how we use this information in a clinical environment—It unlocks a deeper level of thinking that appeals to me a lot more'.* |

## 5.3 Discussion

The effects on learning outcomes of several of the author's collaborative educational innovations centring on anatomy have been presented. Notwithstanding the methodological shortcomings of the present studies, it can be concluded that students appreciated the effort put into improving their learning experience, but the effects of these innovations on learning outcomes were mixed. Moreover, interpretation of results within—and between—studies was hampered by multiple confounding variables, which is similar to the situation seen generally for the evaluation of educational innovation (Fragkandreas 2017; Serdyukov 2017). This may be a consequence of the natural history surrounding the development of many educational innovations, which are often action research that is inspired by student feedback and modified by unexpected changes to the richness of teaching and learning resulting from the innovation. Action Research has been defined as being '... a cyclical or spiral process that includes problem identification (by a teacher, a group of teachers or administrators) taking action and fact-finding about the result of the action taken' (Lufungulo et al. 2021). It is likely, therefore, that the investigative approach itself introduces variables that are difficult to identify and control. Overall, we are presented with a type of problem that has been defined by (Rittel and Webber 1973) as a 'wicked problem' in that it is 'difficult or impossible to solve because of incomplete, contradictory and changing requirements that are often difficult to recognize'. As a result, we may need to accept that in the absence of carefully controlled experimental designs, the best we can do is to identify trends and associations that allow us to draw inferences about the effects of educational innovation on learning outcomes.

**Limitations**

A major limitation for most of the studies described here is that different cohorts of students have been compared. For example, exam results in 1 year that included an educational innovation have been compared with exam results for a different (usually a preceding-) year without the educational innovation. In these circumstances, the ability and attitudes of the cohorts to learning may be different, and teaching staff and the availability of resources may change. It is also possible, despite close attention to standard-setting and the efforts of exam moderation committees, that assessments in different years may be set at different academic levels. One way to control for different cohorts and the associated variables mentioned above would be to use a parallel study design where students from the same cohort are randomly allocated to the treatment or test group. Unfortunately, this can disadvantage one or other group, which is an important consideration where the study includes summative assessments that can determine course progression for students. A crossover design has the potential to avoid disadvantaging students. Here, students are randomly allocated to treatment or control groups and then at a halfway point during the course are swapped into the other group. This assumes the course is long enough to allow meaningful use of the educational innovation before the crossover. It also assumes no sharing of resources by students in both groups and that having the educational innovation in one half of the course compared to another half does not give students an educational advantage.

In all the studies described here, an educational innovation was introduced in response to an educational need perceived by lecturers. On occasion, it was informed by comments by students in course surveys, but there remains the possibility that what lecturers think students want may not reflect what students really want. It would not be surprising, therefore, if learning outcomes were unaffected if students were provided with a resource they did not want or need. This leads to a related limitation where new technologies or pedagogies are introduced to a course primarily to impress academic peers or to avoid feeling left out of fashionable trends. This approach, rather than one based on a desire to improve students' learning experience, is unlikely to be successful.

In the studies described here, it was clear to the students that the educational innovation had been delivered by their course teachers. This raises the possibility that students may have given the innovation an artificially positive response in anonymous opinion polls because (1) they did not want to upset their teachers or (2) they felt a debt of gratitude because their teachers had tried to help them or (3) they feared retribution from their teachers if they criticised the innovation. The extent to which students felt a desire to please—or not displease—the creators of the educational innovations is unknown in the studies described here.

Another limitation of the studies described here was the use of the same assessment for groups with—and without—the educational innovation, without consideration of the new learning that would likely be driven by the innovation. This is clearly the case when an assessment format designed for an older course is used again to assess the cohort with the educational innovation. This problem is discussed further in this article through a review of the author's previous collaborative studies, which were, in any case, a mix of opportunistic trials of new technology, pedagogy or both, and so difficult to compare with each other. In addition, none of the previous studies were designed to test the hypothesis that innovation results in new learnings. This is discussed further in Sect. 5.3.2.

When the results of all the studies described here are compared, improved learning outcomes were more commonly associated with those educational innovations that promoted activities and practices associated with good teaching such as small group activities, emphasising contextual relevance, constructive alignment, assessment for learning, and ensuring new learnings driven by the innovation are assessed. This is indicated qualitatively by a greater prevalence of '+' (moderated improvement in learning outcomes) or '++' (considerable improvement in learning outcomes) in the upper rows highlighted in green in Table 5.11. The non-parametric Spearman rank correlation revealed none of this to be statistically significant, although emphasising contextual relevance and assessing new learning came close at $p = 0.1096$.

Most of the educational changes listed in Table 5.11 were not anticipated or considered at the time the studies were carried out. Instead, they have been added with hindsight, as the author was involved in all the collaborative studies. The possible influence of each educational change on learning outcomes is considered below.

**Technology-Enhanced Learning**

While there has been widespread adoption of digital technologies (Trelease 2008), it is often unclear if learning has actually been enhanced (Kirkwood and Price 2014). Despite this, it was assumed here that new technology would enhance learning. This view was conveniently supported by student opinion surveys, and it is mirrored in the findings for other teaching areas (Asif et al. 2020). However, in the studies described here, learning outcomes showed no clear link to the

**Table 5.11** Relationship of learning outcomes to teaching characteristics driven by different educational innovations (++, considerable; +, moderate; –, none). A loose (non-significant) relationship is seen between improved learning outcomes (highlighted in green) and those educational innovations that promoted activities and practices associated with good teaching. rs = Spearman rank correlation. $p$ = two-tailed probability

| Innovation | Learning outcomes | Technology enhanced learning | Contextual relevance emphasised | Small group activities | Assessment for learning | Constructive alignment | New learning assessed |
|---|---|---|---|---|---|---|---|
| Flipped classroom | Improved | + | ++ | ++ | ++ | ++ | ++ |
| Assessment for learning | Improved | + | ++ | + | ++ | ++ | ++ |
| Reorganised practicals (Y1) | Improved | - | + | ++ | - | + | - |
| Ultrasound simulators | Improved (individual scores) | ++ | ++ | ++ | + | ++ | ++ |
| | No change (class mean) | ++ | ++ | ++ | + | ++ | ++ |
| On-line anatomy | No change | ++ | + | - | + | + | - |
| Cranial nerves tool (bolt on) | No change | ++ | + | - | + | + | - |
| Cranial nerves tool (integrated) | No change | ++ | + | - | + | + | - |
| Reorganised practicals (Y3) | Reduced | - | + | ++ | - | + | - |
| Spearman rank correlation (rs) vs learning outcomes | | rs 0.3878<br>p 0.3023 | rs 0.5692<br>p 0.1096 | rs 0.3061<br>p 0.4229 | rs 0.4596<br>p 0.2132 | rs 0.725<br>p.0271 | rs 0.5692<br>p 0.1096 |

introduction of new technology. A similar conclusion has also been drawn after a meta-analysis of 153 papers on technology-enhanced learning in anatomy-focused teaching (Clunie et al. 2018). One possible reason for this mismatch is that students conflate an improved learning experience with improved learning, and it is difficult to determine where the balance lies in the absence of data on learning outcomes. A cautious conclusion is that the introduction of new technology alone is unlikely to affect learning outcomes.

## Small Group Teaching

Small group teaching has often been advocated for teaching in the health sciences and medicine, primarily because it encourages group discussion and self-evaluation (Jaques 2003; Van Diggele et al. 2020). It can also be an integral part of an assessment for learning approach to teaching (Tormey 2015). Students generally like it (Burgess et al. 2020), but evidence that it improves learning outcomes is sparse. A meta-analysis found that small group teaching improved learning outcomes in schoolchildren (Dietrichson et al. 2017), and a study of 80 final year medical students also found it improved post-test scores compared to lectures (Singh et al. 2016). The introduction of small group teaching in radiology has also been found to improve anatomy scores (Chew et al. 2020). As several educational innovations described here involved the anatomy laboratory, where students traditionally work in small groups, it is unsurprising that small group teaching is a feature common to all the innovations with improved learning outcomes. One anatomy laboratory activity, however, included a considerable amount of small group activity but was associated with reduced learning outcomes. This serves as a reminder that good teaching is subject to multiple variables and influences.

**Contextual Relevance**

Contextual relevance has been described as something that relates to—and is appropriate to—an idea or statement (Doak 2019). At its root is contextual learning theory, where information is delivered in such a way that students can construct meaning which is relevant to their own experiences and circumstances (Imel 2000; Jensen 2005). Contextual relevance is often introduced in anatomy by emphasizing its clinical relevance through clinical problems and practicals (Clough and Lehr 1996). In the present series of educational innovations, this was done by working through clinically applied questions, the use of simulators and physical examinations. All these approaches can be considered as educational approaches that promote situated cognition in learning (Ataizi 2012), where students are encouraged to do something concerned with their learning. This idea is not new: The educator John Dewey said: '*Give the pupils something to do, not something to learn; and the doing is of such a nature as to demand thinking; learning naturally happens*' (Dewey 1916). This process of an activity demanding thinking, and this in turn driving learning, shares much with the idea of '*constructed understanding*' where students build upon previous superficial and deep learning to create their own constructs (Hattie 2008). Unfortunately, for the innovations described here, student opinions were sought primarily on the utility of the innovation and not on educational techniques associated with them. Nevertheless, it appears that students were aware of educational methods (based on some of their free comments), and it may therefore be useful in the future to include questions for scoping students' opinions of **specific** educational methods driven by educational innovations. Another educational approach that formed part of the educational innovations considered here is the theory of constructive alignment and this is considered below.

**Constructive Alignment**

Constructive Alignment is a theory that combines two educational ideas: (1) constructivism (Duffy and Jonassen 1992), whereby the learner uses their own experiences to construct meaning from the learning activity and (2) instructional design (Gagne et al. 2004), where attention is paid to the nature and delivery of information so that it can be assessed effectively. It has been described (Biggs 1996; Biggs 2014) as '*a design for teaching in which what it is intended students should learn, and how they should express their learning, is clearly stated before teaching takes place. Teaching is then designed to engage students in learning activities that optimise their chances of achieving those outcomes, and assessment tasks are designed to enable clear judgments as to how well those outcomes have been attained*'. For some of the later educational innovations described here, there was a deliberate effort made to facilitate constructivism in learning. Examples of this are:

- Pairs of students by applying their knowledge of anatomy to conduct physical examinations and comment on each other's performance.
- Laboratory manuals rewritten with an emphasis on activities that consolidated clinical applications enhanced by numerous applied questions that encouraged deep learning, e.g. after identifying the ligamentum teres of the liver, students were asked to explain its role in the development of a caput medusa in portal hypertension.
- Aligning learning objectives with assessment and regular testing with feedback on those learning objectives.
- Using the same assessment format for both regular formative assessments and higher stakes summative assessments, so that students had little doubt about what would be assessed and how.

The last two educational approaches might generally be described as facilitating constructive alignment, but the increased emphasis on assessment with feedback adopted in later innovations has much in common with an 'assessment for learning' approach. This is discussed below.

**Assessment for Learning**

Assessment for learning has been described as '*an approach to teaching and learning that creates feedback which is then used to improve students' performance. Students become more involved in the learning process and from this gain confidence in what they are expected to learn and to what standard*' (Cambridge 2023; Broadfoot and Black 2004). Increased examination scores have been reported with this approach (Mcdonald and Boud 2003). At face value, a 'no surprises' approach to what learning is expected and how it will be assessed will be of clear benefit to students. This must be balanced, however, with the total assessment of students. As such, it also requires supplementing with tutorial-style teaching to explore nuances and broader issues so that it does not degenerate into 'rubric-driven instruction', a term originally coined for portfolio assessment (Stetcher 1998; Stetcher and Mitchell 1995), but which translates into many unimaginative teaching practices. In the later educational innovations described here (Sect. 5.2.5 on the flipped classroom and Sect. 5.2.6 on assessment for learning), attention was given to devising assessment questions where the intellectual process required to arrive at the answer was more important than the answer itself. For example, in the clinical anatomy and medical imaging unit, students would not be expected simply to memorise the fact that mitral valve stenosis was linked to pulmonary oedema. Instead, they would be asked to explain, with reference to blood flow, blood vessels and Starling forces, why this was so. In doing so, it would then be easy for students to apply this understanding to describe likely sites of oedema with stenosis or incompetence of valves of the left—or right—side of heart. Providing contextual relevance gives students another 'hook' on which to hang their knowledge and highlights the inadequacy of anatomy teaching that focuses solely on the ability to identify structures.

### 5.3.1 Innovation Facilitates Good Teaching

One conclusion from the above discussion is that a correlation between improved learning outcomes and the introduction of educational innovation does not prove a causal relationship. Instead, improved learning outcomes are more likely to be due to the extent to which an educational innovation improves pedagogy. Thus, educational innovation may simply be one of many tools that facilitate good teaching. A corollary is that educational innovation without any improvement in pedagogy will not affect learning outcomes. This could explain the failure of expensive technology to improve learning, which has been suspected for some time '*Available evidence suggests not that technology creates educational improvement but rather that educational improvement comes about through coherent instruction and assessment that supports high-quality student learning*' (Lawless and Pellegrino 2007).

### 5.3.2 Innovation Results in New Learnings

The studies presented here also highlight a role for the assessment of new learning in the evaluation of educational innovation. This insight is not new. It has been proposed that data sources other than those used so far and in the development of the survey instruments should also be employed to measure the effects of educational innovation (Vincent-Lancrin et al. 2019), a view that echoes Barrow's comment that '*The assessment of an innovation should be designed before the innovation is put into practice to allow for appropriate observations and the acquisition of other data needed before, during and after the innovation to be carried out*' (Barrows 2008). At first sight, it may seem impossible to assess new learnings before they are delivered via the innovation. It may, however, be possible to anticipate some of the expected new learnings, such as improved problem-solving, improved data mining, application of theoretical knowledge to real-life scenarios, and better teamwork. These anticipated new learnings could then be assessed before—and after—the introduction of an educational innovation. If the 'new learnings' assessment tasks are clearly identifiable to staff after the assessment has been taken, they can be omitted from inclusion in the students' final grade so as not to disadvantage them (Fig. 5.5).

## 5.4 Conclusion

This review, informed by the author's personal recollections of the studies, has identified educational innovation as one of many ways to drive good teaching, which in turn results in improved learning outcomes. The evaluation of the effect of educational innovation on learning outcomes is hampered by the presence of multiple confounding variables, but a schema to anticipate and evaluate new learning is proposed, which may provide some insights into the effects of the educational innovation on learning. Student opinions of educational innovation were largely positive in this study, reflecting what is found throughout the literature. It might safely be assumed, therefore, that any educational innovation that is clearly identified as such will be warmly received by students, regardless of whether it improves learning outcomes or not. Care must therefore be used when using student opinion alone to evaluate educational innovations, bringing to mind a quote attributed to the musician Wynton Marsalis: '*Those who play for the applause... that's all they get*' (AZquotes 2023).

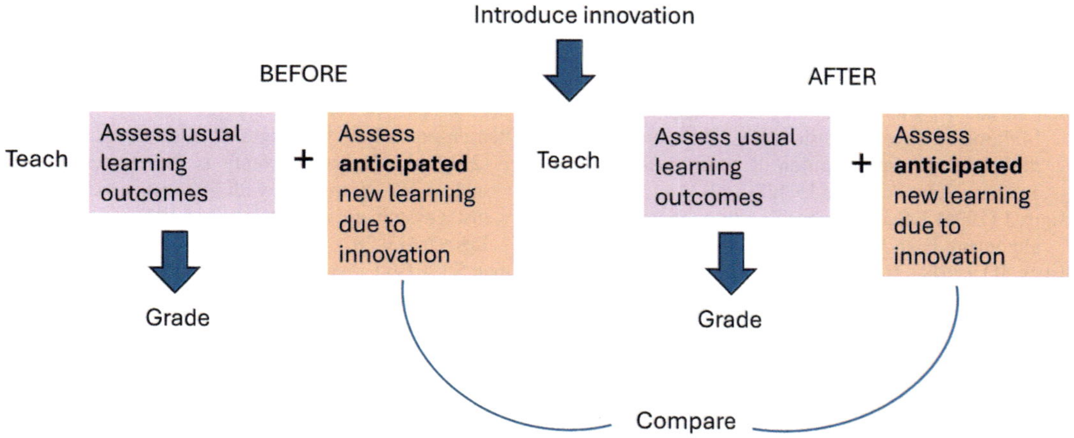


**Fig. 5.5** Schema suggesting how anticipated new learnings due to the educational innovation might be assessed before—and after—the introduction of the innovation. The questions assessing new learning can be omitted when calculating the final mark to avoid students being disadvantaged

**Acknowledgement** The author thanks all collaborators in the studies described in this review for their valued contributions. These studies could not have been completed without the generous assistance provided by The University of Adelaide, The University of South Australia and Macquarie University, Sydney.

## References

Abdullah E, Lone M, Cray JJ, Dvoracek P, Balta JY (2021) Medical students' opinions of anatomy teaching resources and their role in achieving learning outcomes. Med Sci Educ 31:1903–1910

AIHW (2022) Indigenous health and wellbeing. Australian Institute of Health and Wellbeing. https://www.aihw.gov.au/reports/australias-health/indigenous-health-and-wellbeing:

Akçayir G, Akçayir M (2018) The flipped classroom: a review of its advantages and challenges. Comput Educ 126:334–345

Anbu D, Robson A, Kurn O, Taylor C, Dean O, Payne D, Nagy E, Harrison C, Hall S, Border S (2021) Evaluating the efficacy and optimisation of the peer-led flipped model using TEL resources within neuroanatomy. Adv Exp Med Biol 1334:1–22

Asif M, Edirisingha P, Ali R, Shehzad S (2020) Teachers' practices in blended learning environment: perception of students at secondary education level. J Educ Educ Dev 7:286–306

Ataizi M (2012) Situated cognition. In: Seel NM (ed) Encyclopedia of the sciences of learning. Springer US, Boston, MA

AZquotes (2023). https://www.azquotes.com/quote/811888. Accessed 22 June 2023

Backhouse M, Fitzpatrick M, Hutchinson J, Thandi CS, Keenan ID (2017) Improvements in anatomy knowledge when utilizing a novel cyclical "Observe-Reflect-Draw-Edit-Repeat" learning process. Anat Sci Educ 10:7–22

Barrows HS (2008) Innovations without appropriate assessment are of limited usefulness. Teach Learn Med 20:287

Ben Awadh A, Clark J, Clowry G, Keenan ID (2022) Multimodal three-dimensional visualization enhances novice learner interpretation of basic cross-sectional anatomy. Anat Sci Educ 15:127–142

Biggs J (1996) Enhancing teaching through constructive alignment. High Educ 32:347–364

Biggs JD (2014) Constructive alignment in university teaching. HERDSA Rev High Educ 1:5–22

Broadfoot P, Black P (2004) Redefining assessment? The first ten years of assessment in education. Assess Educ Princ Policy Pract 11:7–26

Brown JS, Collins A, Duguid P (1989) Situated cognition and the culture of learning. Educ Res 18:32–42

Burgess A, Van Diggele C, Roberts C, Mellis C (2020) Facilitating small group learning in the health professions. BMC Med Educ 20:457

Cambridge (2023) Getting started with assessment for learning [Online]. https://cambridge-community.org.uk/professional-development/gswafl/index.html. Accessed 20 June 2023

Canty DJ, Hayes JA, Story DA, Royse CF (2015) Ultrasound simulator-assisted teaching of cardiac anatomy to preclinical anatomy students: a pilot randomized trial of a three-hour learning exposure. Anat Sci Educ 8:21–30

Chen F, Lui AM, Martinelli SM (2017a) A systematic review of the effectiveness of flipped classrooms in medical education. Med Educ 51:585–597

Chen S-W, Yang C-H, Huang K-S, Fu S-L (2017b) Digital games for learning energy conservation: a study of impacts on motivation, attention, and learning outcomes. Innov Educ Teach Int 56:66–76

Chew C, O'Dwyer PJ, Young D, Gracie JA (2020) Radiology teaching improves anatomy scores for medical students. Br J Radiol 93:20200463

Choi-Lundberg DL, Low TF, Patman P, Turner P, Sinha SN (2016) Medical student preferences for self-directed study resources in gross anatomy. Anat Sci Educ 9:150–160

Clough RW, Lehr RP (1996) Testing knowledge of human gross anatomy in medical school: an applied contextual-learning theory method. Clin Anat 9:263–268

Clunie L, Morris NP, Joynes VCT, Pickering JD (2018) How comprehensive are research studies investigating the efficacy of technology-enhanced learning resources in anatomy education? A systematic review. Anat Sci Educ 11:303–319

Collins-Praino L, Burton J, Johnson IP (2015) Evaluation of the utility of an in-house three-dimensional teaching tool for undergraduate neuroanatomy. Clin Anat 28:941

Collins-Praino L, Burton JL, Johnson IP (2017) Integration of a novel three-dimensional teaching tool for undergraduate neuroanatomy: student opinions and learning outcomes. J Anat 230:365

Crosscombe N (2018) Innovation. Brock Educ. J. 27:48–52

Dewey J (1916) Thinking in education. Democracy and education: an introduction to the philosophy of education. The Free Press, New York

Dietrichson J, Bog M, Filges T, Klint Jørgensen A-M (2017) Academic interventions for elementary and middle school students with low socioeconomic status: a systematic review and meta-analysis. Rev Educ Res 87:243–282

Doak S (2019) Contextual relevance in schools [Online]. https://www.drshannondoak.com/2019/09/06/contextual-relevance-in-schools/. Accessed 26 June 2023

Duffy TM, Jonassen DH (1992) New implications for instructional technology. New York, Taylor and Francis

Fatmi M, Hartling L, Hillier T, Campbell S, Oswald AE (2013) The effectiveness of team-based learning on learning outcomes in health professions education: BEME guide no. 30. Med Teach 35:e1608–e1624

Flexner A (1910) Medical education in the United States and Canada: a report to the Carnegie Foundation for the Advancement of Teaching. The Carnegie Foundation for the Advancement of Teaching, New York

Fragkandreas T (2017) Innovation paradoxes: a review and typology of explanations. Prometheus 35:267–290

Gagne RM, Wagner WW, Golas KC, Keller JM (2004) Principles of instructional design. California, Wadworth

Gewin V (2020) Five tips for moving teaching online as COVID19 takes hold. Nature 580:295–296

Gladman M, Johnson IP (2017) Blended learning in health: are nurses ready to be flipped? In: Higher Education Group of Adelaide Conference: bend or break: flexible institutions, teachers and students, September 2017

Greenhalgh T (2001) Computer-assisted learning in undergraduate medical education. BMJ 322:40–44

H5P (n.d.) HTML5 package for the collaborative creation of interactive web content. https://h5p.org/. Accessed 21 June 2023

Hall S, Stephens J, Parton W, Myers M, Harrison C, Almansouri A, Lowry A, Border S (2018) Identifying medical student perceptions on the difficulty of learning different topics of the undergraduate anatomy curriculum. Med. Sci. Educ. 28:469–472

Hattie J (2008) Visible learning: a synthesis of over 800 meta-analyses relating to achievement. Routledge, London

Hillier MK, Wijenayake N (2020) E-examinations: the impact of technology problems on student experience. In: Brinda T, Passey D, Keane T (eds) Empowering teaching for digital equity and agency. OCCE 2020. IFIP advances in information and communication technology. Springer

Imel S (2000) Contextual learning in adult education. In: University OS (ed) Practice Application brief no. 12. ERIC Clearinghouse on adult, career, and vocational education. Center on Education and Training for Employment, Columbus

Jaques D (2003) Teaching small groups. BMJ 326:492–494

Jensen PE (2005) A contextual theory of learning and the learning organisation. Knowl Process Manag 12:53–64

Johnson IP (2016) They liked it, but did they learn anything? Challenges of measuring student outcomes associated with innovations in anatomy teaching. In: Higher education academy surveys conference, July 2016, Birmingham, UK

Johnson IP (2019) Innovation to improve learning: what are we missing? In: Australian conference on science and mathematics education, 2019 Adelaide, Australia, pp 47–48

Johnson IP, Palmer E, Burton J, Brockhouse M (2013) Online learning resources in anatomy: what do students think? Clin Anat 26:556–563

Keenan ID, Powell M (2020) Interdimensional travel: visualisation of 3D-2D transitions in anatomy learning. Adv Exp Med Biol 1235:103–116

Kirkwood A, Price L (2014) Technology-enhanced learning and teaching in higher education: what is 'enhanced' and how do we know? A critical literature review. Learn Media Technol 39:6–36

Knowles M (1970) Andragogy: an emerging technology for adult learning. In: Knowles M (ed) The modern practice of adult education: from pedagogy to andragogy. Cambridge Book Company, Cambridge

Lawless KA, Pellegrino JW (2007) Professional development in integrating technology into teaching and learning: unknowns and ways to pursue better questions and answers. Rev Educ Res 77:575–614

Lee Y-J (2011) A study on the effect of teaching innovation on learning effectiveness with learning satisfaction as a mediator. World Trans. Eng. Technol. Educ. 9:92–101

Lufungulo E, Mambwe R, Kalinde B (2021) The meaning and role of action research in education. Multidiscip J Lang Soc Sci Educ 4:115–128

Mahler SA, Wolcott CJ, Swoboda TK, Wang H, Arnold TC (2011) Techniques for teaching electrocardiogram interpretation: self-directed learning is less effective than a workshop or lecture. Med Educ 45:347–353

Massey-Westropp N, Parangee N, Johnson IP (2015) Ultrasound simulators as aids to learning anatomy. Cinical Anatomy 28:946

Mcdonald B, Boud D (2003) The impact of self-assessment on achievement: the effects of self-assessment training on performance in external examinations. Assess Educ 10:209–220

Mione S, Valcke M, Cornelissen M (2013) Evaluation of virtual microscopy in medical histology teaching. Anat Sci Educ 6:307–315

Moust JHC, Van Berkel HJM, Schmidt HG (2005) Signs of erosion: reflections on three decades of problem-based learning at Maastricht University. High Educ 50:665–683

Nilsson M, Bolinder G, Held C, Johansson BL, Fors U, Ostergren J (2008) Evaluation of a web-based ECG-interpretation programme for undergraduate medical students. BMC Med Educ 8:25

Pather N, Blyth P, Chapman JA, Dayal MR, Flack N, Fogg QA, Green RA, Hulme AK, Johnson IP, Meyer AJ, Morley JW, Shortland PJ, Štrkalj G, Štrkalj M, Valter K, Webb AL, Woodley SJ, Lazarus MD (2020) Forced disruption of anatomy education in Australia and New Zealand: an acute response to the Covid-19 pandemic. Anat Sci Educ 13:284–300

Richardson-Hatcher A, Hazzard M, Ramirez-Yanez G (2014) The cranial nerve skywalk: a 3D tutorial of cranial nerves in a virtual platform. Anat Sci Educ 7:469–478

Rittel HW, Webber MM (1973) Dilemas in a general theory of planning. Policy Sci 4:155–169

Rohlfsen CJ, Sayles H, Moore GF, Mikuls TR, O'Dell JR, McBrien S, Johnson T, Fowler ZD, Cannella AC (2020) Innovation in early medical education, no bells or whistles required. BMC Med Educ 20:39

Rubinstein J, Dhoble A, Ferenchick G (2009) Puzzle-based teaching versus traditional instruction in elec-

trocardiogram interpretation for medical students--a pilot study. BMC Med Educ 9:4
Serdyukov P (2017) Innovation in education: what works, what doesn't, and what to do about it? J Res Innov Teach Learn 10:4–33
Singh K, Katyal R, Singh A, Joshi HS, Chandra S (2016) Assessment of effectiveness of small group teaching among medical students. J Contemp Med Educ 4:145–148
Sotgiu MA, Mazzarello V, Bandiera P, Madeddu R, Montella A, Moxham B (2020) Neuroanatomy, the Achille's heel of medical students. A systematic analysis of educational strategies for the teaching of neuroanatomy. Anat Sci Educ 13:107–116
Stetcher B (1998) The local benefits and burdens of large-scale portfolio assessment. Assess Educ 5:335–352
Stetcher BM, Mitchell KJ (1995) Portfolio-driven reform; Vermont teacher's understanding of mathematical problem solving. Final report. CSE Technical Report 400. Los Angeles
Tayyem R, Qandeel H, Qsous G, Badran D, Bani-Hani K (2019) Medical students perception of current undergraduate anatomy teaching. Int J Morphol 37:825–829
Tormey W (2015) Education, learning and assessment: current trends and best practice for medical educators. Ir J Med Sci 184:1–12
Trelease RB (2008) Diffusion of innovations: smartphones and wireless anatomy learning resources. Anat Sci Educ 1:233–239
Turning Technologies (2012) Youngstown, Ohio
Twyman JS (2021) The evidence is in the design. Perspect Behav Sci 44:195–223
Van Diggele C, Burgess A, Mellis C (2020) Planning, preparing and structuring a small group teaching session. BMC Med Educ 20:462
Vincent-Lancrin S, Urgel J, Kar S, Jacotin G (2019) Measuring innovation in education 2019. What has changed in the classroom? OECD, Paris

**Ian P. Johnson** Ph.D., F.R.M.S., F.H.E.A., F.R.S.B., C. Biol is an Anatomy Education Consultant based in Adelaide, South Australia, having retired in 2023 as Professor of Anatomy at Macquarie Medical School, Sydney. He was previously Head of Anatomy and Pathology at the University of Adelaide, South Australia and Senior Lecturer in Anatomy at University College London (UCL), UK. His research is on the effects of ageing on the survival of neurons and on the effectiveness of educational innovation in learning human anatomy.

# Introduction of a Combined Digital and Cadaveric Anatomy Learning Approach for Dental Students: A Cross-Institutional Pilot Study from the Universities of Padua and Trieste

6

Andrea Porzionato, Vanessa Nicolin, Veronica Macchi, Elena Stocco, Stefano Sivolella, Edoardo Stellini, Roberto Di Lenarda, and Raffaele De Caro

**Abstract**

The Universities of Padua and Trieste have recently signed an agreement to deliver a cross-institutional anatomical education programme for students of a dentistry and orthodontics course. According to this agreement, the students of the two universities are jointly involved in anatomical education activities. The University of Trieste has recently implemented a programme of digital anatomy, which aims to provide teaching with detailed three-dimensional views of anatomical structures. Firstly, dental students at both universities attend digital anatomy sessions on this programme. The Institute of Human Anatomy at the University of Padua holds the Reference Centre of the Veneto Region for Body Donation (Decree of 8th March, 2019), which is also one of the few Italian Reference Centres recognised by the Health Ministry (Decree of 21st November, 2022). As such, the University of Padua is one of the few Italian institutions that is permitted to deliver anatomy teaching using cadaveric resources. Following the sessions of digital anatomy in Trieste, dissections are performed at the Institute of Human Anatomy in Padua, with both sessions delivered for the students at both universities. For both the digital and cadaveric sessions, dental specialists contribute to delivery by emphasising the clinical implications of the anatomy being taught, and by highlighting surgical approaches to anatomical dissection. Here, a pilot evaluation of a combined digital-cadaveric approach to anatomy education is

A. Porzionato (✉) · V. Macchi · R. De Caro
Section of Human Anatomy, Department of Neuroscience, University of Padua, Padua, Italy
e-mail: andrea.porzionato@unipd.it

V. Nicolin · R. Di Lenarda
Department of Medical, Surgical, and Health Sciences, University of Trieste, Trieste, Italy

E. Stocco
Section of Human Anatomy, Department of Neuroscience, University of Padua, Padua, Italy

Department of Women's and Children's Health, University of Padua, Padua, Italy

Department of Surgery, Oncology and Gastroenterology, University of Padua, Padua, Italy

S. Sivolella · E. Stellini
School of Dentistry, Department of Neuroscience, University of Padua, Padua, Italy



I. D. Keenan et al. (eds.), *International Anatomical Education*, Biomedical Visualization 9,
https://doi.org/10.1007/978-3-031-91849-0_6

presented, which shows highly positive student perceptions of the format across several areas, including the content, resources, educators and facilities. The valuable cross-institutional collaborative interactions between anatomists, dental specialists and dental students who are facilitated by this combined approach are also described.

**Keywords**

Anatomical education · Forensic clinical anatomy · Virtual anatomy · Anatomage · Body donation programme · Dental school

## 6.1 Introduction

Human anatomy is a fundamental subject in the education and training of medical and health professional students. Learners are expected to identify relevant structures and understand their spatial relationships. However, acquiring an adequate understanding of three-dimensional (3D) anatomy from images remains a challenge (Zhao et al. 2020).

For centuries, gross anatomy has been considered the cornerstone of medical education and has been regarded as a highly clinically relevant discipline compared to other subjects (Pabst et al. 2001). Additionally, cadaveric dissection has long been the traditional method for learning anatomy by exposing and examining tissues and structures within the body. Anatomical knowledge is essential for patient examination and diagnosis formulation, and is crucial for surgeons and anyone performing invasive, emergency or diagnostic procedures (Turney 2007; Sugand et al. 2010). Cadaveric dissection continues to play a critical role in the traditional medical curriculum. Through this approach, students encounter the realities of life, morbidity, mortality and the responsibilities of being a physician (Kalthur et al. 2022). Furthermore, dissection can support learners in their development of observational skills and confidence in their anatomical knowledge (Lempp 2005; Burgess et al. 2016). Dissection also encourages small group discussion, which is important for the development of communication skills and introduces students to a wide array of new terminology (Lempp 2005; Burgess et al. 2012). Therefore, there is a consensus among medical professionals that cadaveric dissection supports anatomy learning while also providing relevant clinical correlations (Ghosh 2017). The cadaver can be considered as the ‘first patient’ for medical students, and for many, it represents their first encounter with death. The opportunity to touch a cadaver is not only valuable for improving the knowledge of gross anatomy, but also helps students reflect on the value of human life (Collett and Lester 1969). Dissection plays a role in the formation of professional identity by reinforcing compassion, respectful attitudes and patient-centred professionalism, as supported by qualitative research (Ong et al. 2023). By providing hands-on experiences with the human body, dissection enhances student appreciation and respect, an essential part of the process of professional identity formation (Ong et al. 2023). However, despite its benefits, there are several limitations to cadaveric dissection, including restricted access to cadaveric material, high costs and increasing numbers of students. These challenges, along with the advent of modern technologies in learning (audiovisual aids, YouTube videos and virtual/augmented reality tools), have prompted changes in how human anatomy is taught (Bartoletti-Stella et al. 2021; Fyfe et al. 2018). Moreover, the student profile has changed, and modern students are now more familiar with technology-based teaching methodologies than previous generations (Border 2019).

Currently, the effectiveness of modern methods in meeting the needs of practical anatomical education is still under discussion, and many studies have explored the role of virtual and digital approaches in anatomical education (Willett et al. 2022; Santos et al. 2022; Owolabi et al. 2022; Stone et al. 2022; Massalou et al. 2022). Additionally, in the specific field of dental and maxillary surgery, various digital resources have been progressively incorporated into dental schools, raising a debate about their specific roles in anatomical education. The use of digital anatomy visualisation tables for teaching the anatomy of the temporomandibular joint showed that, while this specific approach may support understanding of this particular topic, it may not be directly applicable to the investigation of other anatomical regions (Keenan and Ben Awadh 2019; Ben Awadh et al. 2022; da Silveira et al. 2022). A recent editorial article highlighted that three-dimensional (3D) and digital technologies in anatomical education may be more effective for learning than traditional didactic teaching methods but are unlikely to entirely replace cadaveric teaching (Chytas et al. 2021). Most authors agree that an approach combining both digital virtual and cadaveric methods is the optimal strategy for anatomical education (Darras et al. 2019; Quek 2024; Şişu et al. 2024). A recent study on the potential role of virtual classes in anatomy education indicated that dental students experience additional challenges when studying cadaveric anatomy and embryology in virtual classes: they miss every aspect of anatomy education, including cadaveric lab but also face-to-face lectures which support this study. Therefore, entirely digital approaches may be considered as an option, but not a permanent solution, for dental anatomical education (Makaju and Rai 2021).

In our experience, neuroanatomy (cranial nerves, ventricles, cerebellum), abdominal anatomy (orientation of peritoneal cavity, peritoneal reflections, relations of viscera), head and neck anatomy, osteology of skull, structural relations of head and neck, upper limb anatomy (course and relations of vessels, hand, blood supply), pelvis (peritoneal cavity, peritoneal reflections, relations of viscera) and lower limb anatomy, all present significant challenges. This is due to the complexity of structures, spatial relationships, and intricate functional systems. Traditional teaching methods, such as dissection and laboratory work, may be complemented by digital tools like 3D models, interactive software and virtual dissections in order to address these challenges.

Here, we present a combined approach using both digital and cadaveric anatomy for fourth-year dental students, in a format of joint delivery by the Universities of Padua and Trieste. We incorporated the facilities and resources of the two universities in a joint experience across two types of teaching sessions (digital dissection and practical dissection), to investigate student perceptions of our own combined approach for teaching and studying human anatomy. The sessions focused on the floor of the mouth, oral cavity, salivary glands and the main branches of the trigeminal and facial nerves. In both session types, dental specialists highlighted the clinical implications of, and surgical approaches to, anatomical dissections.

### 6.1.1 Study Goals

This study aimed to evaluate the effectiveness of combining digital and cadaveric learning for enhancing the learning and experience of challenging anatomical topics among dentistry and orthodontics students. To address this aim, this study has explored the following research question:

*To what extent do dental students perceive value in elements of a combined practical and digital learning approach for head and neck anatomy education?*

We hypothesise that by providing an integrated, dynamic, flexible and engaging learning experience, our combined approach will yield positive student perceptions relating to the alignment with course objectives, their involvement in activities, the quality of specimens used, educator facilitation and the facilities.

## 6.2 Materials and Methods

### 6.2.1 Educational Context

#### 6.2.1.1 The Body Donation Programme at the University of Padua

The development of body donation programs, grounded in ethical principles, is essential for the use of cadavers in medical and healthcare professional education. Additionally, surgical simulations based on donor cadavers have enhanced surgical practice and improved outcomes (Zdilla and Balta 2023). As such, human body donors are fundamental to the advancement of medical education and practice. However, the limited number of body donations, often due to misinformation on the topic, i.e. a lack of accurate knowledge about body donation procedures and legal frameworks (Figueiredo and Filho 2023); and cultural beliefs and societal norms (Singh et al. 2011); combined with the increasing demand for training courses and research activities; has led to a global shortage of cadavers (De Caro et al. 2021).

To meet the increasing demand for cadavers, the expansion of body donation programmes is essential. As a result, several countries, including Italy, introduced new regulations to govern this activity, with Italy implementing changes in 2020 (Boscolo-Berto et al. 2023). The University of Padua has retained a Body Donation Program, which accepts both whole bodies and anatomical parts, for over 30 years (Macchi et al. 2011). Since 2011, the University of Padua Body Donation Program has been subject to annual certification in accordance with ISO 9001 standards (Porzionato et al. 2012). This external certification aims to enhance the efficiency of the body donation process, in turn also affecting the quality and outcomes of medical education. Key aspects of the certification process include the collection of body donations, management of death certificates, handling of personal data, procurement of body parts from living donors, transportation and identification of cadavers, and the handling of bodies or body parts. Additionally, it covers the delivery of anatomy training sessions and the periodic testing and updating of equipment (De Caro et al. 2009; Boscolo-Berto et al. 2023). Furthermore, the programme undergoes annual internal and external quality audits aimed at both reviewing outcomes and planning for continuous improvement. Specifically, the number of donors and donated body parts per year, the number of training sessions conducted, the number of learners and donor satisfaction with the programme are monitored through questionnaires. These evaluations help identify potential critical issues in the body donation process and related activities (Boscolo-Berto et al. 2023).

In 2019, the Padua Body Donation Programme was recognised as the Reference Centre for the Veneto Region for the 'Use and Conservation of Donated Bodies' (Decree of the Veneto Region 8th March, 2019). The following year, Italian Law No. 10/2020 established the legal, ethical and technical guidelines for the donation of tissues and bodies post-mortem for purposes of study, training and scientific research (De Caro et al. 2021; Boscolo-Berto et al. 2023). In 2021, the Padua Body Donation Program was further recognised as one of the seven National Centres for Use and Conservation of Dead Bodies (Decree Health Ministry, 21st November 2022). The program ensures that from the arrival of a donor to the return of their body to relatives, strict protocols are followed to manage chemical, biological and carcinogenic risks. Key requirements for the reference centre include designated rooms for receiving the deceased, conducting preliminary external inspections, and performing embalming procedures, along with appropriate storage for coffins, technological equipment and personal protective equipment. Facilities are required to include cold rooms with automated or semi-automated systems for handling of bodies, as well as clearly defined dirty, clean and transit areas, all with surfaces, floors, and walls that are resistant to embalming chemicals and are easy to clean and sanitise. The facility must include changing rooms, a controlled mechanical ventilation system for dissecting rooms with separate air expulsion channels (which must be equipped with special filters) and aspirated anatomy dissection tables for waste management, including dedicated tanks for liquid decontamination. Safety cabinets for chemical/flammable products, controlled access to the areas and a remote alarm system for detecting system malfunctions are also essential. Additional spaces, such as dissecting rooms and classrooms, should be available for staff training, while the centre must be staffed with medical, technical and administrative personnel to ensure smooth operations. These requirements comply with the guidelines set by the General Directorate of Health Prevention, Ministry of Health, Italy (2021) and the Italian

Republic (2017) to ensure the safe and hygienic handling of deceased bodies.

The *Interdepartmental Centre for Continuous Training on Cadaver Andrea Vesalio* was established from the convergence of interests between the Departments of Neurosciences, Surgical Oncological and Gastroenterological Sciences and Cardio-Thoraco-Vascular Sciences and Public Health of the University of Padua. The Interdepartmental Centre, working in close coordination with the promoting and affiliated departments, encourages, supports and coordinates research activities focused on human anatomy, surgical and topographical anatomy and clinical anatomy. It also supports translational research activities in fields such as regenerative medicine and tissue engineering, provides specialised training in surgery, and promotes communication and collaboration between scholars from different scientific backgrounds. As part of an anatomical education initiative, the centre has partnered with the University of Trieste to collaboratively develop a combined programme for dental students. For this purpose, the dissection rooms (*Vesalio* and *Aquapendente* rooms), which were specially equipped in recent years for practical dissection courses, have been used by medical and dental students, and by dental residents and specialists from various surgical fields (Macchi et al. 2014).

#### 6.2.1.2 Digital Technologies at the University of Trieste

The Institute of Human Anatomy of the University of Trieste utilises advanced software for digital and topographic Anatomy, specifically the Visible Body Software (https://www.visiblebody.com/) and the Anatomage Table (Anatomage, Inc., Santa Clara, California) (https://anatomage.com/).

The Visible Body is an interactive human anatomy atlas that includes over 5000 anatomical structures across regional and systemic anatomy. It allows the learner to explore both macroscopic and microscopic anatomy through high-definition, 3D interactive models. The models were developed by a team of medical illustrators, biomedical visualisation experts and animators. The database includes programs focused on circulatory, muscular and skeletal systems, offering in-depth information about any anatomical structure. Users can rotate, zoom and dissect the models for a more immersive learning experience (Williams and George 2013; Lewis et al. 2014).

The Anatomage Table, developed by California-based company Anatomage Inc. in partnership with the Stanford Clinical Anatomy Department, is an advanced anatomy visualisation system (Baratz et al. 2019; Raja et al. 2022; Alasmari 2021). This virtual table provides a life-size digital representation of the human body, which is displayed on a screen measuring 2.13 × 0.67 m. Together with cadaver visualisation, it allows for the manipulation and virtual dissection of complex anatomical structures in 3D space. Learners can progressively deconstruct the digital body, removing the skin to reveal structural layers including the muscles, skeleton, internal organs, nerves and blood vessels. Each structure can be viewed in varying layers, scales and sizes, also focusing on the relationships between different body parts and internal organs. Furthermore, the system is enhanced by the integration of advanced imaging technologies including computed tomography (CT), X-ray, ultrasound and Magnetic Resonance Imaging (MRI), all accessible through a user-friendly touchscreen interface. Both male and female anatomy can be analysed, as the imagery featured on the platform is derived and licensed from the male and female bodies included in the Visible Human Project and Visible Korean Human (Martín 2018; Kavvadia et al. 2023). The platform provides attention to detail, the ability to observe any section and the magnification of anatomical structures.

#### 6.2.1.3 Forensic Implications of Clinical Anatomy

We have previously highlighted the importance of anatomical and surgical education using donor cadavers, as well as the medico-legal implications of anatomical education and surgical training with cadavers (Pirri et al. 2021). As such, a key focus of our combined approach is to emphasise the clinical and medico-legal implications of illustrating anatomical structures through digital resources and dissection techniques, particularly within what we have previously defined as 'forensic clinical anatomy' in the dental field (Porzionato et al. 2017a, b, 2019, 2022; Pirri et al. 2021). Forensic clinical anatomy is defined as the practical application of anatomical knowledge and methods (ranging from ultrastructural to macroscopic anatomy) within clinical and surgical contexts, in order to address medico-legal issues (Porzionato et al. 2017a, b), In our approach to forensic clinical anatomy, we emphasised the concept of 'individual anatomy', which refers to the anatomical features of a specific patient at any given moment within a clinical, surgical or forensic context.

When considering individual anatomy, we account not only for normal anatomical structures and their variations, but also for factors such as sex, age, development and the anatomical changes resulting from diseases, surgery or para-physiological conditions (e.g. pregnancy or physical training). These factors can considerably alter the topographic relationships of otherwise normal anatomical structures, e.g. a volumetric increase in organs due to a neoplasm may alter the topographic anatomy of the region, with subsequent clinical implications. Additionally, in forensic examinations, post-mortem or artifact-induced modifications must be considered, and we consequently emphasise this in our teaching.

## 6.3 Study Design

### 6.3.1 Population and Sampling

A cohort of fourth-year undergraduate students from our dentistry and orthodontics course ($n$ = 40) participated in the joint educational activity in 2023. The group consisted of students from the University of Trieste ($n$ = 20) and the University of Padua ($n$ = 20).

### 6.3.2 Organisation of the Anatomical Education Activities

The collaborative meetings between Padua and Trieste began in 2021, featuring cadaveric dissections in Padua and digital anatomy sessions via web link. In 2022 and 2023, other improved editions of the activity were proposed. The events expanded to 2 days, with the first day in Trieste focused on using Anatomage and other digital anatomy tools, while the second day in Padua involved hands-on dissection. Students were in-person present both in Trieste and in Padua. Both Padua and Trieste students engaged in these activities, with discussions on forensic implications such as iatrogenic lesions and professional liability based on relevant scientific literature.

The topics covered in the sessions included the surgical anatomy of the inferior alveolar nerve; the maxillary sinus; the surgical and topographic anatomy of the parotid gland and duct of parotid duct; the topographic anatomy of the neck and submandibular gland.

The learning approaches described were designed for fourth-year students, who can understand clinical-surgical and medico-legal implications of various anatomical structures. Students were encouraged to use the instrumentation, fostering direct interaction between teachers and dental students. The dissections were also recorded, and videos were shared with other students (students of the degree courses in Medicine and Surgery, Dental Hygienist and students of the Specialisation Schools in Paediatric Dentistry and Orthodontia).

#### 6.3.2.1 Learning Activity 1: Surgical Anatomy of the Inferior Alveolar Nerve

Within the combined learning experience, we first described the topographic anatomy of the inferior alveolar nerve using anatomical software in Trieste, with a particular focus on its course within the mandibular canal and the emergence of the mental branch. A series of cone-beam computed tomography images were also used to show the anatomical variability of the inferior alveolar nerve canal, especially in relation to the apices of the teeth. The significance of the anatomical variability of the mandibular canal and inferior alveolar nerve was also discussed, specifically considering studies that correlate the anatomy of these structures with the risk of iatrogenic lesions (Gupta et al. 2021; Kumar et al. 2021; Chua et al. 2022; Srivastava et al. 2022; Al-Haj Husain et al. 2022).

Students need to understand the dimensions of the mental nerve. The emergence of the nerve was first identified through sectioning of the inferior lip and inferior labial fornix, followed by dissection, in Padua, of the soft tissues from the bone surface of the mandible. The gingival layers and the alveolar mucosa were cut to reach the bony plane and identify the mental nerve (Fig. 6.1a–c). The mandibular canal was opened using a piezoelectric bone cutter, allowing for direct observation of the calibre, topographical relationships and radicular branches of the inferior alveolar nerve (Fig. 6.1d–h).

Subsequently, the clinical and forensic implications of iatrogenic injuries related to individual anatomical structures were discussed with the students in Padua. For instance, the inferior alveolar nerve can be injured during a range of dental procedures, including the administration of local anaesthetic, implant bed preparation, and placement endodontic procedures, and third molar extraction (Renton 2010a, b; Suomalainen et al. 2013; Cassetta et al. 2014; Leite et al. 2014; Bürklein et al. 2015; Li et al. 2017; Kavarthapu and Thamaraiselvan 2018; Ali et al. 2018; Vidya et al. 2019; Oliveira et al. 2019; Diakonoff and Moreau 2022; Pääsky et al. 2022; Sodnom-Ish et al. 2022). Injuries to the inferior alveolar nerve are recognised as a severe clinical problem with devastating consequences for the quality of life of patients. Such lesions affect nerve function, leading to temporary or permanent paraesthesia. Moreover, they may be responsible for alterations in speech, taste, chewing, tooth brushing and the ability to maintain lip competence with an impact on food retention and swallowing. Patients suffering from inferior alveolar nerve lesions may suffer from social, personal and psychological disabilities (Matus et al. 2023). Most iatrogenic injuries to the inferior alveolar nerve can be prevented by careful patient assessment and planning, based on consistent evaluation of anatomical risk factors. Furthermore, if iatrogenic injuries occur, there are various approaches to facilitate recovery in the early post-surgical phase (Renton 2010a, b).

#### 6.3.2.2 Learning Activity 2: Surgical Anatomy of the Maxillary Sinus

Preliminarily, the maxillary sinus was evaluated at the University of Trieste using anatomical software. This site is important as it undergoes modification (a progressive thinning) and ageing. Then, in Padua, surgical access to the maxillary sinus was anatomically demonstrated. The superior lip was cut in the sagittal plane to show the layers (Fig. 6.2a); then the alveolar mucosa and the gingiva were cut and cranially detached from the bone surface to show the anterior aspect of the maxillary bone (Fig. 6.2b, c). The infraorbital foramen and emergence of the infraorbital nerve were also identified, as students need to identify and measure the nerve diameter and terminal nasal, labial and eyelid branches, in order to appreciate mechanisms of iatrogenic nerve damage. Students also observed variations in the thickness of the anterior bony wall of the maxillary sinus. The maxillary sinus was opened through incision of the anterior wall to illustrate the space, volume and characteristics of the bony walls and mucosal layer (Fig. 6.2d–f). In this procedure, the mucosal layer may maintain integrity or may detach from the bone (Fig. 6.2g, h). This allows students to understand the anatomical basis for surgical procedures, e.g. maxillary sinus augmentation lifting, in addition to the topography of key structures.

During dissection, the implications of anatomical variations were discussed with students, including reference to surgical procedures, iatrogenic risks and medico-legal implications, e.g. age-related progressive thinning of the maxillary sinus walls; variations in the prevalence of septa within the maxillary sinus (Underwood 1910; Krennmair et al. 1999; Kim et al. 2006) when

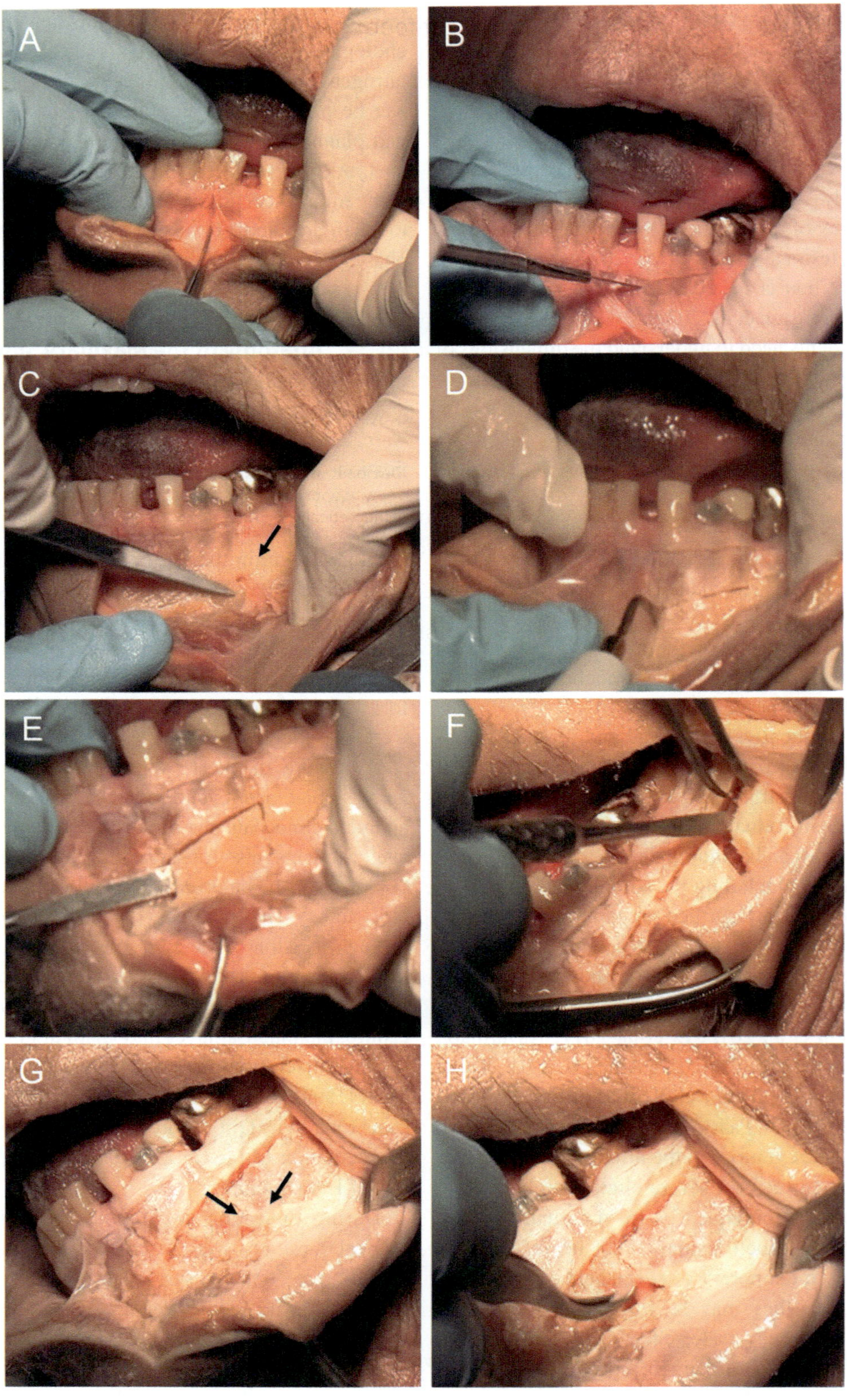


**Fig. 6.1** Surgical anatomy of the inferior alveolar nerve. Section of the inferior lip (**a**) and detachment of gingiva and alveolar mucosa (**b**) to reach the mandible bone surface and the emergence of the mental branch (**c**). Opening of the mandibular canal through piezoelectric bone cutter (**d–f**). General view of the course of the inferior alveolar nerve (arrows) in the mandibular canal (**g**, **h**)

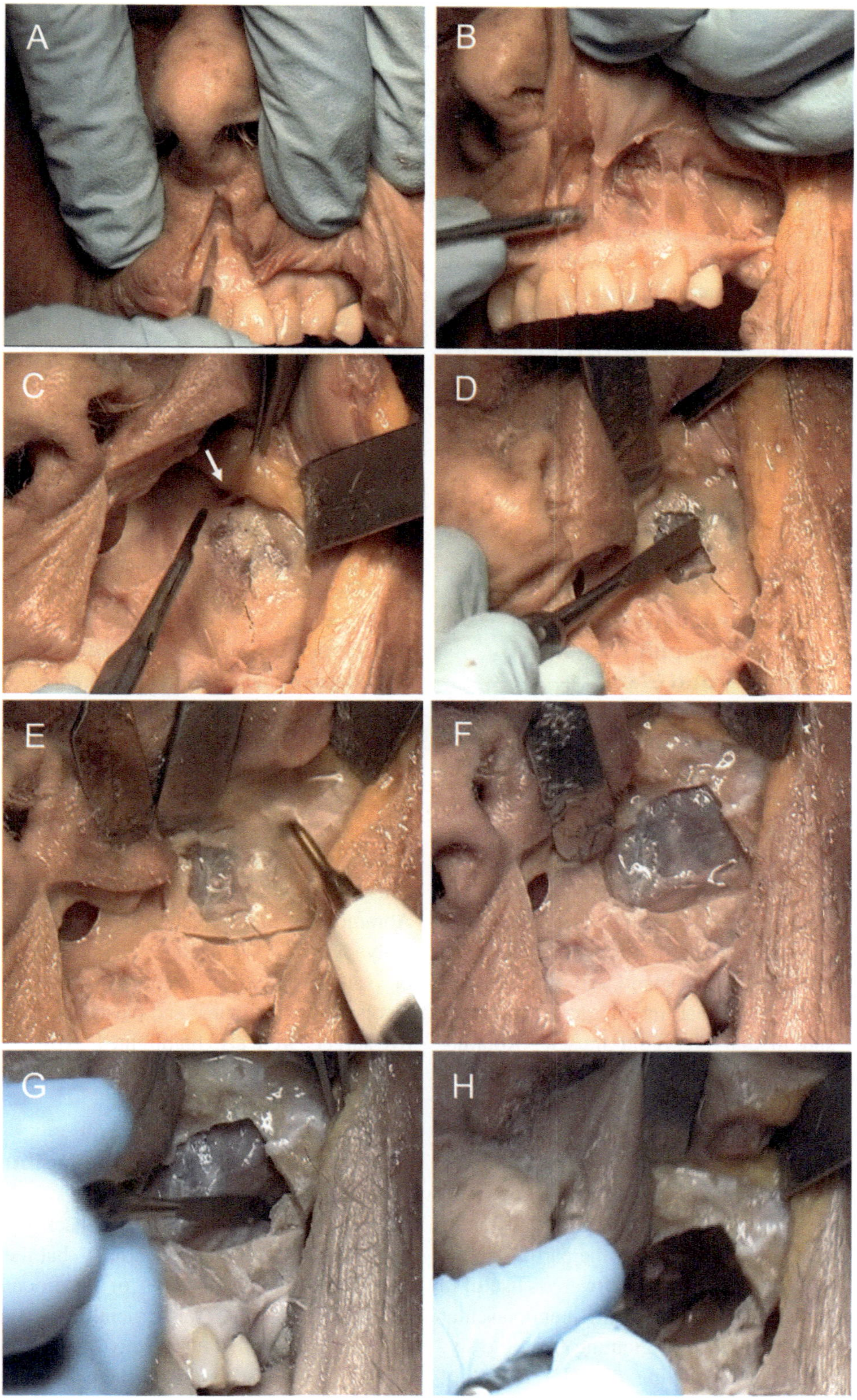


**Fig. 6.2** Surgical anatomy of the maxillary sinus. Section of the superior lip (**a**) and detachment of gingiva and alveolar mucosa (**b**) to reach the maxillary bone surface and the emergence of the infraorbital branch (arrow) (**c**). Note that the anterior wall of the maxillary sinus is particularly thin in this case. Progressive opening and removing of the anterior wall of the maxillary sinus, maintaining the integrity of the maxillary mucosal membrane (**d**–**f**). Detachment of the maxillary sinus membrane from the inferior wall of the maxillary sinus (**g**, **h**)

compared to complete separation (Krennmair et al. 1999; Lee et al. 2010; Maestre-Ferrín et al. 2011; Iwanaga et al. 2020). The presence of septa is used to predict perforation of the olfactory mucosa during sinus floor augmentation procedures (Pommer et al. 2012; Von Arx et al. 2014; Jung et al. 2019).

Finally, the upper topographical relationships of the maxillary sinus with the orbital cavity were demonstrated, showing the thickness of the superior wall (orbital cavity floor), which is important for understanding the risk of orbital and periorbital emphysema after maxillary sinus floor elevation (Farina et al. 2016), particularly as the aetiology of emphysematous complications of dental surgery are largely iatrogenic (Heyman and Babayof 1995; McKenzie and Rosenberg 2009; Farina et al. 2016).

#### 6.3.2.3 Learning Activity 3: Surgical and Topographic Anatomy of the Parotid Gland and Parotid Duct

The topographic anatomy of the parotid gland and its course was considered, particularly referring to associated structures that have direct topographical relationships.

Firstly, in Trieste, a series of images from the Anatomage were used to provide a three-dimensional perspective of relevant anatomical structures, e.g. the anterior topographical relationships of the parotid gland with the masseter muscle, posterior margin of the mandibular ramus and medial pterygoid muscle. Rotating and moving the images in three spatial dimensions enables a more effective demonstration of topographical relationships, which is challenging to achieve during cadaveric dissection due to the deep medial extension of the parotid gland. With reference to the deep topographical relationships of the pharyngeal extension of the parotid gland, 3D anatomical visualisation software is useful to demonstrate relationships with neurovascular structures and muscles originating from the styloid process, and enabling digital removal of structures to enhance visualisation of other topographical relationships.

The parotid duct was shown in the 3D software, particularly in its course on the masseter and buccinator muscles. These structures were also shown on MRIs (magnetic resonance images) to emphasise their appearance on diagnostic images, in different sections and at different levels, showing the topographic relationships of the parotid gland and duct with bone, muscular and neurovascular structures.

The above digital demonstrations were combined with anatomical dissection of the parotid gland and duct (Fig. 6.3) in Padua, including the internal topographical relationships of the parotid gland with the external carotid artery, internal maxillary and superficial temporal arteries, venous network and the auriculotemporal and facial nerves. The course and branches of the facial nerve were discussed with students, with reference to risks of iatrogenic injury during parotid gland surgery. Facial nerve injury represents the most severe complication of parotid gland surgery, as it can lead to transient facial nerve dysfunction after a parotidectomy, and a temporary or absolute facial nerve palsy, severely affecting the quality of life of patients (Chiesa-Estomba et al. 2023). Clinical and forensic perspectives were also discussed with respect to both digital and cadaveric resources, e.g. emphasising the risk of post-traumatic (Landau and Stewart 1985; Parekh et al. 1989) and iatrogenic (Barron et al. 2001; Capaccio et al. 2004; Abramova et al. 2008; Lawson 3rd et al. 2012; Kulyapina et al. 2014; Rajeev et al. 2016) fistulae or sialoceles. The potential iatrogenic risks involved in oncological and plastic or reconstructive surgery were also highlighted.

#### 6.3.2.4 Learning Activity 4: Topographic Anatomy of the Neck and Submandibular Gland

The topographic anatomy of the neck is important for dental students to learn, so that they may understand the organisation of the floor of the oral cavity. It is therefore crucial for students to view all muscles that insert into the hyoid bone in three dimensions, to enable them to understand the role of this bone and supra-hyoid muscles in forming the floor of the oral cavity.

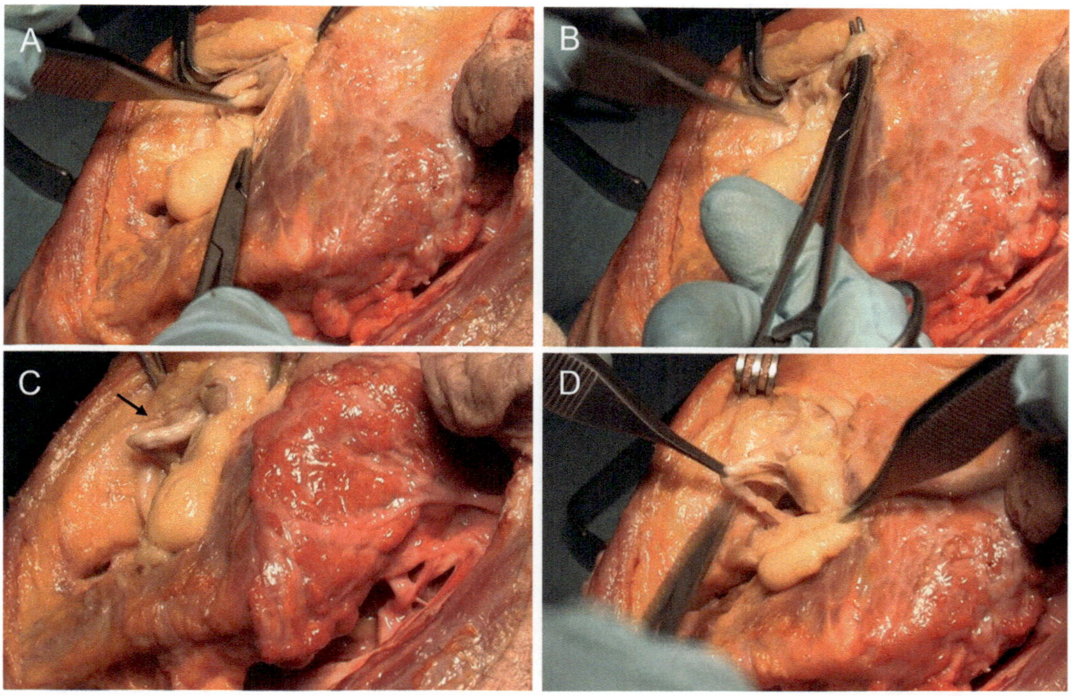


**Fig. 6.3** Topographic anatomy of the parotid gland and parotid duct. Different phases of dissection of the parotid duct (arrow) with respect to buccal fat pad (**a–d**). Note the topographical relationships of the parotid gland with the masseter muscle

In Anatomage sessions, in Trieste, educators highlighted important and relevant anatomical structures for dental students. Firstly, the sub- and supra-hyoid muscles, and their origins, insertions and topographical relationships with neurovascular structures were demonstrated. The topographical relationships of the submandibular and sublingual glands with respect to the supra-hyoid muscles were also shown. Finally, the course of neurovascular structures, e.g. the lingual artery, vein and nerve and the hypoglossal nerve, which are less challenging to identify in labelled Anatomage images or videos than in cadaveric dissections, was demonstrated.

Cadaveric dissection of the neck was then performed, in Padua, to emphasise the methodological approach of dissection by layers. The first superficial dissection allowed identification of the thin platysma muscle, the superficial jugular network and the deeper sternocleidomastoid muscle (Fig. 6.4a, b). The sternal and clavicular insertions of the sternocleidomastoid muscle were cut, and the muscles cranially and laterally moved to permit direct access to the deeper sub-hyoid muscles (Fig. 6.4c). The sternocleidomastoid and sub-hyoid muscles were also described with reference to the superficial and middle cervical fasciae. The course of the cervical neurovascular bundle, and the fascial specialisations were also considered, with particular emphasis on the relevance of the course of the external carotid artery and origin of its collateral branches.

After the specific identification of the sub-hyoid muscles, the anatomy of the visible supra-hyoid (digastric, stylohyoid and mylohyoid) muscles was demonstrated. The submandibular gland was also shown in strict topographic relationships with the tendons of the supra-hyoid muscles (Fig. 6.4d–f), and with the posterior margin of the mylohyoid muscle and its continuity with the uncinate process of the submandibular gland.

Educators combined anatomical descriptions and cadaveric dissections with forensic images

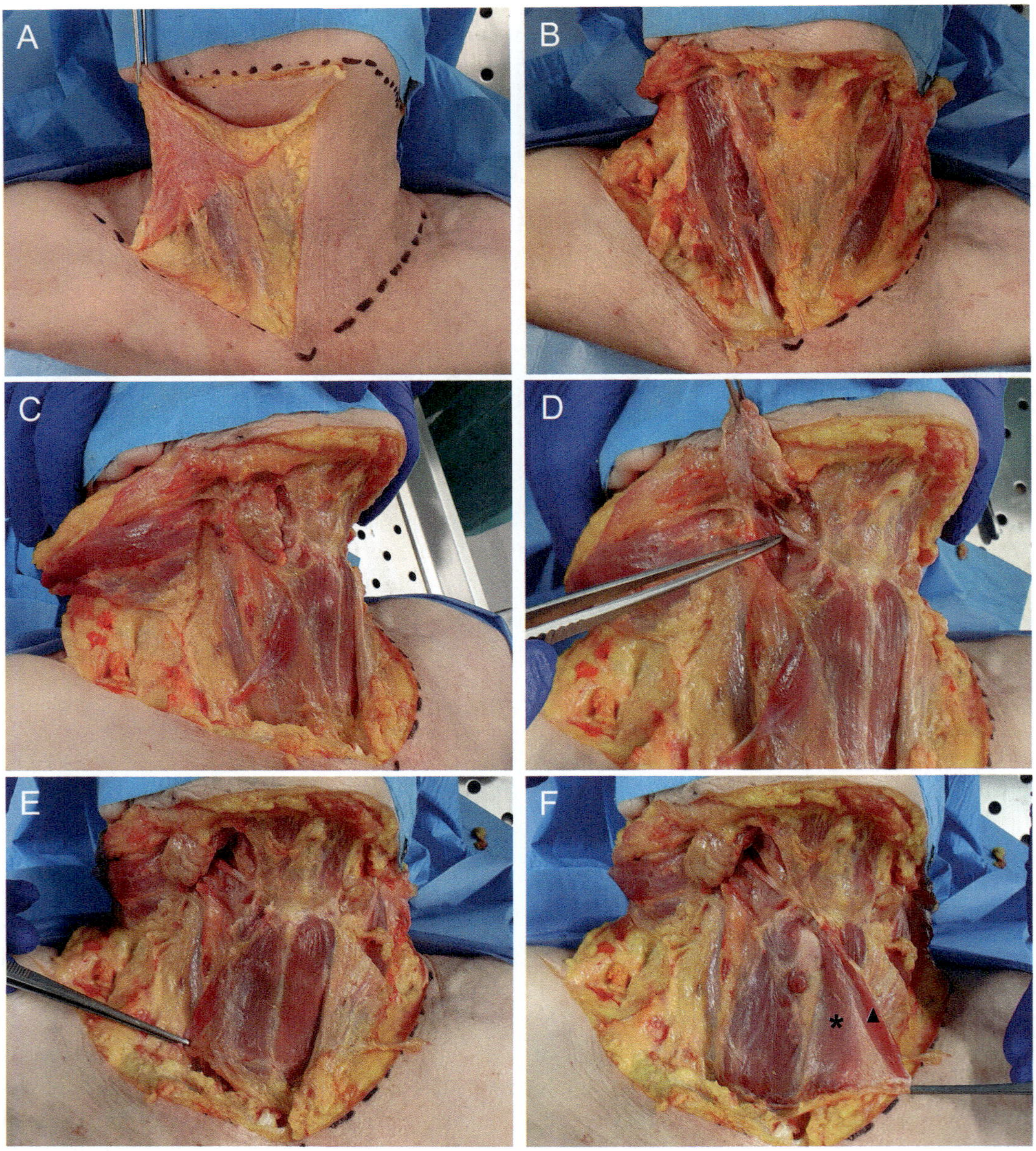


**Fig. 6.4** Topographic anatomy of the neck and submandibular gland. Raising of the skin and platysma on the right side of the neck (**a**). General view of the right sternocleidomastoid muscle and submandibular gland (asterisk) (**b**). Detachment of the caudal insertions of the right sternocleidomastoid muscle, which is also latero-posteriorly moved, with direct view of the right omohyoid muscle (**c**). Dissection of the right submandibular gland which is superiorly moved in order to show its deep topographical relationships, with the intermediate tendon of the digastric muscle (passing through the stylohyoid muscle) and mylohyoid muscle (**d**). Section of the inferior muscle belly of the omohyoid muscle (**e**) and medial movement of the above muscle to show the sternothyroid muscle (**f**). In (**f**), the black asterisk shows the superior belly of the right omohyoid muscle, whereas the black arrow shows the right sternohyoid muscle

showing traumatic injuries of specific anatomical structures, asking students to consider which anatomical structure had been injured in each case, and discussed the anatomical basis of iatrogenic injuries to the submandibular gland (Papadiochos et al. 2021; Dediol et al. 2022).

### 6.3.3 Questionnaire Instrument

Following the 'hands-on' experiences (edition 2023), all participating students ($n$ = 40) were asked to anonymously report their perceptions and experiences of the learning approaches by responding to ten-point Likert-type scale questionnaire items. The questionnaire was designed to evaluate the activities performed at the University of Padua, only. It was adapted from the questionnaire routinely administered to participants in dissection courses, following the quality management procedures of the Body Donation Programme at the University of Padua (Porzionato et al. 2012). Learners responded to the following questionnaire items:

(a) *How would you evaluate the alignment of the procedural content with the objectives of the course?*
(b) *How would you assess your level of involvement in the training activity?*
(c) *What is your evaluation of the anatomical specimen used?*
(d) *How would you assess the Professor's management of the procedure?*
(e) *How would you evaluate the adequacy of the facilities and equipment?*

### 6.3.4 Data Analysis

The mean and standard deviation (SD) of student responses to each item were calculated.

### 6.3.5 Ethical Assessment

All dissections were performed on donated bodies of the Reference Centre of the Veneto Region for Body Donation, in accordance with Italian and European laws.

The studies involving humans enrolled in the Body Donation Program of the Section of Human Anatomy, University of Padua, followed the European, Italian and regional guidelines. Excision/dissection was further authorised by the Italian law No. 10 of 10 February 2020, entitled 'Rules regarding the disposition of one's body and post-mortem tissues for study, training and scientific research purposes'. The human bodies used in this study were acquired from patients/participants who provided their written informed consent to participate in this study.

Ethical assessment was also performed relating to the use of students as participants in the research, including their consent to participate and the secure and confidential storage of their data.

## 6.4 Results

Mean questionnaire responses of participants ($n$ = 40) to each item were calculated as follows: 9.76 ± 0.22 for the coherence of the procedure contents with the course objectives (*question a*); 9.81 ± 0.09 for involvement in the training activities (*question b*); 9.86 ± 0.07 for the adequacy of the anatomical material (*question c*); 9.81 ± 0.08 for the management of the procedure by the educators (*question d*) and 9.86 ± 0.1 for the facilities and equipment (*question e*) (Fig. 6.5).

In addition to the questionnaire responses they provided, students were given the opportunity to provide written free-text feedback and to suggest topics for future activities. Some students recommended increasing cadaver dissection, highlighting the considerable impact of this approach on their education. In particular, they stated that dissection allowed them to view structures in their natural context; aided them in retaining theoretical knowledge; taught the attitudes expected of medical professionals; triggered a range of emotional responses, from curiosity to respect. Other students ($n$ = 5) emphasised the impact of virtual learning on the practical 'hands-on' experience; interestingly, they referred that virtual environment helped them focus on the activities; made learning feel much more exciting; reinforced their knowledge.

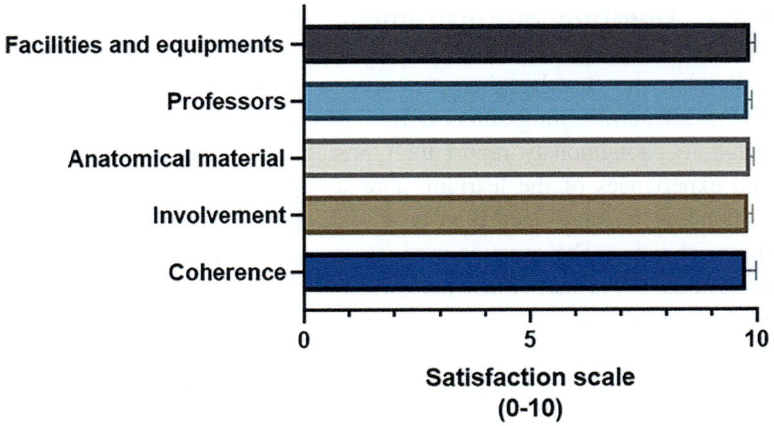


**Fig. 6.5** Questionnaire results. The graph shows the total score given by the students for each of the evaluation questions submitted to them, related to the hands-on experiences

## 6.5 Discussion

In this evaluation, students reported overwhelmingly positive perceptions of the combined digital-cadaveric learning activities. Aligning dissection activities with course objectives is crucial for achieving effective learning outcomes in medical education. This coherence ensures that students are prepared not only to understand theoretical concepts but also to apply them in real-world contexts, enhancing their grasp of anatomical structures. When dissection procedures are thoughtfully integrated with educational goals, students gain a deeper, more practical understanding of anatomy (Flack and Nicholson 2018). Such alignment also promotes attention to detail, problem-solving and technical proficiency that are essential critical skills for future medical practice (Cotofana and Lachman 2020). As students engage with dissections that are purposefully linked to learning objectives, they refine their ability to observe, analyse and interpret anatomical features with precision, which can be applied to clinical practice in the future. Furthermore, this also allows students to assess their own progress more effectively. They can identify areas of strength and weakness, leading to targeted improvements and greater confidence in their knowledge (Abdellatif 2020). This self-assessment not only motivates students but also enhances long-term retention of key concepts.

In summary, harmonizing dissection activities with course objectives is fundamental for creating a structured learning experience that maximises student success and prepares them for the challenges of medical practice. This approach ultimately leads to improved educational outcomes and better-prepared healthcare professionals (Nikolenko et al. 2019).

Involvement in training activities, such as anatomical dissection, plays a crucial role in enhancing learners' visual-spatial awareness. This cognitive skill is vital for students to mentally construct and manipulate three-dimensional anatomical structures. As Estai and Bunt (2016) suggest, dissection activities not only help improve this skill but also foster a deeper understanding of the complex relationships between structures within the human body. By physically engaging with anatomical specimens, students gain a more tangible and intuitive sense of spatial relationships that may be difficult to replicate through textbooks or digital models alone.

Moreover, these hands-on experiences provide an invaluable opportunity for students to observe and practice professional behaviours and non-technical skills (Cahill et al. 2002). Dissection activities often occur in collaborative settings, where students must engage in communication, teamwork and problem-solving, fostering a collaborative learning environment (Huitt et al. 2015). Importantly, students also develop key interpersonal skills such as empathy, respect, compassion, and self-regulation as they work together to handle delicate specimens, learn to respect the human body and engage in ethical decision-making (Weeks et al. 1995; Abrams et al. 2021). These experiences are pivotal for cultivating the professional identity of students, as they are encouraged to reflect on the broader implications of their work in healthcare and their future roles as practitioners (Ghosh 2020; Robertson et al. 2024).

In addition to technical and professional development, dissection also provides students with insights into potential career aspirations. For many, the dissection process serves as a formative experience that may influence their interest in specific fields of medicine, surgery or research. By immersing themselves in the anatomical world, students can develop a clearer understanding of their personal interests and career trajectories within the broader field of health professions. Thus, dissection not only aids in academic learning but also nurtures the personal growth of students, helping them align their educational experiences with future professional aspirations (MacPherson and Lisk 2022). Certainly, all these achievements are also dependent on the adequacy of the anatomical material used and facilities/equipment available. The authors believe that the use of high-quality cadaveric specimens in dissection is crucial for effective medical education and surgical training, offering to students a unique opportunity to engage with real human anatomy also providing a direct connection between theoretical knowledge and its practical application; they can observe, palpate and interact with anatomical structures in ways that cannot be replicated by textbooks, 3D models or virtual simulations.

Thus, adequate cadaveric specimens provide students with essential hands-on experience (Wong et al. 2023), bridging the gap between theoretical knowledge and practical application. It facilitates a realistic learning environment, ultimately leading to better surgical outcomes.

While the focus on technical skills is essential, the emotional and psychological aspects of dissection education are equally important. The preparation of educators in dissection is crucial for fostering both technical and emotional competencies in medical students. Effective preparation not only enhances students' understanding of anatomy but also mitigates anxiety associated with cadaver dissection, ultimately shaping their professional attitudes and skills (Arráez-Aybar et al. 2008).

In the evaluation presented here, student perceptions that increased cadaveric teaching would benefit their learning may be due to the tangible experience of dissection. The process helps human anatomy understanding in a manner that digital or virtual tools cannot replicate. Moreover, through dissection, students can gain comprehension of the complexity and interconnectivity of the human body with the opportunity to recognise anatomical variations that might not be reported in textbooks or virtual models (Brenner 2011; Morris and Jacques 2018; Khorsandi et al. 2021).

Students also reported that they valued the practical 'hands-on' elements of the combined activity. Virtual learning can provide high-resolution images with accurate views of anatomical structures; moreover, virtual tools can be used anytime, making learning easier and more flexible. Given that anatomy requires hands-on experience with cadavers, the development of technologies has provided new ways for active engagement and visualisation of structures in anatomy education (Pettersson et al. 2024). One such tool is the Anatomage Table, an advanced anatomy visualisation system, which several studies have suggested can augment learning experiences for students in anatomy courses (Chaudhry et al. 2023). Modern teachers of human anatomy are expected not only to teach but also to facilitate student learning by integrating innovative technologies. To do so effectively, anatomy educators must stay updated on the latest advancements in medical education (Keenan and Ben Awadh 2019; Patra et al. 2022), as digital resources offer alternative learning modes and serve as a valuable complement to traditional teaching methods (Pettersson et al. 2024).

In the view of the authors, based on personal experiences and observations, the anatomical education approach developed by the Institutes of Human Anatomy of the Universities of Padua

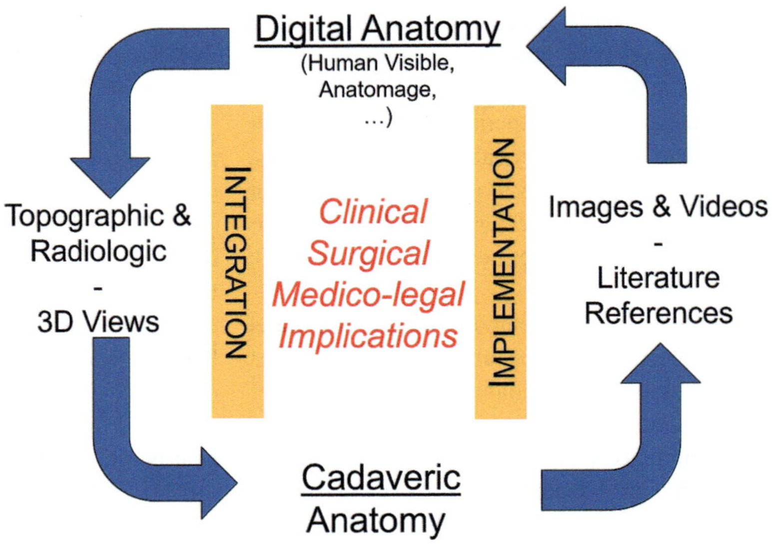


**Fig. 6.6** Reciprocal integration and implementation between digital and cadaveric anatomy. Digital anatomy can be combined with cadaveric anatomy through topographic and radiologic 3D views; in turn, cadaveric anatomy can be implemented by resources including images, videos and literature

and Trieste has successfully combined digital and dissection-based anatomy, optimising the resources available at both institutions. The University of Trieste provided access to its anatomical-digital technologies (i.e. Visible Body Software, and the Anatomage Table), which were used for preliminary demonstrations of the three-dimensional appearance of the anatomical structures. Meanwhile, the University of Padua provided anatomical dissection, working on cadavers from its Body Donation Program.

Through this experience, students had the opportunity to compare digital anatomy with 'real' appearance of tissues and anatomical structures, in situ, enabling them to envision scenarios closely resembling clinical practice.

From a methodological perspective, the combined implementation of digital and cadaveric anatomy was achieved using topographic and radiological images, as well as three-dimensional views of different anatomical structures, both before and during cadaveric dissection. While some anatomical details are difficult to fully demonstrate on the dissection table, these could be integrated by virtual images and videos (Darras et al. 2019; Koney et al. 2024). However, a series of additional images and videos derived from cadaveric dissections were used in the following educational events to permit continuous implementation of the anatomical educational project (Fig. 6.6).

Implementation was also supported by adding new anatomical, surgical and medico-legal references to discussions. Students were encouraged to identify and present the clinical, surgical, and forensic implications of the different anatomical structures they studied (Pirri et al. 2021). This is particularly important in the context of rising litigation related to alleged malpractice in dentistry and oral surgery. A focus on professional responsibility and liability for iatrogenic injuries is also expanding, as demonstrated in recent literature (Wood 2014; Story 2015; Selaković et al. 2016; Sanderson 2017; Nassani 2017; Stilwell 2017; Di Lorenzo et al. 2018; Alsaeed et al. 2022; Johnstone and Evans 2023).

## 6.6 Limitations

Refining feedback collection methods will offer a more comprehensive understanding of the educational impact achieved and help improve teaching approaches to better address the needs of students and enhance the overall educational experience.

The joint cross-institutional learning experience involved only 40 students from the dentistry and orthodontics course. To better evaluate the impact of the combined teaching approaches, it would have been beneficial to gather not only student perceptions of the hands-on experience, but also their overall impressions of the activity. Feedback on the combination of digital and cadaveric teaching methods was collected from the questionnaire's free-text comment section. It would have been useful to include more specific questions regarding the perceived effectiveness of each teaching method (e.g. theoretical lessons, interactive discussions, cadaveric dissection), and to have more deeply explored student insights through the implementation of focus groups and/or semi-structured interviews, in order to collect rich qualitative data, which could have then been analysed thematically. This would have helped to identify how, why and which approach(es) students found valuable and the benefits and limitations that they perceived when the methods were combined.

Another area for improvement would be to track student progress over time, and potential impacts on their formative and/or summative assessment performance. A pre- and post-program survey could have provided a clearer picture of student knowledge gains and practical skills development throughout the course. It would be interesting to assess whether students gained a deeper understanding of anatomical structures and felt more confident in applying their practical skills. Additionally, tracking knowledge retention over time would provide valuable data on the long-term effectiveness of the programme.

Incorporating students from other degree programmes, such as medicine, physiotherapy or nursing, would also provide a comparative perspective on anatomical knowledge, enabling an interdisciplinary learning experience. Comparing outcomes across different groups could help identify areas for improvement and refine the teaching approach to better meet the needs of diverse student populations.

To further enhance the learning experience, broadening the scope of anatomical content covered would be beneficial too. While the focus was on dental anatomy, additional areas such as the musculoskeletal, nervous or circulatory systems would provide students with a more comprehensive understanding of human anatomy. This would enable them to integrate knowledge across different systems.

## 6.7 Conclusions and Future Perspectives

This pilot study could pave the way for broader research on the role of digital tools in anatomy education, with the potential to reshape how challenging anatomical topics are taught and learned across medical disciplines.

For the future, suggested topics include the dissection of the tongue and the floor of the oral cavity from the superior aspect. However, it would be beneficial to first assess student needs related to specific topics. This approach could help tailor the curriculum to focus on those concepts that might be more challenging to understand.

Looking to the future, the authors plan to further enhance the integration of digital and cadaveric anatomy, incorporating topographic, radiological and three-dimensional imaging. This approach builds upon the work of Erkonen et al. in 1990, who suggested integrating imaging anatomy within an existing course of gross anatomy, with a particular emphasis on cross-sectional anatomy. According to the two-year follow-up, this integration provided significant improvement in the students' ability to identify anatomic structures on diagnostic images (Erkonen et al. 1992; Keenan and Ben Awadh 2019; Ben Awadh et al. 2022). Moreover, digital anatomy will be expanded with new images and videos from dissection. Specific (preliminary or subsequent) sections dedicated to discussing literature about clinical, surgical and medico-legal implications could be further implemented, particularly reference to iatrogenic injuries and professional liability.

As Vogel (1993) discussed, what may seem irrelevant as a student can later become crucial in medical practice. In consideration of this, it is fundamental to balance clinical relevance and fundamental basic science concepts in the preclinical curricula (Huppatz 1996). To make basic science concepts more relevant for students, a clinically integrated and problem-based learning approach could be particularly effective (Ozan et al. 2005). Certainly, educating students on the prevalence of anatomical variations and their clinical/surgical implications (increased risk of iatrogenic lesions, need for preoperative diagnostic procedures) is fundamental in the education of future health professionals (Porzionato et al. 2022). However, to achieve this purpose, a solid knowledge of human anatomy is mandatory, and integration of different methodological approaches in teaching is the key strategy.

## References


Abdellatif H (2020) Time spent in practicing dissection correlated with improvement in anatomical knowledge of students: experimental study in an integrated learning program. Cureus 12(4):e7558. https://doi.org/10.7759/cureus.7558. PMID: 32269892; PMCID: PMC7138458

Abramova L, Mann M, Hessler J, Sengelmann RD (2008) Iatrogenic parotid sialocele after excision of malignant melanoma of the cheek. Dermatologic Surg 34(11):1584–1588

Abrams MP, Eckert T, Topping D, Daly KD (2021) Reflective writing on the cadaveric dissection experience: an effective tool to assess the impact of dissection on learning of anatomy, humanism, empathy, well-being, and professional identity formation in medical students. Anat Sci Educ 14(5):658–665. https://doi.org/10.1002/ase.2025

Alasmari WA (2021) Medical students' feedback of applying the virtual dissection table (anatomage) in learning anatomy: a cross-sectional descriptive study. Adv Med Educ Pract 12:1303–1307. https://doi.org/10.2147/AMEP.S324520

Al-Haj Husain A, Schönegg D, Valdec S, Stadlinger B, Gander T, Essig H, Piccirelli M, Winklhofer S (2022) Visualization of inferior alveolar and lingual nerve pathology by 3D double-echo steady-state MRI: two case reports with literature review. J Imaging 8(3):75

Ali AS, Benton JA, Yates JM (2018) Risk of inferior alveolar nerve injury with coronectomy vs surgical extraction of mandibular third molars-a comparison of two techniques and review of the literature. J Oral Rehabil 45(3):250–257

Alsaeed S, Aljarallah S, Alarjani A, Alghunaim G, Alanizy A (2022) Dental malpractice lawsuit cases in Saudi Arabia: a national study. Saudi Dent J 34(8):763–771

Arráez-Aybar LA, Castaño-Collado G, Casado-Morales MI (2008) Dissection as a modulator of emotional attitudes and reactions of future health professionals. Med Educ 42(6):563–571

Baratz G, Wilson-Delfosse AL, Singelyn BM, Allan KC, Rieth GE, Ratnaparkhi R, Jenks BP, Carlton C, Freeman BK, Wish-Baratz S (2019) Evaluating the anatomage table compared to cadaveric dissection as a learning modality for gross anatomy. Med Sci Educ 29(2):499–506. https://doi.org/10.1007/s40670-019-00719-z

Barron R, Margulis A, Icekson M, Zeltser R, Eldad A, Nahlieli O (2001) Iatrogenic parotid sialocele following rhytidectomy: diagnosis and treatment. Plast Reconstr Surg 108(6):1782–1786

Bartoletti-Stella A, Gatta V, Mariani GA, Gobbi P, Falconi M, Manzoli L, Faenza I, Salucci S (2021) Three-dimensional virtual anatomy as a new approach for medical student's learning. Int J Environ Res Public Health 18(24):13247

Ben Awadh A, Clark J, Clowry G, Keenan ID (2022) Multimodal three-dimensional visualization enhances novice learner interpretation of basic cross-sectional anatomy. Anat Sci Educ 15(1):127–142

Border S (2019) Assessing the role of screencasting and video use in anatomy education. Adv Exp Med Biol 1171:1–13

Boscolo-Berto R, Porzionato A, Stecco C, Macchi V, De Caro R (2023) Reference centers for tissue and body donations: compulsory requirements in Italy. Clin Anat 36(3):465–470

Brenner E (2011) "Thoughts on human variations" by Ronald A. Bergman. Clin Anat 24(8):941

Burgess A, Ramsey-Stewart G, Mellis C (2016) Anatomy by whole body dissection: a new Masters of Surgery course. ANZ J Surg 86(9):665–669

Burgess AW, Ramsey-Stewart G, May J, Mellis C (2012) Team-based learning methods in teaching topographical anatomy by dissection. ANZ J Surg 82(6):457–460

Bürklein S, Grund C, Schäfer E (2015) Relationship between root apices and the mandibular canal: a cone-beam computed tomographic analysis in a German population. J Endodont 41(10):1696–1700

Cahill DR, Leonard RJ, Weiglein AH, von Lüdinghausen M (2002) Viewpoint: unrecognized values of dissection considered. Surg Radiol Anat 24(3–4):137–139. https://doi.org/10.1007/s00276-002-0053-2

Capaccio P, Paglia M, Minorati D, Manzo R, Ottaviani F (2004) Diagnosis and therapeutic management of iatrogenic parotid sialocele. Ann Otol Rhinol Laryngol 113(7):562–564

Cassetta M, Pranno N, Barchetti F, Sorrentino V, Lo Mele L (2014) 3.0 Tesla MRI in the early evaluation of inferior alveolar nerve neurological complications after mandibular third molar extraction: a prospective study. Dentomaxillofac Radiol 43(7):20140152

Chaudhry H, Rana S, Bhatti MI, Al-Ansari N, Al Theyab A, Almutairi T, Kazani B, Almasri M, Sadiq Z, Hussein R, Kim D, Chung D, Khalil O, Alroobi H,

Aly A, Raoof A (2023) Utility of the anatomage virtual dissection table in creating clinical anatomy and radiology learning modules. Adv Med Educ Pract 14:973–981

Chiesa-Estomba CM, González-García JA, Larruscain E, Sistiaga Suarez JA, Quer M, León X, Martínez-Ruiz de Apodaca P, López-Mollá C, Mayo-Yanez M, Medela A (2023) Facial nerve palsy following parotid gland surgery: a machine learning prediction outcome approach. World J Otorhinolaryngol Head Neck Surg 9(4):271–279

Chua MKW, Koh WJ, Nimbalkar S, Patil PG (2022) CBCT evaluation of buccolingual orientation of inferior alveolar canal in mandibular posterior region for implant planning. Int J Dent 2022:4682105

Chytas D, Piagkou M, Salmas M, Johnson EO (2021) Three-dimensional digital technologies in anatomy education: better than traditional methods, but are they better than cadaveric dissection? Clin Anat 34(8):1122–1123

Collett LJ, Lester D (1969) The fear of death and the fear of dying. J Psychol 72(2):179–181

Cotofana S, Lachman N (2020) The superficial face dissection as an example for integrating clinical approaches, authentic learning, and changing perspectives in anatomy dissection. Anat Sci Educ 13(1):117–121. https://doi.org/10.1002/ase.1881

Darras KE, Spouge R, Hatala R, Nicolaou S, Hu J, Worthington A, Krebs C, Forster BB (2019) Integrated virtual and cadaveric dissection laboratories enhance first year medical students' anatomy experience: a pilot study. BMC Med Educ 19(1):366

De Caro R, Boscolo-Berto R, Artico M, Bertelli E, Cannas M, Cappello F, Carpino G, Castorina S, Cataldi A, Cavaletti GA, Cinti S, Cocco LI, Cremona O, Crivellato E, De Luca A, Falconi M, Familiari G, Ferri GL, Fornai F, Gesi M, Geuna S, Gibelli DM, Giordano A, Gobbi P, Guerra G, Gulisano M, Macchi V, Macchiarelli G, Manzoli L, Michetti F, Miscia S, Montagnani S, Montella ACM, Morini S, Onori P, Palumbo C, Papa M, Porzionato A, Quacci DE, Raspanti M, Rende M, Rezzani R, Ribatti D, Ripani M, Rodella LF, Rossi P, Sbarbati A, Secchiero P, Sforza C, Stecco C, Toni R, Vercelli A, Vitale M, Zancanaro C, Zauli G, Zecchi S, Anastasi GP, Gaudio E (2021) The Italian law on body donation: a position paper of the Italian College of Anatomists. Ann Anat 238:151761

De Caro R, Macchi V, Porzionato A (2009) Promotion of body donation and use of cadavers in anatomical education at the University of Padova. Anat Sci Educ 2(2):91–92

Dediol E, Harmicar D, Borumandi F (2022) Sialocele of submandibular gland, an unreported complication of bilateral sagittal split osteotomy. J Craniofac Surg 33(5):e493–e495

Di Lorenzo P, Casella C, Capasso E, Delbon P, Fedeli P, Policino F, Niola M (2018) Aesthetic dental procedures: legal and medico-legal implications. Open Med (Wars) 13:96–100

Diakonoff H, Moreau N (2022) Inferior alveolar nerve injury following dental implant placement: a medico-legal analysis of French liability lawsuits. J Stomatol Oral Maxillofac Surg 123(2):158–162

Erkonen WE, Albanese MA, Smith WL, Pantazis NJ (1990) Gross anatomy instruction with diagnostic images. Investig Radiol 25(3):292–294

Erkonen WE, Albanese MA, Smith WL, Pantazis NJ (1992) Effectiveness of teaching radiologic image interpretation in gross anatomy. A long-term follow-up. Invest Radiol 27(3):264–266

Estai M, Bunt S (2016) Best teaching practices in anatomy education: a critical review. Ann Anat 208:151–157. https://doi.org/10.1016/j.aanat.2016.02.010

Farina R, Zaetta A, Minenna L, Trombelli L (2016) Orbital and periorbital emphysema following maxillary sinus floor elevation: a case report and literature review. J Oral Maxillofac Surg 74(11):2192.e1–2192.e7

Figueiredo RR, Filho OG (2023) Perspectives of a university community about body donation for human anatomy study. Revista Brasileira de Educação Médica 47(1). https://doi.org/10.1590/1981-5271v47.1-20220309.ing

Flack NAMS, Nicholson HD (2018) What do medical students learn from dissection? Anat Sci Educ 11(4):325–335

Fyfe S, Fyfe G, Dye D, Radley-Crabb H (2018) The Anatomage table: differences in student ratings between initial implementation and established use. FoHPE 19(2):41–52

General Directorate of Health Prevention, Ministry of Health, Italy (2021) Criteria and requirements whose possession makes universities, high-specialty hospitals, and the Scientific Institute for Research, Hospitalization and Healthcare (IRCCS) suitable to be appointed as reference centers as per law 10/2020. Directorial decree, 31 May 2021

Ghosh SK (2017) Cadaveric dissection as an educational tool for anatomical sciences in the 21st century. Anat Sci Educ 10(3):286–299

Ghosh SK (2020) The practice of ethics in the context of human dissection: setting standards for future physicians. Ann Anat 232:151577. https://doi.org/10.1016/j.aanat.2020.151577

Gupta A, Kumar S, Singh SK, Kumar A, Gupta A, Mehta P (2021) Assessment of anterior loop of inferior alveolar nerve and its anatomic variations with age, gender, and dentition status in Indian population: a CBCT study. Int J Dent 2021:1813603

Heyman SN, Babayof I (1995) Emphysematous complications in dentistry, 1960-1993: an illustrative case and review of the literature. Quintessence Int 26:535

Huitt TW, Killins A, Brooks WS (2015) Team-based learning in the gross anatomy laboratory improves academic performance and students' attitudes toward teamwork. Anat Sci Educ 8(2):95–103. https://doi.org/10.1002/ase.1460

Huppatz C (1996) The essential role of the student in curriculum planning. Med Educ 30(1):9–13

Italian Republic (2017) Guidelines for the prevention of biological risk in the field of necropsy, autopsy and funeral services. In: State-regions conference, November 14th

Iwanaga J, Tanaka T, Ibaragi S, Okui T, Hamaguchi J, Min S, Tubbs RS (2020) Revisiting major anatomical risk factors of maxillary sinus lift and soft tissue graft harvesting for dental implant surgeons. Surg Radiol Anat 42(9):1025–1031

Johnstone M, Evans M (2023) Clinical and medico-legal considerations in endodontics. Aust Dent J 68:S153

Jung J, Hwang BY, Kim BS, Lee JW (2019) Floating septum technique: easy and safe method maxillary sinus septa in sinus lifting procedure. Maxillofac Plast Reconstr Surg 41:54

Kalthur SG, Pandey AK, Prabhath S (2022) Benefits and pitfalls of learning anatomy using the dissection module in an Indian medical school: a millennial learner's perspective. Transl Res Anat 26:100159. https://doi.org/10.1016/j.tria.2021.100159

Kavarthapu A, Thamaraiselvan M (2018) Assessing the variation in course and position of inferior alveolar nerve among south Indian population: a cone beam computed tomographic study. Indian J Dent Res 29(4):405–409

Kavvadia EM, Katsoula I, Angelis S, Filippou D (2023) The anatomage table: a promising alternative in anatomy education. Cureus 15(8):e43047

Keenan ID, Ben Awadh A (2019) Integrating 3D visualisation technologies in undergraduate anatomy education. Adv Exp Med Biol 1120:39–53

Khorsandi M, Vakilian K, Fadavi M, Bayat P (2021) Ethical insight and safety principles of working with cadaver among the first-year medical school students-an educational research. Open Public Health J 14:98–103

Kim MJ, Jung UW, Kim CS, Kim KD, Choi SH, Kim CK, Cho KS (2006) Maxillary sinus septa: prevalence, height, location, and morphology: a reformatted computed tomography scan analysis. J Periodontol 77:903–908

Koney NK, Ansah AO, Asaku BNA, Ahenkorah J, Hottor BA, Adutwum-Ofosu K, Abdul-Rahman M, Arko-Boham B (2024) Anatomage virtual dissection versus traditional human body dissection in anatomy pedagogy: insights from Ghanaian medical students. BMC Med Educ 24(1):1059. https://doi.org/10.1186/s12909-024-06029-2

Krennmair G, Ulm CW, Lugmayr H, Solar P (1999) The incidence, location, and height of maxillary sinus septa in the edentulous and dentate maxilla. J Oral Maxillofac Surg 57:667–671

Kulyapina A, Lopez-de-Atalaya J, Ochandiano-Caicoya S, Navarro-Cuellar C, Navarro-Vila C (2014) Iatrogenic salivary duct injury in head and neck cancer patients: report of four cases and review of the literature. J Clin Exp Dent 6(3):e291–e294

Kumar D, Brar R, Ahmad T, Narad C, Sodhi SPS, Kaur A (2021) Assessment of the anterior and caudal extent of inferior alveolar nerve canal, location of inferior alveolar canal and mental foramen, and the depth of submandibular fossa using computed tomography. Natl J Maxillofac Surg 12(3):380–386

Landau R, Stewart M (1985) Conservative management of post-traumatic parotid fistulae and sialoceles: a prospective study. Br J Surg 72(1):42–44

Lawson GA 3rd, Kreymerman P, Nahai F (2012) An unusual complication following rhytidectomy: iatrogenic parotid injury resulting in parotid fistula/sialocele. Aesthet Surg J 32(7):814–821. Erratum in: Aesthet Surg J 2012;32(8):1040

Lee WJ, Lee SJ, Kim HS (2010) Analysis of location and prevalence of maxillary sinus septa. J Periodontal Implant Sci 40:56–60

Leite GM, Lana JP, de Carvalho MV, Manzi FR, Souza PE, Horta MC (2014) Anatomic variations and lesions of the mandibular canal detected by cone beam computed tomography. Surg Radiol Anat 36(8):795–804

Lempp HK (2005) Perceptions of dissection by students in one medical school: beyond learning about anatomy. A qualitative study. Med Educ 39(3):318–325

Lewis TL, Burnett B, Tunstall RG, Abrahams PH (2014) Complementing anatomy education using three-dimensional anatomy mobile software applications on tablet computers. Clin Anat 27(3):313–320. https://doi.org/10.1002/ca.22256

Li B, Wang L, Zhao Z, Qian H, Cui H, Li L (2017) Surgical approach to foreign material within the mandibular canal due to iatrogenic factors. J Craniofac Surg 28(6):1514–1516

Macchi V, Porzionato A, Stecco C, De Caro R (2014) Evolution of the anatomical theatre in Padova. Anat. Sci Educ 7(6):487–493

Macchi V, Porzionato A, Stecco C, Tiengo C, Parenti A, Cestrone A, De Caro R (2011) Body parts removed during surgery: a useful training source. Anat Sci Educ 4(3):151–156

MacPherson E, Lisk K (2022) The value of in-person undergraduate dissection in anatomical education in the time of Covid-19. Anat Sci Educ 15(4):797–802. https://doi.org/10.1002/ase.2186

Maestre-Ferrín L, Carrillo-García C, Galán-Gil S, Peñarrocha-Diago M, Peñarrocha-Diago M (2011) Prevalence, location, and size of maxillary sinus septa: panoramic radiograph versus computed tomography scan. J Oral Maxillofac Surg 69:507–511

Makaju S, Rai CK (2021) Virtual anatomy classes among the first and second year medical and dental students of a medical college: a descriptive cross-sectional study. JNMA J Nepal Med Assoc 59(240):767–770

Martín JG (2018) Possibilities for the use of anatomage (the anatomical real body-size table) for teaching and learning anatomy with the students. Biomed J Sci Tech Res 4(4):4080–4083

Massalou D, Bronsard N, Hekayem L, Baqué P, Camuzard O (2022) Modern and synchronized clinical anatomy teaching based on the BDIE method (board-digital dissection-imaging-evaluation). Surg Radiol Anat 44(5):803–808

Matus G, Aravena JP, Mariño D, Niklander SE (2023) Decellularized allografts as an alternative for reconstruction of large inferior alveolar nerve defects: a

systematic review. Med Oral Patol Oral Cir Bucal 28(2):e183–e190
McKenzie WS, Rosenberg M (2009) Iatrogenic subcutaneous emphysema of dental and surgical origin: a literature review. J Oral Maxillofac Surg 67:1265
Morris H, Jacques S (2018) Dissection should remain a key part of medical education. ANZ J Surg 88(1–2):10–11
Nassani MZ (2017) Aspects of malpractice in prosthodontics. J Prosthodont 26(8):672–681
Nikolenko VN, Oganesyan MV, Zakirov F, Kudryashova VA, Rizaeva N, Valiullina P (2019) Dissection course in anatomy as stimulus to independent research and to a real step into medicine. J Morphol Sci 36(02):063–066. https://doi.org/10.1055/S-0039-1685222
Oliveira ACS, Candeiro GTM, Pacheco da Costa FFN, Gazzaneo ID, Alves FRF, Marques FV (2019) Distance and bone density between the root apex and the mandibular canal: a cone-beam study of 9202 roots from a Brazilian population. J Endodont 45(5):538–542.e2
Ong CX, Foo YY, Compton S (2023) The impact of human cadaveric dissection on professional identity formation in medical students. BMC Med Educ 23(1):970
Owolabi J, Ojiambo R, Seifu D, Nishimwe A, Masimbi O, Okorie CE, Ineza D, Bekele A (2022) African medical educators and anatomy teachers' perceptions and acceptance of the anatomage table as an EdTech and innovation: a qualitative study. Adv Med Educ Pract 13:595–607
Ozan S, Karademir S, Gursel Y, Taskiran HC, Musal B (2005) First graduates' perceptions on a problem-based and task-based learning curriculum. Educ Health (Abingdon) 18(2):256–271
Pääsky E, Suomalainen A, Ventä I (2022) Are women more susceptible than men to iatrogenic inferior alveolar nerve injury in dental implant surgery? Int J Oral Maxillofac Surg 51(2):251–256
Pabst R, Nave H, Rothkötter HJ, Tschernig T (2001) Evaluation of the medical curriculum: why, when, by whom and for whom should questionnaires be used. Eur J Morphol 39(4):237–239
Papadiochos I, Papadiochou S, Jajimos AR (2021) Iatrogenic penetration of the mouth floor during mandibular molar extraction: a case of protracted bleeding in an emergency department and clinico-anatomical considerations. Stomatologija 23(2):51–55
Parekh D, Glezerson G, Stewart M, Esser J, Lawson HH (1989) Post-traumatic parotid fistulae and sialoceles. A prospective study of conservative management in 51 cases. Ann Surg 209(1):105–111
Patra A, Asghar A, Chaudhary P, Ravi KS (2022) Integration of innovative educational technologies in anatomy teaching: new normal in anatomy education. Surg Radiol Anat 44(1):25–32
Pettersson A, Karlgren K, Hjelmqvist H, Meister B, Silén C (2024) An exploration of students' use of digital resources for self-study in anatomy: a survey study. BMC Med Educ 24(1):45
Pirri C, Stecco C, Porzionato A, Boscolo-Berto R, Fortelny RH, Macchi V, Konschake M, Merigliano S, De Caro R (2021) Forensic implications of anatomical education and surgical training with cadavers. Front Surg 8:641581. https://doi.org/10.3389/fsurg.2021.641581
Pommer B, Ulm C, Lorenzoni M, Palmer R, Watzek G, Zechner W (2012) Prevalence, location and morphology of maxillary sinus septa: systematic review and meta-analysis. J Clin Periodontol 39:769–773
Porzionato A, Macchi V, De Caro R (2019) Forensic clinical anatomy of the spleen in medical malpractice. Forensic Sci Int 304:109772
Porzionato A, Macchi V, Loukas M, De Caro R (2017b) Forensic clinical anatomy—definitions, methods and fields. In: Ferrara SD (ed) P5 medicine and justice. Innovation, unitariness and evidence. Springer, Cham, pp 377–394
Porzionato A, Macchi V, Stecco C, Boscolo-Berto R, Loukas M, Tubbs RS, De Caro R (2022) Clinical anatomy and medical malpractice-a narrative review with methodological implications. Healthcare (Basel) 10(10):1915
Porzionato A, Macchi V, Stecco C, Loukas M, Tubbs RS, De Caro R (2017a) Forensic clinical anatomy: a new field of study with application to medicolegal issues. Clin Anat 30(1):2–5. https://doi.org/10.1002/ca.22796
Porzionato A, Macchi V, Stecco C, Mazzi A, Rambaldo A, Sarasin G, Parenti A, Scipioni A, De Caro R (2012) Quality management of body donation programme at the University of Padova. Anat Sci Educ 5(5):264–272
Quek FF (2024) Revolutionising anatomy education: the current role of digital technologies in enhancing anatomy learning for undergraduate medical students. Cureus 16(12):e75919
Raja BS, Chandra A, Azam MQ, Das S, Agarwal A (2022) Anatomage - the virtual dissection tool and its uses: a narrative review. J Postgrad Med 68(3):156–161. https://doi.org/10.4103/jpgm.jpgm_1210_21
Rajeev R, Sajesh S, Jose M, Kumar ND (2016) Sialocele: a rare sequlae of transparotid approach in subcondylar fracture management. Natl. J Maxillofac Surg 7(2):201–204
Renton T (2010a) Prevention of iatrogenic inferior alveolar nerve injuries in relation to dental procedures. SADJ 65(8):342–344, 346–348, 350–351
Renton T (2010b) Prevention of iatrogenic inferior alveolar nerve injuries in relation to dental procedures. Dent Update 37(6):350–352, 354–356, 358–360 passim
Robertson KA, Organ JM, Yard M, Byram JN (2024) First patient project: engaging pathology through the donor dissection experience and its role in professionalism. Anat. Sci Educ 17(1):199–212. https://doi.org/10.1002/ase.2341
Sanderson S (2017) Medico-legal considerations in providing emergency dental care in practice. Prim Dent J. 6(2):20–25
Santos VA, Barreira MP, Saad KR (2022) Technological resources for teaching and learning about human anatomy in the medical course: systematic review of literature. Anat Sci Educ 15(2):403–419
Selaković S, Mirković S, Bajkin B, Sarcev I, Tadić A, Selaković J (2016) Medico-legal expertise of pain in dental trauma. Vojnosanit Pregl 73(1):66–72

da Silveira CR, Miamoto Dias PE, Oenning AC, de Brito Junior RB, Turssi CP, Oliveira LB (2022) Digital anatomy table in teaching-learning process of the temporomandibular joint anatomy. Eur J Dent Educ 26(1):131–137

Singh AK, Sharma RC, Sharma RK, Musmade DM (2011) Challenges in cadaver availability for learning and research in medical sciences. Int J Med Clin Res 2(2):67–71

Şişu AM, Stoicescu ER, Bolintineanu SL, Faur AC, Iacob R, Ghenciu DM, Dănilă A-I, Haţegan OA (2024) Blending tradition and innovation: student opinions on modern anatomy education. Educ Sci 14(11):1150

Sodnom-Ish B, Seo MH, Yang HJ, Kim SM (2022) Legal liability of a dentist: inferior alveolar nerve damage after calcium hydroxide extrusion. J Craniofac Surg 33(3):e314–e316

Srivastava S, Alharbi HM, Alharbi AS, Soliman M, Eldwakhly E, Abdelhafeez MM (2022) Assessment of the proximity of the inferior alveolar canal with the mandibular root apices and cortical plates-a retrospective cone beam computed tomographic analysis. J Pers Med 12(11):1784

Stilwell C (2017) Mandibular implant overdentures: treatment and medico-legal considerations. Prim Dent J 6(4):28–35

Stone D, Hennessy CM, Smith CF (2022) Teaching with cadavers outside of the dissection room using cadaveric videos. Adv Exp Med Biol 1356:281–297

Story RD (2015) Medico-legal aspects of dental treatment of the ageing and aged patient. Aust Dent J 60(Suppl 1):64–70

Sugand K, Abrahams P, Khurana A (2010) The anatomy of anatomy: a review for its modernization. Anat Sci Educ 3(2):83–93

Suomalainen A, Apajalahti S, Vehmas T, Ventä I (2013) Availability of CBCT and iatrogenic alveolar nerve injuries. Acta Odontol Scand 71(1):151–156

Turney BW (2007) Anatomy in a modern medical curriculum. Ann R Coll Surg Engl 89(2):104–107

Underwood AS (1910) An inquiry into the anatomy and pathology of the maxillary sinus. J Anat Physiol 44:354–369

Vidya KC, Pathi J, Rout S, Sethi A, Sangamesh NC (2019) Inferior alveolar nerve canal position in relation to mandibular molars: a cone-beam computed tomography study. Natl J Maxillofac Surg 10(2):168–174

Vogel WH (1993) Relevance of "irrelevant" facts in medical education: the value of basic science teaching for later medical practice. Acad Med 68(2 Suppl):S27–S28

Von Arx T, Fodich I, Bornstein MM, Jensen SS (2014) Perforation of the sinus membrane during sinus floor elevation: a retrospective study of frequency and possible risk factors. Int J Oral Maxillofac Implants 29:718–726

Weeks SE, Harris EE, Kinzey WG (1995) Human gross anatomy: a crucial time to encourage respect and compassion in students. Clin Anat 8(1):69–79. https://doi.org/10.1002/ca.980080113

Willett GM, Forbes A, Spagna J, O'Kane B, Weston M, Jergenson M, Norton NS (2022) A preliminary investigation of digital media supplements for head and neck anatomy: does production format matter? Clin Anat 35(6):773–779

Williams SE, George DR (2013) 3D virtual anatomy technology in the voice studio: a pilot study to evaluate the functionality and limitations of visible body®. J Sing 69(4):437

Wong AWJ, Kok YO, Chew KY, Tan BK (2023) High-fidelity perforator visualization for cadaver dissection in surgical training. Arch Plast Surg 50(6):621–626. https://doi.org/10.1055/s-0043-1771272

Wood GD (2014) Aspects of dento/medico-legal report writing. Dent Update 41(2):161–162, 164–166

Zdilla MJ, Balta JY (2023) Human body donation and surgical training: a narrative review with global perspectives. Anat Sci Int 98(1):1–11

Zhao J, Xu X, Jiang H, Ding Y (2020) The effectiveness of virtual reality-based technology on anatomy teaching: a meta-analysis of randomized controlled studies. BMC Med Educ 20(1):127

**Andrea Porzionato** Full Professor of Human Anatomy at the Department of Neurosciences, Section of Human Anatomy, University of Padua, Italy. At the University of Padua, he received the Medical Degree, a specialisation in Legal Medicine and then a Ph.D. in Medical, Clinical and Experimental Sciences. He is active in many research projects in the field of neuroanatomy, clinical anatomy and surgical anatomy, with a particular focus on the anatomical and functional relationships of the head and neck regions and vascular-associated structures. His studies often aim to bridge the gap between basic anatomical research and clinical applications. He is a member of the 'Body Donation Programme' of the University of Padua (Reference Centre in Veneto Region, Italy).

**Vanessa Nicolin** Associate Professor of Human Anatomy since, at the Department of Medical Sciences, Surgery and Health, University of Trieste. She is graduated in Trieste, and she obtained the title of Doctor of Research in Human and Molecular Morphology at the University of Bologna. She attended several research laboratories in Italy and the USA for long periods, learning different techniques of microscopic analysis and clinical anatomy. She is responsible for the Bone-Lab Research Laboratory, afferent to the clinic of Maxillofacial Surgery and Odontostomatology in Trieste. Her research focuses on the development of dental materials, anatomy and bone remodelling.

**Veronica Macchi** Full Professor in Human Anatomy, Department of Neurosciences, Section of Human Anatomy, University of Padua, Italy. At the same Institution, she received the Medical Degree, a specialisation in Radiology and a Ph.D. in Medical, Clinical and Experimental Sciences. Her scientific activity focuses on

Clinical and Radiological Anatomy and Topographical Anatomy of the head and neck regions and the knee. She is a member of the 'Body Donation Programme' of the University of Padua (Reference Centre in Veneto Region, Italy); she is Director of the 'Interdepartmental Research Centre for Training on Cadaver—Andrea Vesalio'.

**Elena Stocco** Researcher of Human Anatomy at the Department of Neurosciences, Section of Human Anatomy, University of Padua, Italy. She received a degree in Pharmacy and a PhD in Biology and Regenerative Medicine at the University of Padua. Her research interests deal with Clinical Anatomy and also include the regeneration of peripheral nerves and cartilage through Tissue Engineering approaches.

**Stefano Sivolella** Graduated in Dentistry at the University of Padua in 1994. Specialisation in Odontostomatological Surgery in 1999 (University of Florence). Since 1998, various teaching positions at the Dental School and the Course in Dental Hygiene of the University of Padua. From 2000 to 2016, clinical assistant at the Dental Clinic of the Padua Hospital. Research Doctor (PhD) since 2015 (University of Ferrara). Since 2016, Researcher at the University of Padua, with assistance activities. Since 2016, teaching of Oral Surgery at the Dental School, and Director of the Postgraduate Course in Oral Surgery, University of Padua. Since 2019, Associate Professor at the University of Padua. Author of several publications in international journals.

**Edoardo Stellini** Full Professor. Director of the UOC of the Dental Clinic of the Hospital of Padua. Director of the UOC of Paediatric Dentistry/Pedodontics ULSS n.8—Veneto Region until 2014. He has many years of personal professional experience in the field of dentistry, with particular reference to the field of paediatric dentistry and prosthodontics, where he has carried out extensive scientific activity. President of the Degree Course in Dentistry from October 2013 to 30 September 2021. President of the Degree Course in Dental Hygiene at the Faculty of Medicine and Surgery of the University of Padua from 1 October 2021. Director of the Department of Neurosciences from 1 October 2023.

**Roberto Di Lenarda** Full Professor of Odontostomatological diseases at the University of Trieste. Director of the Clinic of Maxillofacial Surgery and Odontostomatology at the Azienda Sanitaria Universitaria Giuliano Isontina (ASUGI). Director of the Clinical Assistance Department of Specialised Surgeries at the ASUGI since 2016. President of the degree course in Dental Hygiene from 2001 to 2006. Executive Secretary of the National Dental Hygiene Committee from 2001 to 2010. President of the degree course of Dentistry and Dental Prostheses from 2006 to 2012. Secretary and Vice-President of the Board of University Professors of Odontostomatological Disciplines (2008–2014). Elected President of the Board of University Professors of Odontostomatological Disciplines since 2019. Rector of the University of Trieste since 2020. His research is focused on the development of dental materials, pathology and oral medicine, orthodontics, periodontology, implantology and regenerative oral surgery.

**Raffaele De Caro** Full Professor of Human Anatomy at the Department of Neurosciences, Section of Human Anatomy, University of Padua, Italy. He received a Medical Degree, Specialisation in Pathology (University of Padua) and Specialisation in Forensic Medicine (University of Trieste, Italy). His research interests deal with Clinical and Topographical Anatomy with a particular focus on the central nervous system. He is the Director of the 'Body Donation Programme' (Reference Centre in Veneto Region, Italy) at the University of Padua.

# 7 Real-Time Volumetric Visualisations of Cone-Beam Computed Tomography Scans as a Simulation Framework for Radiographic Anatomy Learning

Eire Lopetegi and Jon Jatsu Azkue


## Abstract

Learning radiographic anatomy is challenging, as it requires the ability to mentally reconstruct the spatial organisation of anatomical structures from two-dimensional projections. This task is further complicated by the absence of major monocular depth cues that otherwise commonly help the brain interpret spatial relationships in conventional photographs and in the real world.

Three-dimensional (3D) volumetric rendering is based on projecting a virtual light beam through a virtual object represented by a 3D matrix of scalar data. In this regard, volume renditions bear a fundamental similarity with plain radiographs, which are formed by X-ray beams traversing the body. It is thus theoretically possible to reconstruct plain radiograph-like images from computer tomography (CT) scan data, based on information on X-ray absorption across tissues provided by the latter.

The present chapter explores the use of volumetric visualisations of craniofacial cone-beam CT scan data to create real-time interactive simulations of standard intraoral radiographic projections, in the context of teaching and learning radiographic anatomy. We describe the general principles of volumetric rendering and specific visualisation settings required for simulating radiographic images. It is shown that volumetric visualisations can reproduce the radiographic appearance of dental and support tissues, allowing for identification of major gross anatomical landmarks and structures characteristic to intraoral X-ray projections. Changes in X-ray projection angles can easily be simulated in real time by rotating and panning the scene, creating opportunities to experiment with concepts in oral radiography such as the motion parallax principle and the eggshell effect of corticated structures. In addition, simulated X-ray projections can be enriched with 3D annotation tags and sectional CT views to provide additional spatial references and 3D tracking of structures of interest. Strategies for distributing and exchanging volume-rendered scenes for learning practice and their limitations are discussed.



E. Lopetegi
Centre dentaire Bondy Hôtel de Ville – Dentylis, Bondy, France

J. J. Azkue (✉)
Department of Neurosciences, School of Medicine and Nursing - University of the Basque Country UPV/EHU, Leioa, Spain
e-mail: jonjatsu.azkue@ehu.eus




I. D. Keenan et al. (eds.), *International Anatomical Education*, Biomedical Visualization 9,
https://doi.org/10.1007/978-3-031-91849-0_7

**Keywords**

Computer simulation · Medical illustration · Imaging, three-dimensional · Anatomy · Education · Radiography

## 7.1 Depth Cues, or Lack Thereof

Dental students are expected to acquire competence to accurately interpret radiographic images, as radiographic imaging is integral to both clinical diagnosis and patient management in dental care.

As established by regulatory bodies such as the General Dental Council in the United Kingdom, students who graduate from a dental school should be able to 'undertake relevant special investigations and diagnostic procedures, including radiography'. Because many of the terms and concepts used here may be unfamiliar to the reader, a glossary has been provided at the end of this chapter.

A challenging aspect both of learning and interpreting radiographic anatomy is difficulty in mentally reconstructing 3D relations between body structures from two-dimensional (2D) data, as pictorial depth cues that provide valuable information to perceive the third dimension from photographs or paintings are of limited help when interpreting radiographs. In radiographic images, much like in ordinary photographs, the geometric richness of a 3D space is projected onto a flat image and thus intrinsically lost in the process. The perceptual capabilities of the human brain can still build an interpretation of the 3D organisation of the scene based on a variety of depth cues that can be extracted from a 2D image even in the absence of stereopsis or binocular information. The mechanisms of depth perception have been studied extensively, also in the scope of medical imaging (Pizer and ter Haar Romeny 1991; Reinhardt-Rutland 1992; Kim et al. 2016; Welchman 2016). One such mechanism is occlusion. This is a major monocular depth cue occurring when objects of the scene partially block the view of other elements, in such a manner that occluded objects are perceived to be farther away to an observer than the blocking objects. The human mind can use occlusion for ordering objects in depth and rely on amodal completion to predict the shape of parts that are being covered (Michotte et al. 1991; Briscoe 2011). Provided that objects' dimensions are known, those of a larger size and height are perceived as closer to the observer. Linear perspective, an additional major depth cue, provides a notion of depth based on convergence of lines toward a vanishing point in the distance (Howard and Rogers 2012; Grondin 2016; Arampatzakis et al. 2024). Another depth cue is the aerial perspective. This predicts a blurry and less distinct appearance of distant objects, based on the implicit understanding that light is scattered by the atmosphere, and thus more accumulated light scattering is to be expected the farther the object is situated (Howard and Rogers 2012; Grondin 2016; Arampatzakis et al. 2024). The casting of shadows and the effects of colour and luminous intensity represent additional sources of depth information (Cutting and Vishton 1995). A shadow cast by one object on another provides information about their relative distance. Dark and cool colours (i.e. those on the blue side of the colour wheel) tend to recede into the background whereas bright and warmer colours (reds, yellows and oranges) tend to be more apparent at the forefront. Egusa (1983) also suggests that the colour red may determine the object perceived as nearer to the observer in the absence of other monocular cues to depth.

Although ordinary photographs and radiographic images are both 2D projections of a 3D space and thus involve a loss of explicit depth information, there are fundamental differences between the two as to the extent to which depth cues are preserved and hence the quantity of implicit depth information that the viewer is left with to mentally reconstruct the spatial dimension. Radiographs are an absorptive medium where an X-ray beam traverses the body without producing reflection or scattering, and where the final composite image results from the superimposition of semi-transparent contours of anatomical structures. The depth cue of occlusion is not available to the viewer in an X-ray projection, making it difficult to distinguish intuitively which of two superimposed structures is in front of the other when both are being penetrated by the rays. In dental practice, for example, determining the relative depth position of an unerupted canine relative to incisors, which is carried out radiographically if clinical assessment by inspection and palpation is unfruitful, is often challenging due to the lack of occlusion cues (Jacobs 1994; Clark 1910). Another example of difficulty decoding spatial information is the assessment of the location of pathoses in multirooted teeth, due to superimposition of buccal and palatal roots (Huumonen and Ørstavik 2002; Davies et al. 2015). Relying on the size or height of objects to decode proximity or relative positions is also of little avail in a radiographic image, despite its intuitive usefulness when viewing ordinary photographs.

Based on larger magnification of structures lying closer to the beam source, one can theoretically sense the object's orientation. For example, the occiput is displayed with a slight magnification in posterior-anterior skull projections, whereas it is the orbit region which becomes magnified if the X-ray beam traverses the skull in anterior to posterior direction. Such differences are negligible, however, when viewing relatively flat anatomical structures such as the hand and wrist, or tightly packed structures such as intraoral projections. In addition, perception of spatial positions of anatomical structures based on relative sizes can be misleading and give rise to misinterpretations in radiographic images, as shapes can be distorted to varying degrees by magnification, depending on the distances of X-ray source and conventional films or sensors. A further example of difficulty in interpreting spatial relations from radiographic images, both due to superimposition of anatomical structures and unwanted vertical and horizontal magnifications, is the use of panoramic radiography to assess the positions of upper molar root apices relative to the highly variable anatomy of the maxillary sinus floor. Lopes et al. (2016) reported significant differences between panoramic radiography and cone-beam computed tomography when imaging the positioning of the root apices of maxillary molars. Similar deficiencies have also been reported by other authors to varying degrees (Sharan and Madjar 2006; Hassan 2010; Shahbazian et al. 2015).

Neither shadow casting nor colour effects are actually available in radiographic images as an aid for the interpretation of 3D structure and spatial relationships.

## 7.2 Visual-Spatial Ability for Understanding Anatomy

The task of mentally translating 2D graphic data to a 3D reference framework is cognitively demanding as it requires considerable visual-spatial ability in the sense of recognizing, mentally manipulating and reproducing spatial structure (Keenan and Powell 2020). There is recognition that performance in disciplines such as science, technology, engineering and mathematics is related to the ability of mentally processing spatial information (Benbow and Stanley 1982; Hambrick et al. 2012; Uttal and Cohen 2012; Ramey and Uttal 2017; Buckley et al. 2018; Hodgkiss et al. 2018). Comparatively there is scant evidence in literature as to the relevance of visual-spatial ability to medicine, despite calls for further investigation in this area (Corry 2011; Birchall 2015). Early work by Rochford (1985) showed that second-year medical students who scored better in a battery of spatial tasks involving rotation and visualisation of 3D blocks also performed better in an anatomical multiple-choice questionnaire, which required spatial thinking. Spatial ability was defined by the author as the ability to perceive, retain and recognise or reproduce 3D objects in their correct proportions when manipulated in space or verbally described. It is noteworthy that scoring in the spatial task was not related with correct hits in non-spatial anatomical questions (Rochford 1985). A number of subsequent studies have largely confirmed those observations. Garg et al. (2001) found that spatial ability measured using the mental rotation test (MRT) adapted from Vandenberg and Kuse (1978), a frequently used tool to assess visual-spatial ability, significantly affected the process of learning the anatomy of carpal bones. In a study involving first-year medical students, Lufler et al. (2012) recorded the results of a MRT before a gross anatomy course. The authors reported that students scoring in the top quartile of the MRT were 2.2 times as likely as students scoring in the lower quartile to score >90% on all practical examinations. In addition, Guillot et al. (2007) reported strong correlations between performance in anatomy examinations and spatial ability using the MRT. Nguyen et al. (2012) studied the relationship of visualisation ability with performance in a spatial anatomy task including identification of structures on cross-sectional images and mental rotation of anatomical structures, and reported an improvement in performance although the effect varied depending on the way the instructional material was presented. However, weak and weak-to-moderate associations have been found between spatial abilities and scores in anatomical assessment in a group studying a biomedical science subject (Sweeney et al. 2014) and in a group of dental students receiving a gross anatomy course (Sarilita et al. 2022). In addition, a very weak positive correlation has been reported between scores on the MRT and competence for identifying labeled anatomical structures on cross-sectional images in various formats (Ben Awadh et al. 2022).

The above empirical evidence appears to be in line with a model that looks at spatial ability as an innate, fixed trait that predisposes one to perform better during training and in clinical practice. This assumption has guided considerable research into the acquisition of anatomical and surgical skills (Wanzel et al. 2002, 2003; Keehner et al. 2004). The notion that spatial abilities predict training outcome has also served as a criterion for selection procedures in dentistry (Sandow et al. 2002; Ranney et al. 2005). However, an alternative model contemplates visual-spatial cognition as a developing skill, much like language, which can be progressively acquired with appropriate training and experience (Baenninger and Newcombe 1989). Dental students, for example, are reported to improve their performance in the MRT along with their education program, as found by comparing scores from early and late phases of the curriculum (Hegarty et al. 2009). Although few studies have addressed this issue in the domain of gross anatomy, evidence suggests an influence on spatial cognition from various forms of experience and training (Fernandez et al. 2011; Lufler et al. 2012; Vorstenbosch et al. 2013). Fernandez et al. (2011) evaluated possible differences among first-, fourth- and fifth-year medical students and university lecturers with more than 5 years' experience of teaching anatomy as to their abilities to perform in a computerised spatial ability task. The findings, which showed significantly better performance for experts and intermediates in terms of correct spatial judgments and response time, motivated the authors to propose spatial ability sessions to refine spatial skills with the purpose of improving the learning of anatomy (Fernandez et al. 2011). Later studies such as Lufler et al. (2012) have reported benefits in visual-spatial skills as a result of participating in gross anatomy programs. These authors found significant increases in first-year medical students when comparing MRT scores taken during orientation before beginning the anatomy class and just before taking the final examination. In the same vein, Vorstenbosch et al. (2013) reported significant improvements in spatial abilities after a 25-day anatomy course, using the same assessment tool. Interestingly, higher pre-test MRT scores were predictive of better performance in anatomy tests in both studies, suggesting that the two models considering visual-spatial ability either as a trait or a developing skill need not be mutually exclusive. Rather, innate ability may indeed help one to perform well in medical or dental training while also being further developed over the course of training.

As a form of anatomical visualisation, the mental process of radiographic interpretation demands considerable visual and spatial cognition. Published reports appear to confirm a likely reciprocal influence between competence in radiology and visual-spatial ability. By using the Visual Form Reconstruction Test (Form Test) to measure the aptitude for perceiving 3D spatial relations from 2D data, Berbaum et al. (1985) found a high positive correlation between spatial ability and diagnostic performance scores among radiology residents, particularly those in neuroradiology and body CT/sonography subspecialties. In addition, Smoker et al. (1984) assessed spatial ability in a sample of radiology residents using the Form Test. The results showed a positive correlation between scores and the overall radiologic competence and performance derived by averaging all of the faculty ratings for each resident. Studies have also shown an enhancing effect of radiology training on spatial cognition. For example, scores on a cube perspective test (Stumpf and Fay 1983) are reportedly significantly increased in a sample of fourth- and fifth-year medical students following seven radiology seminars using computer-based 3D visualisations (Rengier et al. 2013). In addition, a recent study conducted on veterinary students' performance in radiographic anatomy after a learning program revealed a significant increase in scores on the MRT after training (Hontoir et al. 2023).

In summary, the available evidence supports a central role for visual-spatial ability not only in the cognitive process of understanding and retaining the 3D structure of the body, but also in mentally reconstructing the 3D structure of bone and soft tissues when interpreting radiographic images. The reciprocal advantage of training on both visual-spatial ability and gross and radiographic anatomy may provide novel opportunities in a context like the current one where a transition is taking place from a passive model of teaching to more active learning strategies.

## 7.3 Volumetric Rendering of CT Data

Scientific 3D visualisation methods include two major techniques, namely, volume rendering and surface rendering or visualisation of polygon meshes. Surface rendering provides a view of the surface of an object, most commonly a polygon mesh, by showing its appearance upon external illumination with a virtual light source. The object in question is a hollow structure which, however, conveys the impression of a solid body when viewed externally. Objects visualised as surfaces are described in a file as a collection of points or vertices, each defined by their space coordinates in the X, Y and Z axes, along with the polygonal edges connecting each pair of vertices to form a 3D mesh. This is the case for a standard file format such as the standard tessellation language (*.STL). Other file formats, for example, the Stanford triangle format (*.PLY) or the Wavefront OBJ format can store additional information regarding colour, surface normals, or opacity of the mesh. Both support so-called texture coordinates, i.e. a set of key reference points which allows the rendering software to map a bitmap image onto the polygon mesh. Further file formats are available also supporting additional, more complex features such as light sources, cameras or animation.

The other major modality of 3D visualisation called volume rendering is a technique based on projecting a virtual light beam through a 3D matrix of voxels and displaying the resulting projected image onto a 2D plane (Drebin et al. 1988) (Fig. 7.1). In biomedical visualisation, scalars rendered volumetrically are most often stacks of Computed Tomography (CT) or Magnetic Resonance Imaging scans. The basic definition of any voxel includes its location within the 3D matrix as a set of coordinates in the X, Y and Z axes, as well as the scalar value carried by the voxel, which in CT or MRI data sets describes the level of density of the scan image. The light beam interacts with all voxels in the matrix, i.e. with those representing both the body and the surrounding space, rather than with only its surface.

Stacks of CT or MRI scan data in standard DICOM (Digital Imaging and Communications in Medicine) format are often used for storing and exchanging information in medical imaging, including patient information. Anonymised image data can be stored in a variety of formats. For example, image files in *.RAW format are basically plain 3D arrays of voxel value data. The Analyze file format, which is widely used in the functional neuroimaging field, consists of an *.IMG file and a *.HDR file, the former containing the actual data in binary representation and the latter carrying additional information about the data such as volume dimensions and voxel size.

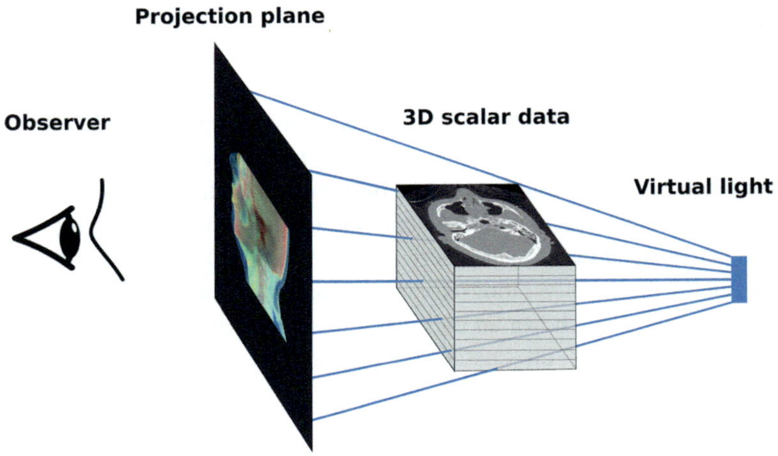


**Fig. 7.1** Basic principle of volume rendering. A virtual light beam traverses a 3D field of scalar data to generate a 2D projection image onto a plane, which is then viewed by the observer

Other available formats such as the Neuroimaging Informatics Technology Initiative (NIfTI) file format or the Visualisation Toolkit format (*. VTK) may include both actual data and additional information such as coordinate definitions, slice ordering, data types and extension data.

The distinction between the types of data used for either surface rendering or volumetric visualisation is relevant as the data type impacts both the rendering process and exchangeability of the rendered scene. Whereas the geometry of a polygon mesh is determined by the underlying polygon data, the shape and appearance of an object rendered volumetrically depends decisively on how light is configured to interact with voxels, and particularly on voxel opacity. In volumetric visualisation, the final graphical rendition is determined by two major user-defined parameters, namely, the colour and the level of opacity with which each voxel will contribute to the final projected image when traversed by the virtual light beam. Scalar values of voxels, which in CT or MRI scans are image density values in a grey scale, are mapped to a colour space to produce a pseudo-colour image. This means that every shade of grey in the original scan data will be represented in the final rendition by a predefined colour. Correspondences between original grey scale values and target colours are defined arbitrarily using a look-up table which can be created numerically by a colour transfer function. The simplest approach consists of setting two different reference colours for representing the highest (near white) and lowest (near black) grey scale values in the original data set and using the resulting colour gradient to represent or map intermediate shades of grey. It is also possible to include additional colours in the look-up table to map intermediate scalar values (Fig. 7.2).

The second major rendering parameter is the mapping of opacity, i.e. the definition of the desired levels of opacity in the final rendition as a function of scalar values of voxels. This is established using an opacity transfer function. A simple opacity transfer function may be a linear one between two reference values assigned to the highest and lowest original scalar values, with voxels carrying intermediate values being mapped by interpolation (Fig. 7.2). Complex functions can be set up by adding intermediate reference points. The opacity transfer function provides the user with crucial control over the final appearance of tissues and structures in the final rendition. For example, voxels representing the space surrounding the body can be assigned zero opacity and thus easily removed from the rendered scene (Fig. 7.2). This is obviously critical to be able to visualise 3D anatomy reconstructed from CT or MRI scan data. Furthermore, and relevant to the present chapter, using specific colour- and opacity transfer function settings provides the opportunity for producing radiograph-like projections from CT data sets.

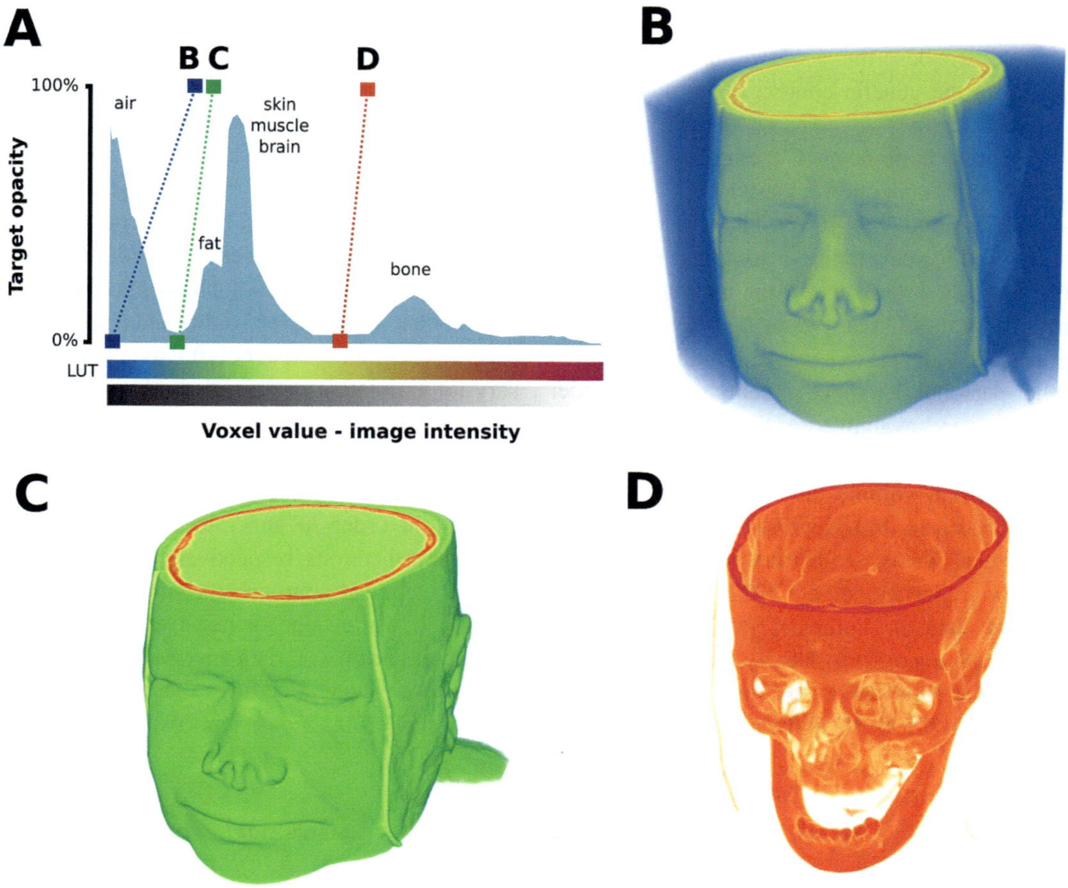


**Fig. 7.2** Example colour and opacity transfer functions to manipulate the appearance of a volume-rendered scene. Head and neck CT data from the Visible Human Female (https://www.nlm.nih.gov/research/visible/visible_human.html) is used as source data, and Paraview software is utilised for rendering. The initial data set has been cropped to remove voxels representing the head vertex, in order to expose subcutaneous tissues, skull and brain in 3D renditions. (**a**) Example opacity transfer functions which result in renditions shown in **b–d**. The distribution of image density values in the source data set is provided for reference, with their approximate correspondences to the represented organic tissues. Both the image densities in the source CT data and the colour look-up table are shown at the bottom. (**b**) Voxels representing air density, which carry the lowest image density values in the original CT data set, are assigned a blue colour and an intermediate opacity level. Using these opacity transfer function settings, the body is rendered opaque and rendered visible through a translucent surrounding air. (**c**) Voxels representing organic tissues are rendered opaque, whereas all those representing air density are set to be rendered entirely transparent and thus removed from the scene. Note the orange colour for bone on the cut surface of the head, in contrast with green for intermediate density voxels representing soft tissues and brain. (**d**) Displacing the opacity transfer function curve to the right sets all voxels representing tissues with intermediate and low densities to be rendered transparent, leaving the skull as the sole visible structure

## 7.4 Simulating Radiographic Imaging

Radiographic- and CT images are both generated by differences in the degree to which X-rays are absorbed or scattered by anatomical structures. However, a radiograph is the result of an X-ray beam traversing internal components with different levels of X-ray absorption, whereas a CT scan provides a 2D map of X-ray attenuation throughout structures and tissues. Theoretically radiograph-like images may therefore be simulated by volumetric visualisation of a 3D field of CT scan data, taking advantage of information on

X-ray absorption contained by CT images. In this section, we describe a basic approach in this regard in the general context of intraoral radiographic imaging.

We used cone-beam CT data sets as source data, from a collection released from a previous study (Cipriano et al. 2022). The data were originally used for developing a computer routine for the automated detection of the mandibular canal in cone-beam CT scans using deep learning techniques. Every data set was anonymised, and no other personal details of patients such as gender or age were available to the present authors. Dental scans (NewTom/NTVGiMK4, 3 mA, 110 kV) had volume dimensions ranging from 312,265,148 to 463,423,178 in the X, Y and Z axes respectively, and intra-slice distance of 300 µm (Cipriano et al. 2022). We converted original data in numpy (*.NPY) file format (https://numpy.org/) to Nearly Raw Raster Data file format (*.NRRD), and data sets were then imported to Slicer 3D software (v. 5.4.0 r31938; https://www.slicer.org/) for volume cropping. In order to reduce computational load, voxels corresponding to anatomical structures irrelevant to the study were removed using the Crop Volume module in 3D Slicer, such as those corresponding to the vertebrae, the skull right above the inferior orbital rim, and the hyoid bone whenever possible. Selected volumes were clipped further to separate discrete sectors such as maxillary and mandibular regions, or right and left halves (Fig. 7.3). Isolating right or left halves was necessary for simulating periapical and bite-wing projections in the premolar and molar regions, which represent dental and facial structures of only one side. Maxillary and mandibular sectors were isolated in order to generate upper and lower occlusal projections, respectively.

The technical approach for simulating a conventional, white bone or negative X-ray image from cone-beam CT data involved establishing first a set of general settings that produce volumetric renderings consistent with the general

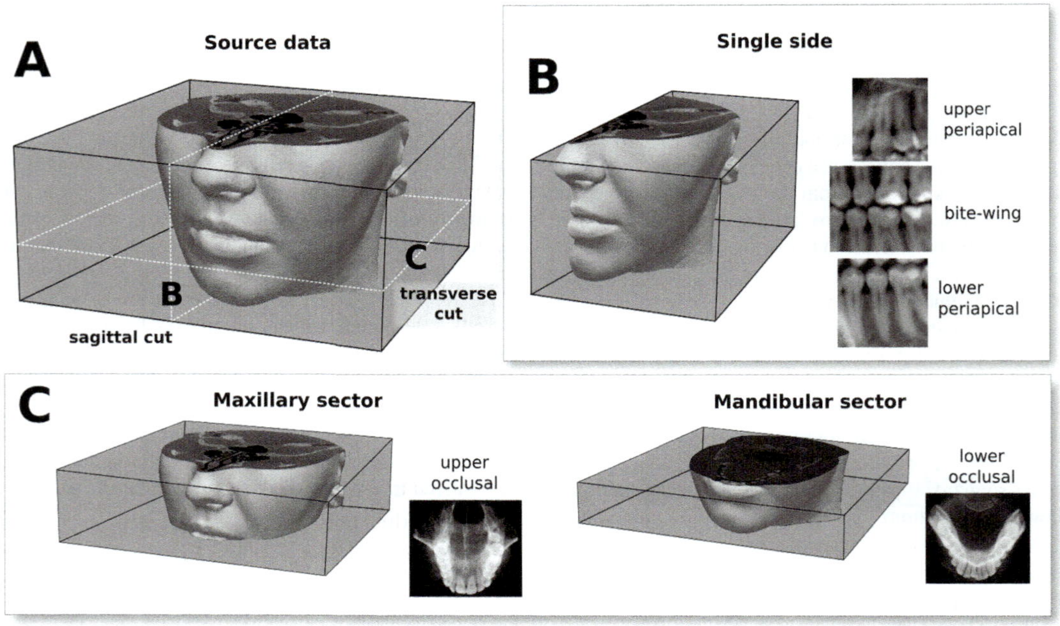


**Fig. 7.3** Clipping of a cone-beam CT volume data set into convenient sectors for X-ray simulations. (**a**) The source craniofacial cone-beam CT data sets utilised here typically represented the portion of the head comprised between the infraorbital and mental regions, which are depicted for reference using a surface rendition of the head. The locations of cuts in the median and occlusal planes, which result in isolation of sectors shown in (**b**) and (**c**), are indicated. (**b**) Cone-beam CT data from only one half of the head provides suitable conditions for simulating projections of the periapical and bite-wing types in the premolar and molar regions. No data clipping is required for projections in the anterior teeth region. (**c**) Maxillary and mandibular sectors are isolated by splitting the volume at the occlusal level, in order to simulate occlusal radiographic projections

appearance of radiographs, and then adjusting specific rendering parameters to refine the desired visual result for each particular data set. General settings are therefore applied equally to all data sets, whereas the specific parameters vary from one data set to another. The general settings consist of a black background colour to the rendered scene and an entirely white, monochromatic look-up table. This means that any non-transparent anatomical structure is set to be rendered white against a black background, consistent with the overall appearance of a radiographic image.

The specific rendering settings proposed here include two parameters in the opacity transfer function (Fig. 7.4). The first parameter defines a scalar value threshold below which all voxels are to be rendered entirely transparent. Conversely, then, all voxels with values exceeding the threshold will be made visible in the rendered scene. Lowering the value of this parameter determines a higher presence of lower density tissues in the

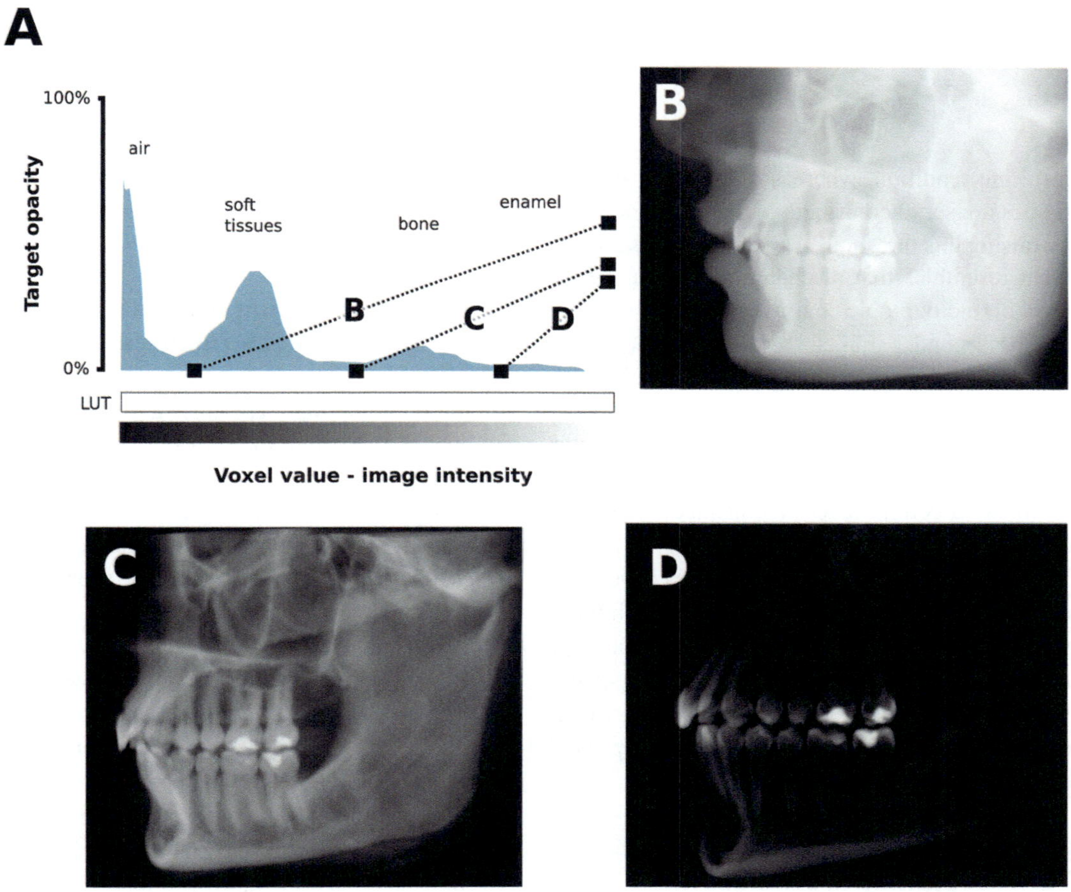


**Fig. 7.4** Example opacity transfer functions for volumetric visualisation within the general conditions for simulating X-ray images. Note that the colour look-up table is monochromatic white. (**a**) A simple approach to setting the opacity transfer function consists of adjusting two parameters, where parameter #1 (nodes on the horizontal axis) determines the threshold above which voxels are set to rendered visible, whereas parameter #2 (nodes on the vertical axis) changes the overall level of opacity of visible voxels. The renditions resulting from the three example opacity transfer functions are shown in B–D. The distribution of image densities in the source data set and their approximate correspondence with tissues are provided for reference. (**b**) Soft tissues including skin, subcutaneous tissue and muscle are all made visible by setting parameter #1 at a lower value. (**c**) Radiograph-like images can be created best by setting parameter #1 right above the level of soft tissue density and always below the level of bone density. (**d**) Displacement of parameter #1 to the right results in removal of most craniofacial anatomy except for higher density components such as enamel and filling material

**Table 7.1** Software tools for volumetric visualisation. Open-source, freely available tools with volume rendering capabilities that can be used for radiographic simulations. The list is not comprehensive but rather a brief guide

| Package | Main purpose | Platform | Web site |
|---|---|---|---|
| Paraview | Scientific visualisation | Windows, macOS, Linux | https://www.paraview.org/ |
| 3D Slicer | 3D visualisation and processing | Windows, macOS, Linux | https://www.slicer.org/ |
| Invesalius 3 | 3D reconstruction | Windows, macOS, Linux | https://invesalius.github.io/ |
| Voreen | Interactive visualisation and analysis | Windows, Linux | https://www.uni-muenster.de/Voreen/ |
| VolView | Radiological viewer | Web-based | https://volview.kitware.com/ |
| ImageVis3D | Volume rendering | Windows, Linux | https://www.sci.utah.edu/cibc-software/imagevis3d.html |
| MeVisLab | Medical image processing and visualisation | Windows, macOS, Linux | https://www.mevislab.de/ |

resulting rendition, whereas only higher density tissues are set to contribute to the rendered scene by raising this threshold. The effect of this adjustment resembles that of changing photon energy level on radiographic contrast.

The second parameter establishes the maximal level of opacity assigned to voxels with higher values, although it also linearly affects the opacities of all visible voxels. This parameter should be set at some level below 100% in order to ensure a certain level of translucency of rendered structures.

All 3D renderings shown in figures have been created using Paraview software (v. 5.8.1; https://www.paraview.org/). A number of additional freely-available, open-source tools supporting volumetric visualisation are also available, which can be used using an analogous approach (Table 7.1).

## 7.5 Volumetric Visualisations for Learning Radiographic Anatomy

The present chapter looks into the potential utility of radiograph-like volumetric visualisations for teaching and learning intraoral radiographic anatomy. As a first prerequisite for qualifying as a valid tool, digital simulations should be capable of reproducing major anatomical structures and landmarks that are characteristic to normal intraoral X-ray projections.

### 7.5.1 Radiodensities of Dental and Support Structures

Volumetric renderings reproduce a range of density shades reflecting variations in X-ray beam attenuation levels across dental and support structures (Fig. 7.5). The higher radiopacity of the enamel is distinguishable, and the dentinoenamel junction is most often clearly seen. Also distinguishable are a range of density shades formed by superimposed profiles of crowns and cusps. The relative radiolucency of the pulp chamber is apparent, and both root canals and pulp horns are distinctly delineated at appropriate projection angles.

In the mandible, the higher density of compact bone can be seen consistently, e.g. at the inferior margin of the body, in contrast to that of cancellous and alveolar bones. Bone densities and textures vary considerably among regions in the maxillary bone and the mandible, and there is substantial variation among cone-beam CT scans as well. Volumetric renderings can reproduce finer anatomical detail in most cases, such as trabecular bone patterns. The appearance of finer anatomical structures is affected by the size, shape, location and projection angle. For exam-

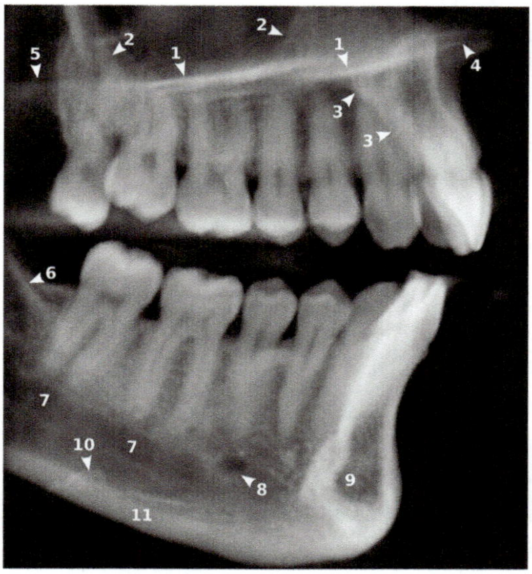


**Fig. 7.5** Overview of the right hemimaxilla and hemimandible in a simulated radiograph based on a volume rendition of cone-beam CT data. The appearance of teeth and support structures closely resembles the contours and shapes commonly found in actual radiographs. **1.** Floor of the nasal cavity. **2.** Walls of the maxillary sinus. **3.** Hard palate. **4.** Anterior nasal spine. **5.** Pyramidal process of palatine bone. **6.** Oblique line. **7.** Mandibular canal. **8.** Mental foramen. **9.** Spongy bone at the sagittal cut plane in mental region. **10.** Superimposed hyoid bone. **11.** Cortical bone of the mandible body

ple, the lamina dura is generally distinguishable in continuity with the alveolar crest and adjacent to the radiolucent line formed by the periodontal ligament space, although both depended critically on the projection angle. This variability in the respective radiopaque and radiolucent lines is also common in actual radiographs, as both are affected not only by variations in the angulation of the X-ray beam but also by the convexity or concavity of proximal tooth surfaces and curvature of the roots (White and Pharoah 2014; Prakash et al. 2015; Hubar 2017).

### 7.5.2 Gross Topographic Landmarks and Anatomical Detail

Landmarks and boundaries demarcating gross radiographic compartments and regions of the maxilla and the mandible are apparent in simulated bite-wing and periapical projections (Fig. 7.6). Radiopaque lines delineating the hard palate, the nasal septum and the floor of both the nasal cavity and maxillary sinus are most often clearly seen. Also distinct is the radiopaque line formed by the inner wall of the maxillary sinus extending forward over the molars and the premolars (Ennis 1937), on occasion as an inverted Y and most often as crossed radiopaque lines, although its morphology varies substantially with the height and concavity of the maxillary sinus. In the mandible, radiopacities formed by the mylohyoid ridge and the oblique line are less consistent and often not simultaneously present. Less conspicuous and prevalent are radiopaque lines formed by the intermaxillary suture or the cortical lining of the mandibular canal. The mandibular canal is a further example of radiographic variability, as its visibility varies considerably not only among individuals but also in different areas of the mandible both in actual periapical and panoramic X-ray projections (Klinge et al. 1989; Lindh et al. 1992; Velvart et al. 2001; Naitoh et al. 2009; Jung and Cho 2014).

The nasal cavity and maxillary sinus are seen as voluminous, markedly radiolucent regions in volumetric renderings. The mental foramen is identifiable most often within the sector housing the premolar teeth. In addition, relatively radiolucent areas such as the ones formed by the submandibular fossa or the incisive foramen can be distinguished in most cases. The incisive foramen, which is considered to be a normal anatomical landmark in diagnostic intraoral periapical radiographs, is actually visible in 40% and 76.7% of cases in bisecting and paralleling radiographs, respectively (Veena et al. 2021). Interocclusal distances between antagonistic teeth in simulated X-ray images do not necessarily coincide with those in actual bite-wing radiographs. This discrepancy cannot be avoided, as the positions of teeth in simulations depend entirely on the source data, and can thus be considered as a potential limitation.

Both the nasal septum and lateral walls of the nasal cavity are delineated by radiopaque lines of variable thickness in simulated maxillary occlusal projections (Fig. 7.7), although obviously also upper teeth occupy a prominent position. The maxillary sinuses are visible as radiolucent areas,

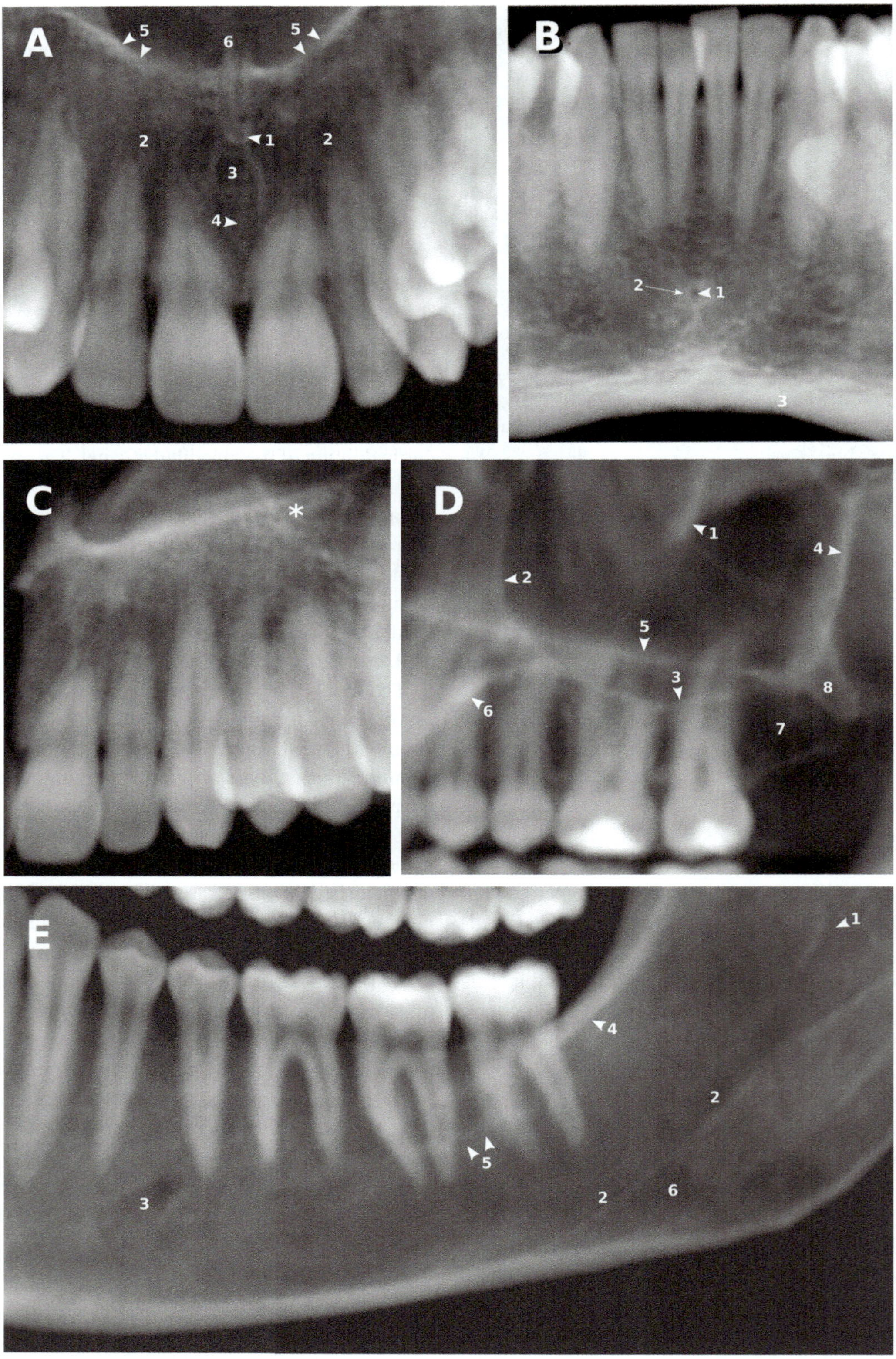


**Fig. 7.6** Major anatomical landmarks and structures characteristic to periapical and bite-wing projections, in volume-rendered simulations. (**a**) Frontal projection of the maxilla and maxillary teeth. **1.** Anterior nasal spine. **2.**

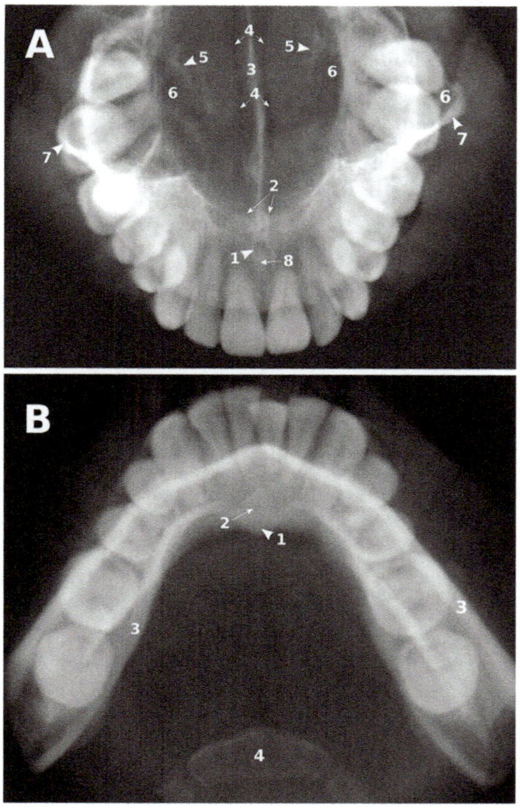


Fig. 7.7 Simulation of occlusal-like projections from cone-beam CT scan data. (**a**) Maxillary projection. **1.** Incisive foramen. **2.** Nasal openings of incisive canal. **3.** Nasal septum. **4.** Inferior nasal concha. **5.** Lateral wall of nasal cavity. **6.** Maxillary sinus. **7.** Zygomatic bone. **8.** Anterior nasal spine. (**b**) Mandibular projection. **1.** Superior genial spine. **2.** Lingual foramen. **3.** Cortical bone of mandible body. **4.** Body of the hyoid bone

with superimposed opacities formed by maxillary molars. All three openings of the incisive canal are distinguishable on occasion. In projections simulating mandibular occlusal radiographs, radiodense silhouettes of the crowns of posterior teeth are superimposed to the dense profile formed by the body of the mandible, and the roots of anterior teeth are visible to varying extents. The higher density of the compact bone is distinctly seen.

Generally speaking, anatomical variability is considerable among cone-beam CT data sets with regard to the occurrence, shape and density of anatomical structures and detail, and there is hardly a single ideal data set that features all anatomical structures of interest. This means that one may want to go through several simulated scenes to cover all relevant structures. It is important to select source data carefully, not only based on technical criteria such as spatial resolution and the absence of motion artifacts but also with consideration of variation in morphology. On the other hand, however, providing a range of real-life examples of anatomical variation can be seen as an opportunity to address an important aspect of anatomical education that is often relatively neglected despite being considered as important conveying to anatomy students (Willan and Humpherson 1999; Sañudo et al. 2003; Alraddadi 2021).

### 7.5.3 Real-Time User Interaction for Learning

In recent decades, there has been a transition in medical teaching from passive learning to more interactive and experiential learning strategies (Prober and Khan 2013; Skochelak and Stack 2017; Atta and AlQahtani 2018; DeWitt 2019), and radiology education poses no exception in this regard (Wertz et al. 2014; Linaker 2015; Shiner 2018; Spence 2019; Tay and McNulty 2023). Computer-based simulations provide new opportunities for acquiring both diagnostic interpretive and procedural skills in projectional radiography in a safe and cost-effective environment (Chetlen et al. 2015; Kasprzak 2016; McNulty

Fig. 7.6 (continued) Radiolucency of lateral fossae. **3.** Incisive foramen. **4.** Intermaxillary suture. **5.** Floor of nasal cavity. **6.** Nasal crest of maxilla and nasal septum. (**b**) Frontal projection of mental region and anterior mandibular teeth. **1.** Lingual foramen. **2.** Superior genial spine. **3.** Cortical bone of the mandible body. (**c**) Oblique projection of the maxillary canine region, where the asterisk shows the radiopaque line formed by the inner wall of the maxillary sinus extending forward over the molars and the premolars. (**d**) Projection at the region of posterior maxillary teeth. **1.** Zygomaticomaxillary suture. **2–4.** Anterior wall, floor and posterior wall of the maxillary sinus. **5.** Floor of the nasal cavity. **6.** Hard palate. **7.** Maxillary tuberosity. **8.** Pyramidal process of palatine bone. (**e**) Lateral projection of the posterior mandibular teeth region. **1.** Mandibular foramen and lingula. **2.** Mandibular canal. **3.** Mental foramen. **4.** Oblique line. **5.** Mylohyoid line. **6.** Submandibular fossa et al. 2021). Volumetric visualisations are fully interactive in most if not all 3D rendering applications, in the sense of providing standard functionalities of spatial navigation such as rotation, panning and zooming in real time. This continuous inspection capability is reminiscent of real-time radioscopy or fluoroscopy, where an X-ray detector such as a fluorescent screen coupled with an image intensifier and a video monitor enables real-time radiographic images to be obtained. This real-time interaction also provides opportunities for acquiring procedural skills, such as proper positioning of the X-ray beam, through deliberate practice (Tung and Awan 2023; Tung and Matalon 2023).

The use of volumetric visualisations of CT data to simulate radiograph-like images has been explored by previous studies in the scope of education and training. Nilsson et al. (2004) created a virtual reality application for simulating multiple oral radiographic projections from a stack of CT scans of a plastic-embedded dry skull. Rather than providing training in radiographic anatomy or diagnostic imaging, however, the goal was to acquire skills to deduce the projection direction of the X-ray beam and the relative depth positions of objects. In that setting, the user was allowed to manipulate the relative position and angle of a virtual X-ray machine relative to a surface model of a patient's head, to which the system was designed to respond by displaying the corresponding synthetic X-ray projection (Nilsson et al. 2004). In addition, Hassan et al. (2007) generated a web-based instructional tool displaying a range of 3D renderings of craniofacial cone-beam CT data in predetermined projection angles, inclusive of static volume renditions presented in QuickTime Virtual Reality® file format. User-interaction included rotation, zooming and panning through predetermined QTVR frames. Although a variety of volume renditions were used in that module, including reconstructions of X-ray views, the focus was primarily on providing training in interpretation of craniofacial cone-beam CT anatomy rather than radiographic anatomy. In a later study, Vuchkova et al. (2011) presented fourth-year dental students with static orthopantomographs reconstructed from cone-beam CT data using curved planar reformation, in conjunction with a 3D view of a skull surface which could be rotated at will. Appreciation of depth in simulated panoramic images was expected to emerge from comparing the location of lesions with 3D views of the skull.

Arguably, a major advantage of the present simulation approach over static radiographs is real-time user interactivity. Interactivity is a desirable attribute of software for 3D anatomy education (Zilverschoon et al. 2019) and a highly regarded feature of e-Learning programs for radiology (den Harder et al. 2016). An analysis of factors in computer visualisations that contribute to spatial anatomy comprehension has shown that learning is facilitated by depicting multiple views of anatomical models, e.g. by rotating models continuously in the X, Y and Z axes (Nguyen et al. 2012). Interactive volumetric visualisations clearly provide an advantage in this regard, as students are allowed to actively create any radiographic view virtually and thereby inspect the desired structures continuously from every direction. Furthermore, it has been suggested that a control interface that allows the manipulations made by the users to be exactly mirrored in the movements of the visualisation may be especially beneficial in helping learners create an integrated spatial mental representation of any object they are viewing (Keehner et al. 2008).

Another likely facilitating factor is visual-spatial ability. Nguyen et al. (2012) found that dynamic 3D visualisations particularly benefited individuals with the ability to manipulate objects mentally, as opposed to those with lower visual-spatial ability who may rather tend to rely on standard, easily recognised views of the model as reference orientations (Bülthoff et al. 1995; Garg et al. 1999). Interestingly, studies suggest that competence in anatomy is facilitated by specific training in mental rotation (Hoyek et al. 2009). In connection with this, it is noteworthy that pattern learning and mental spatial transformations tend to be relatively domain specific, meaning that the benefits of practice mentally rotating one type of shapes are limited primarily to highly similar shapes (Sims and Mayer 2002; National Research Council 2006). Thus, the optimal approach in this direction may be to involve human body and organs as transformation objects during mental rotation training (Jamil et al. 2019).

In the real world, the motion parallax or the motions of objects relative to the observer provide a useful cue to inform of their depth or relative position (Rogers 1993; Hanes et al. 2008; Kim et al. 2016). An example of exploiting the motion parallax in clinical practice is the maneuver of shifting the incidence angle of the X-ray tube in order to determine the relative depth position of an unerupted tooth (Jacobs 1999, 2000) or for refining the localisation of periapical lesions (Davies et al. 2015). With the purpose of improving the skill to interpret the depth locations of unerupted teeth in radiographs utilizing parallax, Nilsson and colleagues (2007) successfully simulated the tube shift technique using pairs of static X-ray projection images exposed with slightly different positions. In this regard, it is likely that interactive volumetric visualisations may provide an intuitive environment for understanding the parallax effect, by enabling the user to follow the relative movement of objects in real time while adjusting the direction of the virtual X-ray beam rotating the scene (Fig. 7.8).

A further task where free mobilisation of the simulated scene may become useful is to understand the eggshell effect of corticated structures (White and Pharoah 2014), which reflects

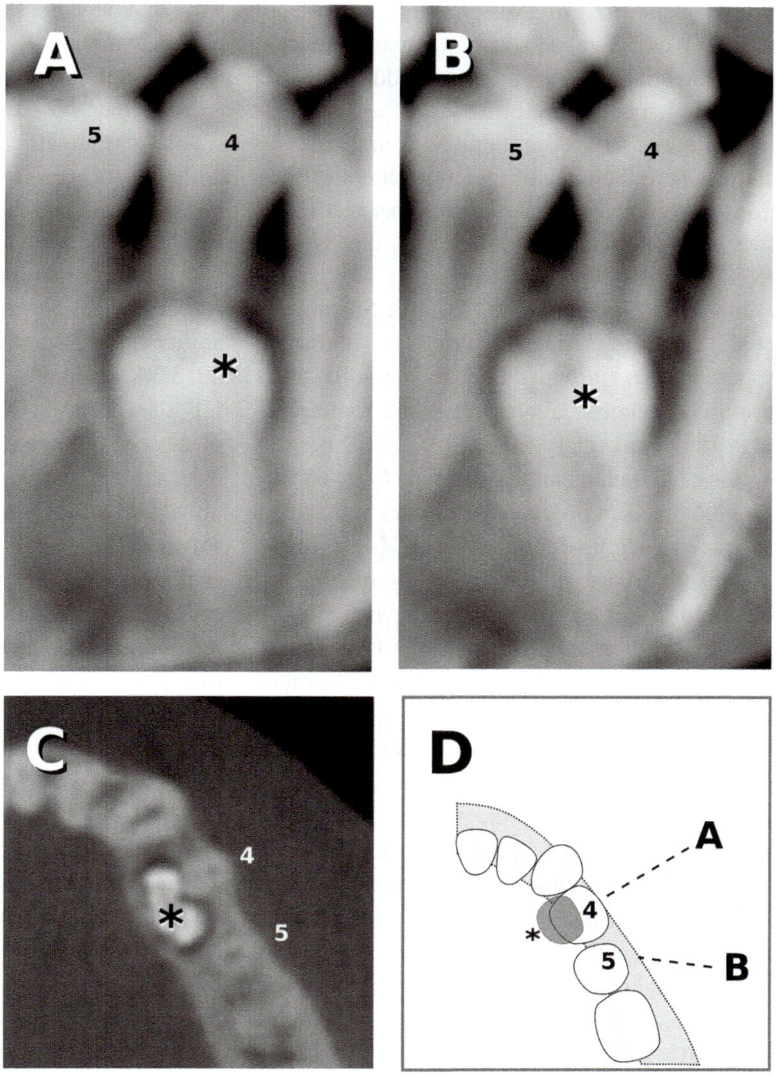


**Fig. 7.8** Identifying the position of an unerupted supernumerary premolar tooth (asterisk) relative to the first (4) and second (5) premolars by slight rotation of the volume-rendered scene, in simulation of the tube-shift method. (**a**) View of the first premolar region from a perpendicular direction to its mesiodistal axis. (**b**) View by shifting the angle distally. (**c**) Axial, cone-beam CT image showing the location of the unerupted premolar relative to the roots of normally erupted premolars (rostral view, matching that in **d**). (**d**) Schematic representation of viewing angles in **a** and **b**

increased attenuation of the X-ray beam due to the longer path followed by photons traversing a structure tangentially. The eggshell effect is commonplace in intraoral radiography and accounts for the radiopacity of a variety of structures such as the lamina dura and the floor of the nasal cavity or the maxillary sinus, among others. By slightly rotating the scene, students have the opportunity to manipulate and actively explore such effect in real time.

## 7.6 Mixed Visualisations

Modern computer-generated 3D graphics allow for simultaneous presentation of multiple actors in a single scene using different rendering techniques. We have made a brief incursion into several potential scenarios where additional 3D graphic elements may provide further information or serve as a visual aid. As a first case we address here the issue of understanding the positions of root apices of posterior upper teeth relative to the maxillary sinus, as this spatial relationship is often challenging to interpret in radiographic images (Hassan 2010; Shahbazian et al. 2015; Eid et al. 2022). We have used small sphere meshes (provided by Paraview software as primitive geometrical shapes) to tag root apices of posterior maxillary teeth (Fig. 7.9). This strategy is not fundamentally different from relying on predefined radiographic contours and shapes as spatial cues for interpreting an X-ray image. Such is the case, for example, of mediastinal lines and stripes used as references when evaluating chest radiographs (Proto 1987; Gibbs et al. 2007), or notional arch-shaped lines traced on an occipitomental view skull radiograph which are known as McGrigor-Campbell lines (McGrigor and Campbell 1950). In the present interactive visualisation approach, the viewer is provided with the opportunity to track the positions of root apices during real-time exploration of the scene, and thus to maintain visual control over their relationships with adjacent structures from every viewing angle.

As a second case, we have sought to combine volumetric visualisation with a surface representation of the maxillary region. For this purpose, we have reconstructed a polygon mesh by segmentation from the same cone-beam CT data used for volumetric visualisation, using ITK Snap (http://www.itksnap.org/) and Blender software (http://www.blender.org/) as described previously (Azkue 2021b). The rationale behind is to present a reference image of the anatomical structure whose radiographic projection is being simulated by volumetric visualisation. It immediately becomes obvious, however, that the surface model obscures the volumetric rendition if both are displayed simultaneously. For the surface representation to serve as a reference, one possible alternative approach is to display volume and surface renditions in separate and adjacent view ports with linked cameras (Fig. 7.9), in such a way that any exploratory mobilisation of one model is automatically accompanied by consistent repositioning of the other.

A further possible strategy to aid in understanding volumetric visualisations is combining with the source cone-beam CT scans. There is reported evidence that training based on analysis of 2D CT views in correlation with volumetric visualisations can improve both diagnostic skills for cross-sectional imaging and visual-spatial ability among medical students (Rengier et al. 2013). A sectional view of the source cone-beam CT data can be reconstructed in orthogonal planes or even in an arbitrary plane as needed. In the example shown in Fig. 7.9, we create sectional images using the Slice filter in Paraview software. Much like in the precedent case, however, sectional images displayed in situ largely cancel the view of the volume rendition, which is much more subtle visually. One possible alternative presentation is again using adjacent and linked view ports. This also provides the opportunity to incorporate surface-rendered elements serving as visual aids (Fig. 7.9).

## 7.7 Distribution and Exchange of Volume-Rendered Content

In contrast with polygon meshes, which can be saved to file carrying important information for visualisation such as materials, textures or even

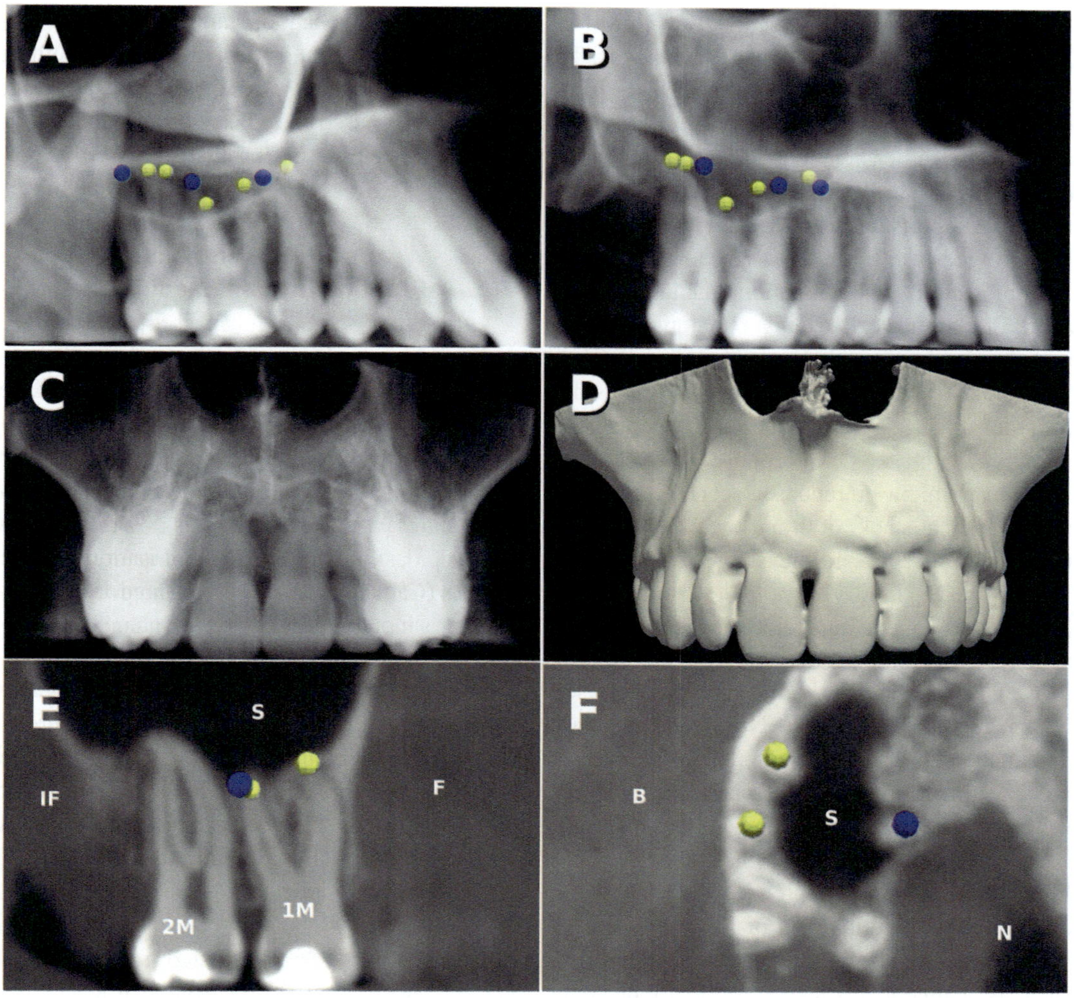


**Fig. 7.9** Parallel volumetric, surface and sectional visualisations of the maxillary region using Paraview software. (**a**, **b**) Simulated X-ray projection in lateral view with slight anterior (**a**) and posterior angulations (**b**), where surface-rendered blue and yellow spheres are used to tag the locations of palatal and buccal apices of posterior maxillary teeth, respectively. (**c**, **d**) A surface view of the maxilla is presented as an anatomical reference in the same viewing angle as the simulated X-ray projection. Both cameras can be linked together so that any exploratory movement of the camera in any of the two view ports results in synchronous movement of the camera in the other. (**e**, **f**) Sectional reconstructions in sagittal (**e**) and transverse planes (**f**) using the Slice filter in Paraview, where blue and yellow surface-rendered spheres tag the root apices of the first molar. *S* maxillary sinus, *N* nasal cavity, *IF* infratemporal fossa, *B* buccal soft tissues, *F* facial soft tissues, *1 M*, *2 M* first and second molars, respectively

illumination, when a data set is rendered, volumetrically no data is generated that could be exported as a model and therefore, there is no simple means to distribute truly interactive volume-rendered scenes. This means that whatever software importing the volume data set would have to be able to volume-render it, and the user will need to configure the rendering settings again. This arguably poses an obstacle to exchanging, sharing and reusing volume-rendered scenes. Several possible workarounds are nevertheless available, some of which will be briefly commented on below.

### 7.7.1 Embedding in a 3D-PDF File

The Portable Document Format (PDF), which is the most widely used digital file format for scholarly and scientific electronic publishing, supports the embedding of 3D surface models in Universal 3D (*.U3D) format. Researchers have advocated for adopting 3D-PDF technology as a de facto standard for publishing the outcomes of biomedical studies (Ruthensteiner and Hess 2008; Barnes et al. 2013; Newe and Becker 2018). The PDF not only supports textured surface renderings, but also provides a Javascript interface for enhanced user interactivity and animation, making the PDF a versatile framework to deliver anatomical information and knowledge (Azkue 2021a). Unfortunately, the 3D-PDF technology does not provide direct support for volume rendering graphics natively. One proposed workaround seeks to mimic a prerendered scene by converting it into a stack of translucent 2D images (Barnes et al. 2006; Ruthensteiner et al. 2010). The S2PLOT library (available from https://astronomy.swin.edu.au/s2plot) implements this approach by creating three stacks of 2D images in *.TGA format from the volumetric data set, one orthogonal to each axis. During display, only the image stack most orthogonal to the current viewing direction is chosen for display, and the set of images changes automatically as the camera is moved around. Although the S2PLOT library was originally intended for displaying astronomic data, utilisation of this approach for volumetric visualisation of clinical imaging data such as magnetic resonance angiographic data has been demonstrated (Barnes et al. 2013). In addition, Newe et al. (2018) developed a module to MeVisLab software (https://www.mevislab.de/), which can be used to create model files in U3D format containing all the data needed to embed volume renderings into PDF files. MeVisLab is a medical imaging platform and visual development environment, where algorithms or libraries wrapped into modules can be connected using a graphical user interface to create complex procedures without programming.

Volume-rendered scenes can be contained and exchanged in 3D-PDF files as learning objects, i.e. specialised pieces of educational software where content is broken down into compact, self-contained chunks that can be repurposed for different educational applications (Wiley 2002; Churchill 2007). Metadata or data about the contained data, an essential component of learning objects, can be incorporated to PDF files using the built-in Extensible Metadata Platform (XMP), a powerful XML format-based metadata standard created by Adobe (Adobe Systems Inc. 2012a, b, c).

### 7.7.2 Wrapping the Visualisation Pipeline into an Executable File

Creating an interactive volume-rendered scene entails reading the source data file, setting up the rendering parameters and eventually performing the rendition proper. Using modern scripting languages and basic templates, it is possible to create basic desktop applications that automate this process (Caban et al. 2007). This approach limits the user's input to providing a file name of source data and configuring basic rendering parameters. For example, the Visualisation Toolkit scientific visualisation software (https://vtk.org/) provides wrapping functionality to generate an interface layer to scripting languages such as Python or Tcl-Tk among others. Converting the script to a standalone executable file allows for distributing and running the application independently.

### 7.7.3 Web-Based Visualisations

The field of scientific visualisation in the web browser is expected to play an increasingly important role (Mwalongo et al. 2016; Franke and Haehn 2020). Emerging technology such as Web Graphics Library (WebGL) supports 3D renderings and allows GPU-based rendering and acceleration without any third-party browser plugins. This capability provides a cross-browser and cross-platform mechanism to view 3D graphics just by visiting a website. Web-based visualisation is also convenient in that no software installations are required and contents can be easily updated on the server side. Volumetric data can be rendered within the web browser using the JavaScript API WebGL 2.0 (Marrin 2011). Still, a web-based application needs to be preconfigured both to read the source data and to apply the appropriate parameters for visualisation.

In the educational setting, a major advantage of using a web browser for visualisations is that web contents can be packaged and uploaded to Learning Management Systems (LMS), i.e. applications used for the administration and delivery of educational—largely e-Learning—programs. In addition, web-based contents can be imported as assets to authoring tools in order to create educational contents that can then be encapsulated in SCORM format, a de facto standard for e-Learning content.

An additional approach to deliver volume-rendered scenes is to create content in virtual reality modeling language (VRML), a widely used file format for web-based 3D graphics, based on an analogous strategy to the one above implemented by the S2PLOT library (Engel and Ertl 1999). The latter indeed provides a function to export a standard volume rendering to VRML. The Extensible 3D (*.X3D) file format for web-based 3D content also supports volume rendering, including visualisation of clinical imaging data (Arbelaiz et al. 2017). It is however worth noting that web browsers do not support VRML or X3D content natively, which makes it necessary to install a client plug-in to the browser for viewing 3D content in those formats.

## 7.8 Future Research and Implications for Practice

Despite the potentials offered by interactive volumetric visualisations, there are various aspects for the use of this approach for learning radiographic anatomy that merit further exploration. One such aspect is related with delivery and presentation of radiograph-like simulations. By improving the effectiveness and efficiency of visualisations, students are expected to be more successful in retaining visual information required for their understanding of complex aspects of anatomy. Poorly designed instructional materials can inadvertently impose extraneous cognitive load, i.e. ineffective mental effort that does not contribute to consolidation of new learning into long-term memory, according to the cognitive load theory of human cognitive architecture (Sweller 1988; Sweller et al. 2019). The use of visual cues such as arrows, pointers or colour-coding can minimise extraneous cognitive load of visualisations (Thompson et al. 2011; Janssen et al. 2015; Richter et al. 2016; Skulmowski and Rey 2018) and direct learners to elements of crucial importance (Arendasy et al. 2011). However, an excess of 3D tags in a single scene may increase extraneous load and compromise cognitive resources that contribute to actual learning (Khalil et al. 2005; Hadie et al. 2018). This issue needs to be borne in mind in the design process of volume-rendered scenes that incorporate 3D tags and visual cues.

Studies suggest that cross-sectional and multimodal 3D visualisations offer advantages when integrated together as a learning material, synergistically promoting both anatomic understanding, visual-spatial ability and radiological reasoning (Rengier et al. 2013; Ben Awadh et al. 2022). It is shown here that volumetric visualisations simulating radiographic projections can indeed be enriched with both cross-sectional views and surface-rendered models. Further research is needed to validate multimodal approaches using volumetric visualisations with regard to the learning of radiographic anatomy.

Another essential issue to be addressed is whether instructional materials based on interactive volumetric visualisations to simulate radiographic imaging effectively improve learning. A number of studies have so far documented the potentials offered by 3D visualisations for learning anatomy (Yammine and Violato 2015), and there is evidence of volumetric visualisations contributing specifically to a better interpretation of spatial information in radiographs (Nilsson et al. 2007). It is possible, as discussed in Sect. 7.2, that 3D visualisation may also contribute indirectly to anatomic understanding by improving visual spatial ability. Empirical evidence is required to assess this issue in the domain of radiographic anatomy.

As a final consideration, an essential requisite for any potential learning tool to be considered as such is to be valued positively by educators and learners. A widely applied and empirically tested theoretical model to this effect is the technology acceptance model (TAM; Davis 1989), which explains end users' acceptance behaviour based on perceived ease of use and perceived usefulness. Although the TAM has been utilised for evaluating technologies in a broad sense, it is also the most widely used for prediction and explanation of user's behaviour toward acceptance and adoption of educational technology (Abdullah and Ward 2016; Granić and Marangunić 2019). The TAM can provide a conceptual framework also for validating the use of real-time simulation of radiographic imaging as proposed here, in a similar fashion to previous studies that have focused on educational tools for learning anatomy such as virtual reality, mobile apps or simulators (Fang et al. 2014; Harmon et al. 2022; Singer et al. 2024).

## 7.9 Conclusions

This chapter explores the potential of volumetric visualisations of maxillofacial cone-beam CT scan data to simulate intraoral radiographs in the context of learning radiographic anatomy. Volumetric renderings provide adequate graphic detail to recognise major anatomical landmarks and structures that are characteristic to intraoral projections, as a first prerequisite to qualify as a learning tool. Volumetric visualisations can be used to learn standard morphology and normal variation, but an analogous graphic strategy can also be used for the study of pathology by selecting appropriate cone-beam CT data sets. A second requisite is to provide advantages over conventional teaching tools such as collections of static radiographs. In this regard, the main contribution of the present approach is the provision of real-time user interaction and the opportunity to explore radiographic anatomy hands on, with full control of the camera. Furthermore, volume-rendered scenes can be enriched with 3D graphic elements such as polygon meshes and planar views that may contribute additional information or serve as a visual aid.

Although several strategies for exchanging and reusing volume-rendered scenes are indeed available as described above, technical possibilities are still limited at present. Ideally, authoring tools available to content creators should be endowed with volume rendering support, in order to integrate volumetric visualisations into teaching practice via multimedia, dedicated websites or stand-alone desktop applications. Further investigation is warranted both to improve aspects of volumetric rendering such as exchangeability and to inform how to best deliver and present radiographic simulations.

**Acknowledgements** The authors are grateful to Dr. Federico Bolelli for kind assistance in accessing the source cone-beam CT data.

## Glossary

**Amodal completion** Amodal completion is the mental representation of those parts of a perceived object from which we receive no sensory inputs. In visual perception, amodal completion is the task of recognizing and imagining the occluded parts of objects we see.

**Cone-beam CT** Cone-beam computer tomography is a medical imaging technique, in which the X-rays are sent out from the X-ray source using a cone-shaped beam, as opposed to fan-shaped of spiral scans used in traditional computed tomography.

**Curved planar reformation** Curved planar reformation or multiplanar reconstruction is a technique to provide a two-dimensional representation of a curved anatomical structure. In dentistry, curved planar reformation can be used to reconstruct panoramic X-ray views from CT scans.

**Eggshell effect** Tangential X-rays passing through a hollow or tubular structure, e.g. a dental alveolus, traverse a longer path than the radial thickness of the wall, thereby producing a radiographic appearance or pattern that resembles that of an eggshell.

**Motion parallax** The motion parallax refers to the changes in the apparent relative positions of two objects located at different distances from the observer when the viewing angles are varied.

**Panoramic radiography or orthopantomogram** The orthopantomogram, also known as an orthopantomograph, pantomogram, OPG X-ray or OPT, is a composite panoramic X-ray of the mandible, maxilla and teeth, formed by combining images of multiple planes.

**Polygon mesh** A polygon mesh is a mathematical representation of an object's surface as a net of polygons, most commonly triangles or quadrilaterals, connected by lines and vertices.

**Scalar data** Scalar data refer to individual values that can express scale, such as weight, temperature or size, as opposed to aggregate data types expressing spatial coordinates or directions. In volume rendering, scalars are values for some measured or calculated property of the volume, e.g. radiodensity values in a stack of CT scans.

**Scientific visualisation** Scientific visualisation is a branch of computer graphics concerned with the representation and visualisation of scientific data, including but not limited to clinical imaging data.

**SCORM** SCORM, which stands for Sharable Content Object Reference Model, is a set of technical standards and specifications for learning management systems and e-Learning contents to work together.

**Surface normals** In a polygon mesh, surface normals are directional vectors that are perpendicular to polygon faces or to vertices. In surface rendering, normals are crucial for simulating the reflection of light on a 3D object.

**Surface rendering** Surface rendering is one of the two major rendering modes in scientific visualisation. In surface rendering, a 3D object is represented as a set of geometric primitives such as polygon meshes or contours, and rendered for display as a surface illuminated by a virtual light source.

**Volume rendering** Volume rendering is one of the two major rendering modes in scientific visualisation. In volume rendering, the object of interest is sampled into many building blocks called voxels, which are rendered for display by passing a virtual light through them.

**Voxel** Voxel is short for 'volume element', in analogy to picture elements or pixels that make up a 2D bitmap image. It represents any of the discrete elements comprising a three-dimensional entity, such as an image produced by computed tomography imaging.

## References


Abdullah F, Ward R (2016) Developing a general extended technology acceptance model for E-learning (GETAMEL) by analysing commonly used external factors. Comput Hum Behav 56:238–256

Adobe Systems Inc. (2012a) XMP specification Part 1. Data model, serialisation, and core properties. https://wwwimages2.adobe.com/content/dam/acom/en/devnet/xmp/pdfs/XMP%20SDK%20Release%20cc-2016-08/XMPSpecificationPart1.pdf

Adobe Systems Inc. (2012b) XMP specification Part 2. Additional properties. https://wwwimages2.adobe.com/content/dam/acom/en/devnet/xmp/pdfs/XMP%20SDK%20Release%20cc-2014-12/XMPSpecificationPart2.pdf

Adobe Systems Inc. (2012c) XMP specification Part 3. Storage in files. https://wwwimages2.adobe.com/content/dam/acom/en/devnet/xmp/pdfs/XMP%20SDK%20Release%20cc-2016-08/XMPSpecificationPart3.pdf

Alraddadi A (2021) Literature review of anatomical variations: clinical significance, identification approach, and teaching strategies. Cureus 13(4):e14451

Arampatzakis V, Pavlidis G, Mitianoudis N, Papamarkos N (2024) Monocular depth estimation: a thorough review. IEEE Trans Pattern Anal Mach Intell 46(4):2396–2414

Arbelaiz A, Moreno A, Kabongo L, García-Alonso A (2017) Volume visualization tools for medical applications in ubiquitous platforms. In: Giokas K, Bokor L, Hopfgartner F (eds) eHealth 360° - international summit on eHealth, Budapest, Hungary, June 14–16, 2016, Revised selected papers. Springer, Cham, pp 1–8

Arendasy M, Sommer M, Hergovich A, Feldhammer M (2011) Evaluating the impact of depth cue salience in working three-dimensional mental rotation tasks by means of psychometric experiments. Learn Individ Differ 21(4):403–408

Atta IS, AlQahtani FN (2018) Matching medical student achievement to learning objectives and outcomes: a paradigm shift for an implemented teaching module. Adv Med Educ Pract 9:227–233

Azkue JJ (2021a) Embedding interactive, three-dimensional content in portable document format to deliver gross anatomy information and knowledge. Clin Anat 34(6):919–933

Azkue JJ (2021b) External surface anatomy of the postfolding human embryo: computer-aided, three-

dimensional reconstruction of printable digital specimens. J Anat 239(6):1438–1451
Baenninger M, Newcombe N (1989) The role of experience in spatial test performance: a meta-analysis. Sex Roles 20(5/6):327–344
Barnes DG, Fluke CJ, Bourke PD, Parry OT (2006) An advanced, three-dimensional plotting library for astronomy. PASA 23:82–93
Barnes DG, Vidiassov M, Ruthensteiner B, Fluke CJ, Quayle MR, McHenry CR (2013) Embedding and publishing interactive, 3-dimensional, scientific figures in portable document format (PDF) files. PLoS One 8(9):e69446
Ben Awadh A, Clark J, Clowry G, Keenan ID (2022) Multimodal three-dimensional visualization enhances novice learner interpretation of basic cross-sectional anatomy. Anat Sci Educ 15(1):127–142
Benbow CP, Stanley JC (1982) Intellectually talented boys and girls: educational profiles. Gift Child Q 26(2):82–88
Berbaum KS, Smoker WR, Smith WL (1985) Measurement and prediction of diagnostic performance during radiology training. AJR Am J Roentgenol 145(6):1305–1311
Birchall D (2015) Spatial ability in radiologists: a necessary prerequisite? Br J Radiol 88(1049):20140511
Briscoe R (2011) Mental imagery and the varieties of amodal perception. Pac Philos Q 92(2):153–173
Buckley J, Seery N, Canty D (2018) A heuristic framework of spatial ability: A review and synthesis of spatial factor literature to support its translation into STEM education. Educ Psychol Rev 30(3):947–972. https://doi.org/10.1007/s10648-018-9432-z
Bülthoff HH, Edelman SY, Tarr MJ (1995) How are three-dimensional objects represented in the brain? Cereb Cortex 5(3):247–260
Caban JJ, Joshi A, Nagy P (2007) Rapid development of medical imaging tools with open-source libraries. J Digit Imaging 20(Suppl 1):83–93
Chetlen AL, Mendiratta-Lala M, Probyn L, Auffermann WF, DeBenedectis CM, Marko J, Pua BB, Sato TS, Little BP, Dell CM, Sarkany D, Gettle LM (2015) Conventional medical education and the history of simulation in radiology. Acad Radiol 22(10):1252–1267
Churchill D (2007) Towards a useful classification of learning objects. Educ Technol Res Dev 55(5):479–497
Cipriano M, Allegretti S, Bolelli F, Di Bartolomeo M, Pollastri F, Pellacani A, Minafra P, Anesi A, Grana C (2022) Deep segmentation of the mandibular canal: a new 3D annotated dataset of CBCT volumes. IEEE Access 10:11500–11510
Clark CA (1910) A method of ascertaining the relative position of unerupted teeth by means of film radiographs. Proc R Soc Med 3:87–90
Corry CA (2011) The future of recruitment and selection in radiology. Is there a role for assessment of basic visuospatial skills? Clin Radiol 66(5):481–483
Cutting JE, Vishton PM (1995) Perceiving layout and knowing distances: the integration, relative potency, and contextual use of diVerent information about depth. In: Epstein W, Rogers S (eds) Handbook of perception and cognition, perception of space and motion, vol 5. Academic Press, San Diego, pp 69–117
Davies A, Mannocci F, Mitchell P, Andiappan M, Patel S (2015) The detection of periapical pathoses in root filled teeth using single and parallax periapical radiographs versus cone beam computed tomography – a clinical study. Int Endod J 48(6):582–592
Davis FD (1989) Perceived usefulness, perceived ease of use, and user acceptance of information technology. MIS Q 13(3):319–340
DeWitt RD (2019) Planning for active learning in the didactic classroom. J Physician Assist Educ 30(1):41–46
Drebin RA, Carpenter L, Hanrahan P (1988) Volume rendering. ACM SIGGRAPH Comput Graph 22:65
Egusa H (1983) Effects of brightness, hue, and saturation on the perceived depth between adjacent regions in the visual field. Perception 12(2):167–175
Eid EA, El-Badawy FM, Hamed WM (2022) Accuracy of intraoral digital radiography in assessing maxillary Sinus-Root relationship compared to CBCT. Saudi Dent J 34(5):397–403
Engel K, Ertl T (1999) Texture-based volume visualization for multiple users on the world wide web. In: Proceedings of the 5th eurographics workshop on virtual environments, EGVE 1999, Vienna, Austria, May 31 - June 1, 1999. Springer, pp 115–124
Ennis LM (1937) Roentgenographic variations of the maxillary sinus and the nutrient canals of the maxilla and the mandible. Int J Orth Oral Surg 23(2):173–193
Fang T-Y, Wang P-C, Liu C-H, Su M-C, Yeh S-C (2014) Evaluation of a haptics-based virtual reality temporal bone simulator for anatomy and surgery training. Comput Methods Prog Biomed 113(2):674–681
Fernandez R, Dror IE, Smith C (2011) Spatial abilities of expert clinical anatomists: comparison of abilities between novices, intermediates, and experts in anatomy. Anat Sci Educ 4(1):1–8
Franke L, Haehn D (2020) Modern scientific visualizations on the web. Informatics 7(4):37
Garg A, Norman G, Spero L, Maheshwari P (1999) Do virtual computer models hinder anatomy learning? Acad Med 74(suppl):S87–S89
Garg AX, Norman G, Sperotable L (2001) How medical students learn spatial anatomy. Lancet 357(9253):363–364
Gibbs JM, Chandrasekhar CA, Ferguson EC, Oldham SAA (2007) Lines and stripes: where did they go?--from conventional radiography to CT. Radiographics 27(1):33–48
Granić A, Marangunić N (2019) Technology acceptance model in educational context: a systematic literature review. Br J Educ Technol 50(5):2572–2593
Grondin S (2016) Depth perception. In: Psychology of perception. Springer, pp 103–122
Guillot A, Champely S, Batier C, Thiriet P, Collet C (2007) Relationship between spatial abilities, mental rotation and functional anatomy learning. Adv Health Sci Educ Theory Pract 12(4):491–507

Hadie SNH, Sulong HAM, Hassan A, Ismail ZIM, Talip S, Rahim AFA (2018) Creating an engaging and stimulating anatomy lecture environment using the cognitive load theory-based lecture model: students' experiences. J Taibah Univ Med Sci 13(2):162–172

Hambrick DZ, Libarkin JC, Petcovic HL, Baker KM, Elkins J, Callahan CN, Turner SP, Rench TA, Ladue ND (2012) A test of the circumvention-of-limits hypothesis in scientific problem solving: the case of geological bedrock mapping. J Exp Psychol Gen 141(3):397–403

Hanes DA, Keller J, McCollum G (2008) Motion parallax contribution to perception of self-motion and depth. Biol Cybern 98(4):273–293

den Harder AM, Frijlingh M, Ravesloot CJ, Oosterbaan AE, van der Gijp A (2016) The importance of human–computer interaction in radiology e-learning. J Digit Imaging 29(2):195–205

Harmon DJ, Burgoon JM, Kalmar EL (2022) Development and assessment of an integrated anatomy mobile app. Clin Anat 35(5):686–696

Hassan BA (2010) Reliability of periapical radiographs and orthopantomograms in detection of tooth root protrusion in the maxillary sinus: correlation results with cone beam computed tomography. J Oral Maxillofac Res 1(1):e6

Hassan BA, Jacobs R, Scarfe WC, Al-Rawi WT (2007) A web-based instruction module for interpretation of craniofacial cone beam CT anatomy. Dentomaxillofac Radiol 36(6):348–355

Hegarty M, Keehner M, Khooshabeh P, Montello DR (2009) How spatial abilities enhance, and are enhanced by, dental education. Learn Indiv Differ 19:61–70

Hodgkiss A, Gilligan KA, Tolmie AK, Thomas MSC, Farran EK (2018) Spatial cognition and science achievement: the contribution of intrinsic and extrinsic spatial skills from 7 to 11 years. Br J Educ Psychol 88(4):675–697

Hontoir F, Simon V, De Raeve Y, Dumortier L, Dugdale A, Vandeweerd J-M (2023) Can online teaching of radiographic anatomy replace conventional on-site teaching? A randomized controlled study. J Vet Med Educ 50(2):217–227

Howard IP, Rogers BJ (2012) Perceiving in depth, vol 1. Oxford University Press, New York

Hoyek N, Collet C, Rastello O, Fargier P, Thiriet P, Guillot A (2009) Enhancement of mental rotation abilities and its effect on anatomy learning. Teach Learn Med 21(3):201–206

Hubar JS (2017) Fundamentals of oral and maxillofacial radiology. John Wiley & Sons

Huumonen S, Ørstavik D (2002) Radiological aspects of apical periodontitis. Endod Top 1:3–25

Jacobs SG (1994) Localisation of the unerupted maxillary canine: additional observations. Aust Orthod J 13(2):71–75

Jacobs SG (1999) Radiographic localization of unerupted maxillary anterior teeth using the vertical tube shift technique: the history and application of the method with some case reports. Am J Orthod Dentofacial Orthop 116(4):415–423

Jacobs SG (2000) Radiographic localization of unerupted mandibular anterior teeth. Am J Orthod Dentofacial Orthop 118(4):432–438

Jamil Z, Saeed AA, Madhani S, Baig S, Cheema Z, Fatima SS (2019) Three-dimensional visualization software assists learning in students with diverse spatial intelligence in medical education. Anat Sci Educ 12(5):550–560

Janssen A, Shaw T, Goodyear P, Kerfoot BP, Bryce D (2015) A little healthy competition: using mixed methods to pilot a team-based digital game for boosting medical student engagement with anatomy and histology content. BMC Med Educ 15:173

Jung Y-H, Cho B-H (2014) Radiographic evaluation of the course and visibility of the mandibular canal. Imaging Sci Dent 44(4):273–278

Kasprzak T (2016) Technology and radiology education-meeting the needs of millennial learners. Acad Radiol 23(7):844–847

Keehner M, Khooshabeh P, Hegarty M (2008) Individual differences among users: implications for the design of 3D medical visualizations. In: Dong F, Ghinea G, Chen SY (eds) User centered design for medical visualization. IGI Global, pp. 1–24

Keehner MM, Tendick F, Meng MV, Anwar HP, Hegarty M, Stoller ML, Duh QY (2004) Spatial ability, experience, and skill in laparoscopic surgery. Am J Surg 188(1):71–75

Keenan ID, Powell M (2020) Interdimensional travel: visualisation of 3D-2D transitions in anatomy learning. Adv Exp Med Biol 1235:103–116

Khalil MK, Paas F, Johnson TE, Payer AF (2005) Design of interactive and dynamic anatomical visualizations: the implication of cognitive load theory. Anat Rec B New Anat 286(1):15–20

Kim HGR, Angelaki DE, DeAngelis GC (2016) The neural basis of depth perception from motion parallax. Philos Trans R Soc Lond Ser B Biol Sci 371(1697):20150256

Klinge B, Petersson A, Maly P (1989) Location of the mandibular canal: comparison of macroscopic findings, conventional radiography, and computed tomography. Int J Oral Maxillofac Implants 4(4):327–332

Linaker KL (2015) Pedagogical approaches to diagnostic imaging education: a narrative review of the literature. J Chiropr Humanit 22(1):9–16

Lindh C, Petersson A, Klinge B (1992) Visualisation of the mandibular canal by different radiographic techniques. Clin Oral Implants Res 3(2):90–97

Lopes LJ, Gamba TO, Bertinato JVJ, Freitas DQ (2016) Comparison of panoramic radiography and CBCT to identify maxillary posterior roots invading the maxillary sinus. Dentomaxillofac Radiol 45(6):20160043

Lufler RS, Zumwalt AC, Romney CA, Hoagland TM (2012) Effect of visual-spatial ability on medical students' performance in a gross anatomy course. Anat Sci Educ 5(1):3–9

Marrin C (2011) WebGL specification. Khronos WebGL Working Group, Beaverton

McGrigor DB, Campbell W (1950) The radiology of war injuries. VI. Wounds of the face and jaw. Br J Radiol 23(276):685–696

McNulty JP, England A, Shanahan MC (2021) International perspectives on radiography practice education. Radiography (Lond) 27(4):1044–1051

Michotte A, Thines G, Crabbe G (1991) Amodal completion of perceptual structures. In: Thines G, Costall A, Butterworth G (eds) Michotte's experimental phenomenology of perception. Lawrence Erlbaum Associates, New York, pp 140–167

Mwalongo F, Krone M, Reina G, Ertl T (2016) State-of-the-art report in web-based visualization. Comput Graph Forum 35(3):553–575

Naitoh M, Katsumata A, Kubota Y, Hayashi M, Ariji E (2009) Relationship between cancellous bone density and mandibular canal depiction. Implant Dent 18(2):112–118

National Research Council (2006) Learning to think spatially: GIS as a support system in the K-12 curriculum, 1st edn. The National Academies Press, Washington, DC

Newe A, Becker L (2018) Three-dimensional portable document format (3D PDF) in clinical communication and biomedical sciences: systematic review of applications, tools, and protocols. JMIR Med Inform 6(3):e10295

Newe A, Brandner J, Aichinger W, Becker L (2018) An open source tool for creating model files for virtual volume rendering in PDF documents. In: Maier A, Deserno TM, Handels H, Maier-Hein KH, Palm C, Tolxdorff T (eds) Bildverarbeitung für die Medizin 2018, Algorithmen – Systeme – Anwendungen. Proceedings des Workshops vom 11. bis 13. März 2018 in Erlangen. Springer Vieweg Berlin, Heidelberg, pp 133–138

Nguyen N, Nelson AJ, Wilson TD (2012) Computer visualizations: factors that influence spatial anatomy comprehension. Anat Sci Educ 5(2):98–108

Nilsson T, Ahlqvist J, Johansson M, Isberg A (2004) Virtual reality for simulation of radiographic projections: validation of projection geometry. Dentomaxillofac Radiol 33(1):44–50

Nilsson TA, Hedman LR, Ahlqvist JB (2007) A randomized trial of simulation-based versus conventional training of dental student skill at interpreting spatial information in radiographs. Simulation in healthcare: The journal of the society for simulation in healthcare 2(3):164–169. https://doi.org/10.1097/SIH.0b013e31811ec254

Pizer SM, ter Haar Romeny BM (1991) Fundamental properties of medical image perception. J Digit Imaging 4(1):1–20

Prakash N, Karjodkar FR, Sansare K, Sonawane HV, Bansal N, Arwade R (2015) Visibility of lamina dura and periodontal space on periapical radiographs and its comparison with cone beam computed tomography. Contemp Clin Dent 6(1):21–25

Prober CG, Khan S (2013) Medical education reimagined: a call to action. Acad Med 88(10):1407–1410

Proto AV (1987) Mediastinal anatomy: emphasis on conventional images with anatomic and computed tomographic correlations. J Thorac Imaging 2(1):1–48

Ramey KE, Uttal DH (2017) Making sense of space: distributed spatial sensemaking in a middle school summer engineering camp. J Learn Sci 26(2):277–319. https://doi.org/10.1080/10508406.2016.1277226

Ranney RR, Wilson MB, Bennett RB (2005) Evaluation of applicants to predoctoral dental education programs: review of the literature. J Dent Educ 69(10):1095–1106

Reinhardt-Rutland AH (1992) Primary depth cues and background pattern in the portrayal of slant. J Gen Psychol 119(1):29–35

Rengier F, Häfner MF, Unterhinninghofen R, Nawrotzki R, Kirsch J, Kauczor HU, Giesel FL (2013) Integration of interactive three-dimensional image post-processing software into undergraduate radiology education effectively improves diagnostic skills and visual-spatial ability. Eur J Radiol 82(8):1366–1371

Richter J, Scheiter K, Eitel A (2016) Signaling text-picture relations in multimedia learning: a comprehensive meta-analysis. Educ Res Rev 17:19–36

Rochford K (1985) Spatial learning disabilities and underachievement among university anatomy students. Med Educ 19(1):13–26

Rogers BJ (1993) Motion parallax and other dynamic cues for depth in humans. Rev Oculomot Res 5:119–137

Ruthensteiner B, Baeumler N, Barnes DG (2010) Interactive 3D volume rendering in biomedical publications. Micron 41(7):886.e1–886.e17

Ruthensteiner B, Hess M (2008) Embedding 3D models of biological specimens in PDF publications. Microsc Res Tech 71(11):778–786

Sandow PL, Jones AC, Peek CW, Courts FJ, Watson RE (2002) Correlation of admission criteria with dental school performance and attrition. J Dent Educ 66(3):385–392

Sañudo JR, Vázquez R, Puerta J (2003) Meaning and clinical interest of the anatomical variations in the 21st century. Eur J Anat 7(Suppl. 1):1–3

Sarilita E, Lita YA, Firman DR, Wilkinson T, Susilawati S, Saptarini R, Aripin D, Sjamsudin E (2022) Spatial ability and anatomy learning performance among dental students. Korean J Med Educ 34(4):309–318

Shahbazian M, Vandewoude C, Wyatt J, Jacobs R (2015) Comparative assessment of periapical radiography and CBCT imaging for radiodiagnostics in the posterior maxilla. Odontology 103(1):97–104

Sharan A, Madjar D (2006) Correlation between maxillary sinus floor topography and related root position of posterior teeth using panoramic and cross-sectional computed tomography imaging. Oral Surg Oral Med Oral Pathol Oral Radiol Endod 102(3):375–381

Shiner N (2018) Is there a role for simulation based education within conventional diagnostic radiography? A literature review. Radiography (Lond) 24(3):262–271

Sims VK, Mayer RE (2002) Domain specificity of spatial expertise: the case of video game players. Appl Cogn Psychol 16(1):97–115

Singer L, Evans L, Zahra D, Agbeja I, Moyes S (2024) Fostering engagement in virtual anatomy learning for healthcare students. BMC Med Educ 24(1):414

Skochelak SE, Stack SJ (2017) Creating the medical schools of the future. Acad Med 92(1):16–19

Skulmowski A, Rey GD (2018) Realistic details in visualizations require color cues to foster retention. Comput Educ 122:23–31

Smoker WR, Berbaum KS, Luebke NH, Jacoby CG (1984) Spatial perception testing in diagnostic radiology. AJR Am J Roentgenol 143(5):1105–1109

Spence B (2019) Practical applications in radiography education. Radiol Technol 90(4):369–384

Stumpf H, Fay E (1983) Schlauchfiguren – Ein Test zur Beurteilung des räumlichen Vorstellungsvermögens. Verlag für Psychologie Dr. C. J. Hogrefe, Göttingen

Sweeney K, Hayes JA, Chiavaroli N (2014) Does spatial ability help the learning of anatomy in a biomedical science course? Anat Sci Educ 7(4):289–294

Sweller J (1988) Cognitive load during problem solving: effects on learning. Cogn Sci 12(2):275–285

Sweller J, van Merriënboer JJG, Paas F (2019) Cognitive architecture and instructional design: 20 years later. Educ Psychol Rev 31:261–292

Tay YX, McNulty JP (2023) Radiography education in 2022 and beyond - writing the history of the present: a narrative review. Radiography (Lond) 29(2):391–397

Thompson ME, Ford R, Webster A (2011) Effectiveness of interactive, online games in learning neuroscience and students' perception of the games as learning tools: a pre-experimental study. J Allied Health 40(3):150–155

Tung EL, Awan OA (2023) Becoming an expert radiology educator through deliberate practice. Acad Radiol 30(12):3150–3152

Tung EL, Matalon SA (2023) Study smarter: applying the science of learning to radiology. J Am Coll Radiol 20(11):1084–1091

Uttal DH, Cohen CA (2012) Spatial thinking and STEM education: when, why, and how? In: Ross BH (ed) The psychology of learning and motivation. Elsevier Academic Press, pp 147–181

Vandenberg SG, Kuse AR (1978) Mental rotations, a group test of three-dimensional spatial visualization. Percept Mot Skills 47(2):599–604

Veena KM, Razak R, Chatra L, Shenoy P, Prabhu R (2021) Appreciation of incisive foramen in intraoral periapical radiographs - a comparative radiographic study. J Morphol Sci 38:127–129

Velvart P, Hecker H, Tillinger G (2001) Detection of the apical lesion and the mandibular canal in conventional radiography and computed tomography. Oral Surg Oral Med Oral Pathol Oral Radiol Endod 92(6):682–688

Vorstenbosch MA, Klaassen TP, Donders AR, Kooloos JG, Bolhuis SM, Laan RF (2013) Learning anatomy enhances spatial ability. Anat Sci Educ 6(4):257–262

Vuchkova J, Maybury TS, Farah CS (2011) Testing the educational potential of 3D visualization software in oral radiographic interpretation. J Dent Educ 75(11):1417–1425

Wanzel KR, Hamstra SJ, Anastakis DJ, Matsumoto ED, Cusimano MD (2002) Effect of visual-spatial ability on learning of spatially-complex surgical skills. Lancet 359(9302):230–231

Wanzel KR, Hamstra SJ, Caminiti MF, Anastakis DJ, Grober ED, Reznick RK (2003) Visual-spatial ability correlates with efficiency of hand motion and successful surgical performance. Surgery 134(5):750–757

Welchman AE (2016) The human brain in depth: how we see in 3D. Annu Rev Vis Sci 4(2):345–376

Wertz CI, Hobbs DL, Mickelsen W (2014) Integrating technology into radiologic science education. Radiol Technol 86(1):23–31

White SC, Pharoah MJ (2014) Oral radiology. Principles and interpretation, 7th edn. Elsevier, St. Louis

Wiley DA (2002) Connecting learning objects to instructional design theory: a definition, a metaphor, and a taxonomy. In: Wiley D (ed) The instructional use of learning objects. Agency for Instructional Technology: Association for Educational Communications & Technology, Bloomington, IN, pp 3–24

Willan PL, Humpherson JR (1999) Concepts of variation and normality in morphology: important issues at risk of neglect in modern undergraduate medical courses. Clin Anat 12(3):186–190

Yammine K, Violato C (2015) A meta-analysis of the educational effectiveness of three-dimensional visualization technologies in teaching anatomy. Anat Sci Educ 8(6):525–538

Zilverschoon M, Kotte EMG, van Esch B, Ten Cate O, Custers EJ, Bleys RLAW (2019) Comparing the critical features of e-applications for three-dimensional anatomy education. Ann Anat 222:28–39

**Eire Lopetegi** is a General Dentist originally from Azkoitia, in the Basque Province of Gipuzkoa. She graduated from the University of the Basque Country (UPV-EHU) in Bilbao, after a stay at the Dental School at the Università degli Studi di Torino, Italy. She currently works as a general dentist in Bondy and Gennevilliers, France.

**Jon Jatsu Azkue** is a Professor of Human Anatomy and Embryology at the University of the Basque Country (UPV-EHU). He has a background in neuroscience and scientific visualisation, with published contributions on 3D visualisation tools for anatomy teaching and learning.

# 8 Discover Anatomy: A Student-Facilitated Public Engagement Event

Kayleigh Scotcher, Emily Green, Thomas Butts, Evie M. Donald, Kate Dulohery, Munesh P. Khamuani, Harry Miles, Debs Patten, Alice E. M. Roberts, Hannah Swainson, and Joanna Matthan

**Abstract**

Knowledge of the human body is of benefit to all, underpinning our understanding of health and lifestyle choices. Anatomy is the foundation of human biological subjects and is an area of interest for many members of the public. Furthermore, widening participation and outreach is an important role of universities. Anatomists at Newcastle University, North-East England, UK, and the University of Sunderland, North-East England, UK designed and delivered a public engagement event at the Life Science Centre (part of the International Centre for Life), Newcastle-upon-Tyne, UK. This aimed to educate the public and promote anatomy education within the region. The event was split into seven stations covering different topics: (1) overview of the human body, (2) cardiovascular system, (3) respiratory system, (4) nervous system, (5) gastrointestinal system, (6) musculoskeletal system and (7) dental anatomy. The stations were scattered amongst the usual exhibits of the venue. Each station involved hands-on activities for members of the public to undertake. Student volunteers, alongside university staff, were recruited to deliver on these stations and were a crucial part of the success of the event. All student facilitators were provided with tailored training on their assigned stations and general training on public engagement skills. The involvement of students allowed the event to be run on a larger scale, with 520 attendees in total. Feedback from the venue, public and students was very positive. Based on this successful experience, plans are in place for more public engagement events with student involvement in the future. An

K. Scotcher · E. Green (✉) · H. Swainson
School of Medicine, Faculty of Medical Sciences, Newcastle University, Newcastle upon Tyne, UK
e-mail: emily.green@newcastle.ac.uk; hannah.swainson@newcastle.ac.uk

T. Butts · E. M. Donald · K. Dulohery · M. P. Khamuani · H. Miles · D. Patten · A. E. M. Roberts
School of Medicine, Faculty of Health Sciences and Wellbeing, University of Sunderland, Sunderland, UK
e-mail: thomas.butts@sunderland.ac.uk; evie.donald@sunderland.ac.uk; kate.dulohery@sunderland.ac.uk; munesh.khamuani@sunderland.ac.uk; harry.miles@sunderland.ac.uk; debs.patten@sunderland.ac.uk; alice.roberts@sunderland.ac.uk

J. Matthan
Newcastle University Medicine Malaysia (NUMed), Johor, Malaysia
e-mail: joanna.matthan2@newcastle.edu.my



I. D. Keenan et al. (eds.), *International Anatomical Education*, Biomedical Visualization 9,
https://doi.org/10.1007/978-3-031-91849-0_8

interesting potential future project would be to explore the integration of public engagement activities into anatomy curricula, which would help to develop transferable skills and consolidate student anatomical knowledge through teaching.

**Keywords**

Anatomical sciences · World anatomy day · Public engagement · Student involvement · Human body · Outreach

## 8.1 Introduction

### 8.1.1 Public Engagement and Outreach in the Anatomical Sciences

Public engagement and outreach play important roles within higher education institutions (HEIs). Engagement with the public has previously been referred to as the '*third mission*' of universities, elevating its importance to be on par with key teaching and research aims, and demonstrating the emphasis that modern HEIs are placing on social and community engagement (Vargiu 2014). This is not a phenomenon exclusive to the UK; universities across the globe are recognising the importance of their institution actively engaging with society and involving the public in their work (Furco 2010). Despite its widespread presence in the higher education sector, public engagement has a broad and varied definition that may include continuing education initiatives, such as courses or public lectures; research practices that engage and involve the community; or knowledge-transfer activities that seek to educate the public, often making use of university resources such as laboratories and libraries (Ruth et al. 2023). Within all public engagement activities, a common theme is the aim of supporting the embedding of the university within the community, which helps to build public trust and involvement with the institution (Vargiu 2014). Universities are generally considered to have a civic duty to their local community and often have institutional strategies that highlight building partnerships with the public (Oaks et al. 2009).

Where public engagement tends to focus on engaging with local communities and the general public, outreach activities often have more specific target audiences that relate to widening access to higher education for specific student groups (Heaslip et al. 2020). This generally includes those individuals who may otherwise struggle (e.g. financially, physically, mentally) to make successful applications to university. In the modern higher education context of increased competition between HEIs, outreach may also take the form of recruitment, aiming to increase student numbers regardless of background or demographics (Johnson et al. 2019). Outreach, as well as public engagement activities, can drive student recruitment by increasing the visibility of the university involved, acting as a publicity tool for prospective students, not only locally but also internationally using social media (Kuzma and Wright 2013). However, despite these aspirations to increase student recruitment numbers more widely, outreach initiatives more often form a key part of widening access and participation programmes introduced by many HEIs (Chorcora et al. 2023). Educational inequality and barriers to higher education exist for a multitude of reasons beyond the scope of this chapter, but many stem from a lack of parental support, negative perceptions of higher education, and reduced access to information and guidance (Chorcora et al. 2023). Outreach activities seek to dismantle these barriers, increasing the visibility of HEIs and providing a source of information; these have been shown to raise awareness of opportunities and improve the knowledge of attendees (González-Solá and Rosario-Canales 2015; Márquez et al. 2019; Sanders et al. 2022). There may additionally be a 'hidden curriculum' within outreach activities, where the attitudes and demographics of the people involved can imply what 'kind' of people or students may study at that HEI (Finn and Brown 2023). Consequently, careful planning of resources and staff or student involvement is important to ensure the right message is communicated through engagement and outreach activities. In medicine, additional challenges are presented by the fact it is almost always an oversubscribed programme, with more applicants than places available, meaning the emphasis is on selecting rather than recruiting students. However, like many university courses, under-represented student groups are also under-represented in medical school applications (Martin et al. 2018). Fewer students applying for these courses results in fewer students from these groups studying courses such as medicine. Martin et al. (2018) also note that '*medicine, dentistry and veterinary science are the university subjects with the lowest proportion of disadvantaged applicants*', compared to other subjects. These are courses that typically have a strong focus on human anatomy, making the anatomical sciences an excellent choice for outreach projects that aim to widen the pool of potential applicants to health science degrees (Márquez 2013).

As well as being a useful channel through which to spark interest and increase aspirations amongst potential university applicants, human anatomy is also an excellent choice for public engagement as it is a topic that continues to fascinate the lay public. Taylor et al. (2018) showed that there is an 'eagerness' to learn about the human body amongst the public, meaning events incorporating the anatomical sciences are more likely to draw a crowd and connect the public with the work of the local universities that teach it. Anatomical public engagement and outreach activities also play a crucial role in building the understanding, appreciation and awareness of the general public, with respect to the structure and function of the human body. There is a need for public anatomical education; research has shown that public knowledge of the human body is highly variable (Tubbs et al. 2016; Żytkowski et al. 2021). For example, on anatomy knowledge tests, individuals who work in healthcare professions unsurprisingly score higher ($P = <0.001$) than those who do not (Taylor et al. 2018). Improving the anatomical awareness of lay individuals is undoubtedly important, as it could lead to better health outcomes. For example, it could be assumed that those who better understand the functions of a healthy human body may seek advice more quickly for medical concerns when compared to those who lack this understanding. There is evidence that low health literacy is linked to poor health outcomes (Dewalt et al. 2004; Nutbeam 2008). It is important to build health literacy from an early stage, and this can be achieved by targeting anatomy engagement activities for children and young people (Ortug et al. 2021). Anatomy public engagement also provides a space for the layperson to ask questions and discuss topics around healthy anatomy and how to maintain it. Accurate information may otherwise be challenging to access in an era of 'information overload' because of the World Wide Web and social media (Khaleel et al. 2020). Community outreach through science has been used previously to tackle issues with health and science education literacy, with members of the public as well as high school and undergraduate students (Burns 2002; Meyer et al. 2018). Public engagement in human anatomy enables individuals from diverse backgrounds to gain accurate knowledge about their bodies, promoting health literacy and encouraging active participation in healthcare decision-making. Anatomy is, therefore, useful not only as an engagement and outreach tool but also to promote health literacy and awareness amongst the general public.

Fortunately, anatomical science is a highly visual subject, meaning it easily lends itself to the development of innovative and creative events and activities. In previous work, anatomists have developed accessible and interesting approaches, from body painting to dance to other creative methods, for engaging the public in anatomy, which can all be undertaken by a range of ages (Finn 2014, 2019, 2020; Biglino et al. 2022). Additionally, many of the tools used in teaching anatomy to students can be adapted for public engagement and outreach use. The anecdotal experience of the authors suggests that even resources that may seem unstimulating to an experienced anatomy educator, can spark excitement amongst the public. For example, the authors have previously found a simple anatomical model of the trunk with removable viscera to be sufficient to engage the public and stimulate conversation about their bodies. Public engagement is much more straightforward when the subject matter alone can capture the interest of the layperson without the involvement of high-energy activities and versatile resources.

Complex anatomy education tools can, however, also prove useful assets for public engagement and outreach, especially when designing more stimulating activities. For example, recent developments in photogrammetry techniques allow three-dimensional (3D) models of cadaveric specimens to be acquired, which can be used in both medical education and outreach (Jędrzejewski et al. 2020; Gröning et al. 2021). Virtual dissection has also gained popularity over the years, and virtual dissection tables have been used in anatomy outreach activities (Rosario et al. 2019). These virtual dissection and cadaveric modelling tools are particularly useful where ethical or practical barriers prevent the use of real cadaveric material when engaging the public. Nonetheless, work with animal specimens in public engagement (Sanders and Philp 2019) and human cadaveric material for outreach purposes have been successfully used (Meyer et al. 2018; Ruth et al. 2023). Other teaching tools at the cutting edge of anatomy education, such as 3D software or virtual and augmented reality headsets (Venkatesan et al. 2021; Arjomandi Rad et al. 2023; McMenamin 2023), could create further opportunities for dynamic and engaging activities with the public.

Public engagement and outreach are crucial elements of the practice of any university and are of particular importance for healthcare professions-related courses such as medicine, dentistry, pharmacy and nursing. Healthcare graduates will work with hugely diverse people and patients, and consequently, students on these programmes should be drawn from an equally diverse population (NHS England 2023). Engagement and outreach can help ensure diverse populations of potential students are reached. Anatomy lends itself with relative ease to such engagement and outreach activities, is of enduring interest to the public, and also provides opportunities to showcase current teaching and research practices of the HEIs delivering the events.

### 8.1.2 World Anatomy Day

World Anatomy Day (WAD) was established by the International Federation of Associations of Anatomists (IFAA) in 2019. It takes place on 15th October each year, a date which marks the anniversary of the death of Andreas Vesalius. Vesalius is a notable figure in anatomical history, often described as the 'father of modern human anatomy' (McRae 1890; Xiang and Venkatesan 2023), having perpetuated the fundamental idea that accurate human anatomical knowledge could only be gained through observation of dissected human cadavers. Vesalius also authored seminal anatomical texts, most famously '*De Humani Corporis Fabrica Libri Septem*' ('*On the Fabric of the Human Body*') (Arnaldo and Bonar 1996). WAD was established in memory of Vesalius and serves as an opportunity to celebrate the anatomical sciences and raise awareness about the importance of human anatomy in healthcare, education and research.

Events relating to WAD have been limited since its inception. A small number of in-person events took place in 2019, mainly involving lecturers and students of anatomy, but further celebrations were limited to online and virtual events due to the onset of the COVID-19 pandemic (Simok et al. 2022). There have been few events aimed at engaging the public with WAD, and anatomists at Newcastle University and the University of Sunderland saw this as an ideal springboard to develop a series of events to excite and educate children, families and adults of all ages about their bodies.

### 8.1.3 Program of Events in 2022

Newcastle University and the University of Sunderland are two UK HEIs with medical schools that are based in the North-East of England, UK. Both universities deliver cadaveric human anatomy education to their medical student cohort, though the medical curricula do differ between the two (Donald et al. 2023; Keenan et al. 2023). In 2022, anatomists at the two universities designed and delivered a series of public engagement events in the week leading up to WAD. Preparation started a year in advance of the events, and the innovative and collaborative inaugural 2022 program engaged over 700 members of the public in the North-East of England. The program of events includes: (1) a cardiovascular anatomy workshop for local school-age and college students, (2) an adults-only evening event on genitourinary anatomy, called '*Bits and Bobs: The Anatomy of Sex*', (3) a public lecture by Editor-in-Chief of Gray's Anatomy, Professor Susan Standring, and (4) a family-friendly full-day event, called '*Discover Anatomy*'. The series of events aimed to capture the interest of a large range of demographic groups and involved several venues and external institutions throughout the region. The events were also an excellent example of a cross-institutional and inter-departmental working, with colleagues from both universities and several departments including anatomy, physiology, fine art, medicine and dentistry working together to develop and deliver the programme of events.

The locations of each event varied to reach different target audiences. The cardiovascular workshop was planned to take place on-campus at both Newcastle University and the University of Sunderland, involving the universities' widening participation teams to target local schools and colleges to raise aspirations and give these students a taste of studying at university. Professor Susan Standring's public lecture also took place on-campus, as part of Newcastle University's 'Insights' programme of talks available to the public. The evening event, '*Bits & Bobs: The Anatomy of Sex*', took place at the Great North Museum: Hancock, a natural history museum in the centre of Newcastle-upon-Tyne, UK, allowing members of the public not associated with the university to hear about and attend the event. Finally, the family-friendly full-day '*Discover Anatomy*' event took place at a large science centre, the Life Science Centre, also in the city centre of Newcastle-upon-Tyne, UK.

This chapter focuses on the '*Discover Anatomy*' event, as the largest of the four activities. It was designed to be suitable for all ages and demographics. The event engaged 520 members of the public and aimed to (1) educate attendees on a range of body systems, (2) raise aspirations of young people within the region and (3) promote the anatomy education of the institutions involved.

## 8.2 Event Details

### 8.2.1 Planning and Funding

In November 2021, anatomy staff at Newcastle University began brainstorming ideas for public engagement activities for World Anatomy Day in 2022. Before planning these activities in depth, staff approached venues in the local region. This meant that activities could be tailored to the needs of the local community and could be designed based on the space and resources available. The Life Science Centre, Newcastle-Upon-Tyne, UK which is affiliated with Newcastle University, agreed to host the public engagement activities on 15th October 2022, World Anatomy Day. This event was named '*Discover Anatomy*'.

Staff at Newcastle University approached the anatomy team at the University of Sunderland to invite collaboration on the series of events, including '*Discover Anatomy*'. This ensured that there were enough experienced anatomists to design and deliver high-quality anatomy-themed stations at the event. Moreover, it provided an excellent opportunity for cross-institutional working. This meant that the events could reach more of the public within the North-East of England. Future plans involve running the event annually and rotating locations each year, to create a dynamic anatomy community engaging the public across the region. One unintended outcome of working collaboratively between two traditionally 'rival' universities, is that it portrays to the general public that these rivalries and competition for student recruitment do not need to impact on grassroots-level academic contributions to society.

As the Life Science Centre targets a younger audience, the WAD team designed a family-friendly anatomy-themed event. The venue allowed for multiple stations to be strategically positioned throughout the space and interwoven amongst the centre's usual exhibits. It was agreed by the anatomy teams and the venue staff that '*Discover Anatomy*' would be a 'take-over' event, meaning it was a large-scale event that would take place across the whole venue for an entire day. The anatomy staff visited the venue twice before finalising station plans to ensure the size of the space and resources were suitable for the designated stations. Some stations, for example, required electricity, so a plug socket was essential, and others required a space where it would be possible to facilitate potentially messy activities.

The finalised stations each focused on a different region of the human body and involved a range of interactive activities. These were (1) an overview of the human body, followed by specific areas relating to (2) the nervous system, (3) dental anatomy, (4) the gastrointestinal system, (5) the respiratory system, (6) the cardiovascular system, and (7) the musculoskeletal system.

The venue provided approximate visitor numbers for recent weekends in the lead-up to the event, which allowed station leads to better plan the resources and staffing required to deliver their station effectively. Although there were numerous anatomy staff involved in the planning phases, and anatomical resources (such as anatomical models) were available through the affiliated institutions, it was felt that additional resources and support would be required to achieve a high-quality and engaging event. For this reason, funding was sought and acquired through the *Newcastle University FMS Engage Tilly Hale Award* to deliver the series of events for World Anatomy Day in 2022, including '*Discover Anatomy*'. The Tilly Hale Award provides funding to support academics within the Faculty of Medical Sciences (FMS) at Newcastle University in health, wellbeing and social care, to work with the public in innovative and unique ways. The award was set up in dedication to the memory of Tilly Hale, a patient diagnosed with primary biliary cholangitis, who recognised the stigma associated with liver disease and was keen to combat this. She became heavily involved in patient and public involvement and engagement, and took part in many research projects, not only as a participant but also by engaging other patients in the research process and sitting on funding and steering committees. In June 2014, Tilly was invited to give a TEDx talk at Newcastle University, UK, which is available to view online [https://www.youtube.com/watch?v=v1Ml1djOyB8]. Sadly, Tilly passed away in December 2015, and the FMS Tilly Hale Award was set up to continue her legacy in engaging the public with science and research.

To support the delivery of the activities, undergraduate students were recruited. The process of recruitment, training, delivery and evaluation of student involvement is discussed later in this chapter. Additional funding was procured to support student involvement from the British Association of Clinical Anatomists (BACA) [www.baca-anatomy.co.uk/], the Anatomical Society (AS) [www.anatsoc.org.uk/], and Adam, Rouilly (Kent, UK).

Most of the advertising and marketing of the event was managed by the venue. '*Discover Anatomy*' was advertised on the Life Science Centre's website [www.life.org.uk] and social media channels and listings, and the centre also designed and distributed a poster for the event. This poster was used both by the Centre and by the affiliated universities for advertising purposes throughout their institutions, such as on university-wide display screens, mailing lists, and social media pages.

### 8.2.2 Delivery

As described previously, '*Discover Anatomy*' involved seven stations, each focussing on a different aspect of human anatomy: (1) Overview of the Human Body, (2) Nervous System, (3) Dental Anatomy, (4) Gastrointestinal System, (5) Respiratory System, (6) Cardiovascular System and (7) Musculoskeletal System (Table 8.1).

To encourage engagement at all stations, on entry to the event, participants were provided with a 'passport' to take around the venue. The 'passport' was a professionally printed folded booklet that listed each of the stations and where

**Table 8.1** Summary of each station present at the 'Discover Anatomy' event including the overview of activities and the resources, staffing and funding required to run the station

| Station and overview | Resources and funding |
|---|---|
| **Overview of the human body**<br>Attendees were asked to draw an outline of a human body and place the printed organs in the correct location. Staff and students supported this activity and facilitated discussions around organ systems. Anatomical torso models and craft materials were also available as alternative self-directed activities. | ***Materials***<br>Chalkboard vinyl sheets, chalk pens organs printed and laminated in various sizes, Adam, Rouilly (Kent, UK) SOMSO® [www.somso.de/en/] anatomical models of the torso, craft materials, colour pencils and anatomical colouring-in sheets for a range of ages.<br>***Staffing***<br>One staff member and two student volunteers<br>***Approximate funding***<br>All models were provided by Newcastle University. Additional resources and printing totalled approximately £75 GBP. |
| **Nervous system**<br>The station consisted of interactive tasks to educate attendees on the structure and function of the brain. This included building brains using anatomical models, and discussions around the evolution of the brain including matching brains to the correct animal. | ***Materials***<br>Adam, Rouilly (Kent, UK) SOMSO® [www.somso.de/en/] anatomical models of the brain, and printed brain images.<br>***Staffing***<br>Two staff members and two student volunteers<br>***Approximate funding***<br>All models were provided by Newcastle University. Additional resources and printing totalled approximately £5 GBP. |
| **Dental anatomy**<br>The station consisted of activities for children and adults that involved education of basic tooth anatomy and the function of different types of teeth along with awareness of proper brushing techniques to prevent gum disease. | ***Materials***<br>Adam, Rouilly (Kent, UK) SOMSO® [www.somso.de/en/] Oral cavity and teeth models including individual teeth, toothbrushes, modelling clay and plastic (non-sharp) tools.<br>***Staffing***<br>Four staff members and one student volunteer<br>***Approximate funding***<br>All models were provided by Newcastle University. Additional resources and printing totalled approximately £30 GBP. |
| **Gastrointestinal/digestive system**<br>The station consisted of six sections, each dedicated to diverse activities encompassing all facets of the digestive system. The primary objective was to showcase the functionality of its individual components. | ***Materials***<br>Adam, Rouilly (Kent, UK) SOMSO® [www.somso.de/en/] anatomical models of the head, modelling clay, tights, softballs, cardboard tubes, plastic boxes, several one-litre water bottles, printed question/facts cards, plush 'smiling poop emoji', children's sized toilet potty, printed workbooks and pens.<br>***Staffing***<br>One staff member and three student volunteers<br>***Approximate funding***<br>All models were provided by Newcastle University. Additional resources and printing totalled approximately £80 GBP. |
| **Cardiovascular system**<br>The station involved a demonstration of an ultrasound scan of the neck. It also involved participants scanning phantoms to see what was hidden within them. | ***Materials***<br>Vscan Extend™ (GE HealthCare, GE HealthCare Technologies Inc., Chicago, IL, USA) handheld ultrasound machines, and phantoms (jelly-filled tubs containing small objects, to simulate using ultrasound on the body).<br>***Staffing***<br>Two staff members and two student volunteers<br>***Approximate funding***<br>All ultrasound machines and phantoms were provided by the University of Sunderland. Additional resources and printing totalled approximately £5 GBP. |

(continued)

**Table 8.1** (continued)

| Station and overview | Resources and funding |
|---|---|
| **Respiratory system**<br>This station involved a demonstration of respiratory anatomy on sheep lungs using an overhead camera and screens, followed by participants handling and observing sheep hearts. Alternative activity outside the station involves colouring and cutting out alveoli to create a lung tissue collage. | ***Materials***<br>An overhead camera and screen, sheep lungs and hearts (pre-dissected), cool box and ice packs, plastic trays, foot-pump and hose, personal protective equipment (PPE) (including latex-free gloves and aprons), antibacterial hand gel, disinfectant wipes, instruction sheets, alveoli drawings, colouring pencils, safety scissors, whiteboard, Blu-tack, and Adam, Rouilly (Kent, UK) SOMSO® [www.somso.de/en/] anatomical models of the torso, sagittal head and neck, and tracheobronchial tree.<br>***Staffing***<br>One staff member and three student volunteers<br>***Approximate funding***<br>All models were provided by Newcastle University. Additional resources and printing totalled approximately £145 GBP. |
| **Musculoskeletal system**<br>This station involved the sculpting of the muscles of facial expression, using modelling clay, onto a model of the skull. This was guided by a purpose-made booklet that highlighted the positions and shapes of the muscles, as well as their actions. | ***Materials***<br>Modelling clay, Styrofoam skull models, Adam, Rouilly (Kent, UK) SOMSO® [www.somso.de/en/] anatomical models of the skull, and printed (basic and advanced) guidance booklets.<br>***Staffing***<br>Two staff members and three student volunteers<br>***Approximate funding***<br>All models were provided by Newcastle University. Additional resources and printing totalled approximately £70 GBP. |

they were located within the venue. Participants received a sticker for their 'passport' for every station they completed and were awarded a small prize for a fully completed 'passport'. The 'passport' served to create a professional first impression of the event as well as encouraging engagement with all seven stations (Fig. 8.1). The 'passport' was distributed at an information desk at the entrance to the Life Science Centre, which also included general information about the '*Discover Anatomy*' event as well as promotional materials for the universities involved.

#### 8.2.2.1 Overview of the Human Body

This station was positioned close to the entrance of the venue. This naturally encouraged attendees to begin at this station. The station provided attendees with an overview of the different systems and regions of the human body. The aim of the station was to:

- Highlight the main organs of the body as well as their structure, location and function; and
- Spark an interest in the human body and health.

Multiple blackboard vinyl sheets were stuck to the floor of this region of the venue. Attendees that approached the area were invited to draw around one member of their party, or to draw a human body shape on the blackboard vinyl, using chalk pens provided. They were then handed a series of laminated printouts of the organs of the human body to place in the correct position on the body outline (Fig. 8.2). All images were sourced from Servier Medical Art by Servier [smart.servier.com], which are licensed under a Creative Commons Attribution 3.0 Unported Licence. These organ images were printed in various sizes so both adults and children could engage in the activity, and organs could be approximately sized relative to the body outline they had drawn. All printed organs were laminated to ensure they would be durable and withstand frequent use for the duration of the event.

As attendees took part in this task, student volunteers engaged them with information about the organs they were looking at. It was essential that facilitators were present on this station to ensure attendees learned something from the station, and to monitor that organs were being placed in the correct position. Facilitators could then provide

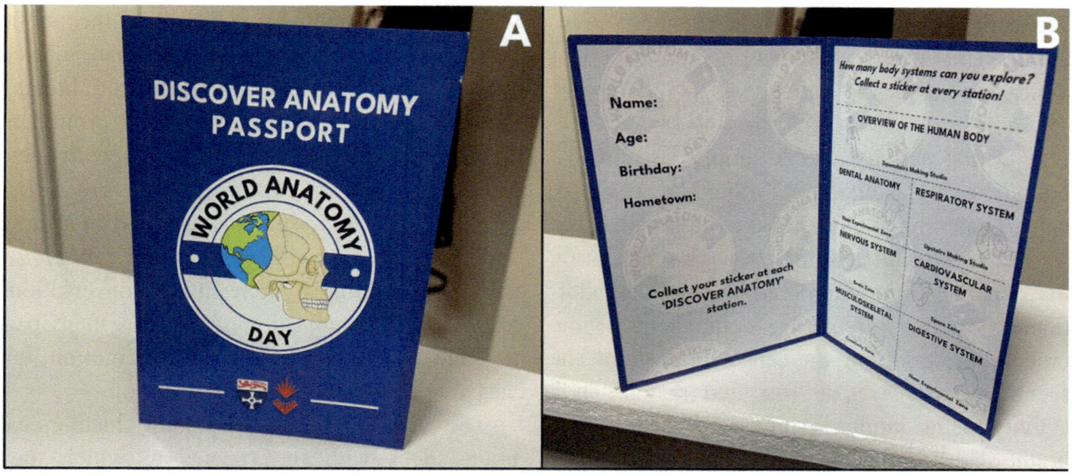


**Fig. 8.1** Photograph of the passport used at the 'Discover Anatomy' event which took place in October 2022 at the Life Science Centre. (**a**) The front page of the passport. (**b**) The internal pages of the passport

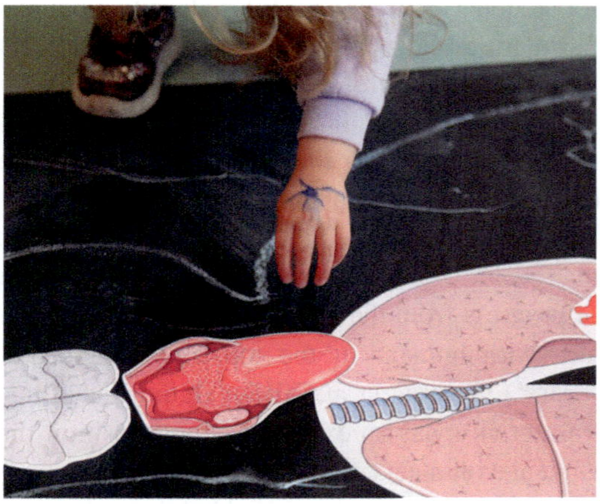

**Fig. 8.2** Photograph from the 'Overview of the human body' station at the 'Discover Anatomy' event in October 2022 at the Life Science Centre. A young attendee placed the printed Servier Medical Art organs on the outline of the human body. The printed organs were generated from Servier Medical Art, provided by Servier, licensed under a Creative Commons Attribution 3.0 unported licence

attendees with a sticker for their '*Discover Anatomy*' passport which indicated that they had completed that station. A total of four student volunteers assisted with this station: two in the morning, and two in the afternoon.

There were also alternative tasks on this station for attendees to engage with; there were SOMSO® [www.somso.de/en/] anatomical models of the torso (Adam, Rouilly, Kent, UK) which attendees could piece together. Student volunteers were present to engage individuals whilst they took part in this. There were also craft activities including colouring-in exercises of different regions of the human body, that were designed to engage younger participants who may have found placing the organs too challenging.

This station was designed to provide a global overview of the systems of the body. Attendees could then explore some of these systems further in the other activities positioned in the six other stations.

#### 8.2.2.2 Nervous System

The nervous system station comprised two activities, each of which was accessible across all age ranges. The aims of the station were to:

- Educate the public regarding basic brain structures and functions; and
- Foster discussion about brain evolution.

Both activities were simple in design, easily portable and served as starting points for engaging conversations with children and adults alike. Despite being a challenging topic for many students, to the extent that a recognised educational term has been coined, i.e. '*neurophobia*' (Jozefowicz 1994; Persons and Sipla 2023), the brain, its function and its evolution consistently excite the popular imagination (Sanders et al. 2022). This made neuroanatomy an excellent topic for engaging attendees at *'Discover Anatomy'*.

The first activity was a simple 'assemble a brain' exercise using the SOMSO® [www.somso.de/en/] anatomical model of the brain with arteries (Adam, Rouilly, Kent, UK). This is a brain that separates into nine parts, including an additional posterior circulation component of the model that runs along the ventral brainstem. The second activity was a 'mix and match' style activity, using laminated photos of histological brain slices and the species to which they belonged.

Anatomical models have long been a mainstay of neuroanatomy teaching (Arantes et al. 2018; Auffret 2024). For students, children and lay members of the public, the first challenge in working with the brain and arteries anatomical model is not an intellectual one: physically assembling it without dropping any of the pieces is not always straightforward and is therefore an excellent task for primary school-aged children (Fig. 8.3). For older participants, the assembly of the model acted as a prompt for interesting discussions. The posterior circulation component of the model is deliberately asymmetrical to represent anatomical variation. For the keen-eyed teen or adult participant, this stimulated discussions about anatomical variation, and why it is so prominent in vasculature. Student volunteers were frequently involved in these discussions,

**Fig. 8.3** Photograph from the 'Nervous system' station at the 'Discover Anatomy' event in October 2022 at the Life Science Centre. This photograph shows a member of staff supporting a young attendee with the SOMSO® brain with arteries anatomical model

which often concerned members of the public with family members or friends who had suffered strokes, highlighting an important health literacy component of this station. The other prominent conversation topic was consciousness, with questions raised such as: '*What bit of the brain controls moving/emotions/memory*?', and in a more sophisticated form: '*Where is my consciousness*?'. Interestingly, in resulting conversations about cortical localisation of function, many participants asked about thought and emotion, but very few about any sensory perception.

The second activity (matching histological brain slices with the species to which they belonged) was again a simple one but served as a jumping-off point for discussions about the evolution of the brain. The evolutionary biology of brain anatomy is a notoriously difficult subject for students (Jozefowicz 1994), despite it being a fascinating and important area. This results in part from the extremely elaborate evolutionary and developmental history of the human brain. More fundamentally though, it results from the consistent Aristotelian misconception that pervades school and medical education in the UK: that evolutionary change is necessarily a process of progress from the simple to the complex (usually with humans at the top of some sort of 'order') (Stearns et al. 2010). Helping facilitate this activity was an excellent opportunity for the student volunteers to solidify their own knowledge of evolutionary brain biology through teaching and discussion with members of the public. In the activity, participants were asked to match up laminated photos of coronal brain sections with the animal species that they belonged to. The species concerned represented a range of different animal species. The combination of species was chosen to include multiple examples that possess either a gyrencephalic (heavily folded) cortex, such as in ferrets, the capybara, elephants or humans. Importantly though, it also included an equal number of lissencephalic (unfolded) cortices representing the same mammalian order: respectively, a bat, a mouse, a manatee and a bushbaby. Thus, participants could equate the degree of brain folding with the levels of cognitive ability of the species concerned but also recognise that there is no difference between humans and other mammals in terms of the structure of brain tissue and folding, only a difference in the degree or amount of folding. The other necessary conclusion was that complex folded brains and enhanced cognitive capacity have evolved independently multiple times. As it has been argued (Conway-Morris 2004), it was not at all inevitable that what we call consciousness would evolve in a bipedal ape—it could have been in an elephant or a dolphin, but it might have been inevitable that it would evolve.

These simple activities served as an excellent method to stimulate and engage participants in discussion about brain anatomy, evolution and function. The simplicity of the activities meant they were cheap, easy to create, and required minimal facilities and equipment. However, these sorts of straightforward engagement activities do require expert staff on hand to engage the public in discussions and answer questions to achieve the station's aims. Student volunteers were invaluable on this station to ensure there were enough facilitators on hand to hold conversations with attendees. In future events, more written information could be provided to participants to reduce the staffing requirement.

#### 8.2.2.3 Dental Anatomy

Several factors, including diet, compromised systemic health and socioeconomic status can lead to common dental issues, such as tooth decay or gum inflammation (Peres et al. 2020). Neglecting and not treating such issues can lead to further periodontal problems, such as periodontitis (Murakami et al. 2018). There is therefore a public health need to ensure the public is educated on maintaining oral hygiene to prevent such diseases. The aim of this station was to:

- Educate the public on maintaining their oral hygiene; and
- Help reduce the risk of developing dental caries or common gum diseases, such as gingivitis.

To fulfil the above aims, this station introduced children and adults to basic tooth anatomy along with the function of different types of teeth. This was followed by guiding them through proper brushing techniques. The resources used by facilitators and attendees during the station included various anatomical dental models, plastic dental tools and laminated activity template sheets with simplified colourful images (Fig. 8.4). Some of the resources were obtained with permission from Newcastle School of Dental Science, others were purchased specifically for the station using the event funding.

**Fig. 8.4** Photograph from the 'Dental Anatomy' station at the 'Discover Anatomy' event in October 2022 at the Life Science Centre. The image shows two of the facilitators (one staff and one student) alongside the resources available at the station

Staff and students facilitating the station were involved in signposting attendees to the activities and models. Facilitators also answered questions and encouraged discussion to raise awareness about common dental problems and how they could be prevented by proper brushing at least twice a day.

The first activity involved attendees observing teeth models and differentiating between different types of teeth (i.e. incisors, canines, premolars and molars), along with learning about their function. The second activity involved attendees observing different parts of a tooth; for children, it was limited to crown and root, whereas for adults, the reference tooth image was labelled with crown, root, enamel, dentine, pulp and cementum. Reference images were obtained from Gray's Anatomy for Students (Drake et al. 2020), Guyton and Hall Textbook of Medical Physiology (Hall and Hall 2020), and BioRender illustrations [www.biorender.com/]. Images were used for educational purposes only based on institutional licences and no images or resources were distributed to attendees. These labelled images were provided in the printed activity sheets for both adults and children to help them with the identification of the models provided. The final activity involved educating attendees on brushing techniques using a giant oral cavity model and toothbrush. This was reinforced by sticking yellow modelling clay onto various surfaces of tooth models. Attendees were given a plastic instrument to remove it as a dentist would during a scaling procedure. Through this activity, attendees were educated and reminded that (1) food accumulation on/between tooth surfaces along with improper tooth brushing leads to deposition of plaque and calculus, (2) calculus cannot be removed with regular brushing once it is deposited, and can only be removed by a dentist using dental equipment, and (3) plaque and calculus build-up can lead to gum problems, such as bleeding and inflammation. This was an important public health education element of this station as it is evident in the literature that effective oral health education programs are effective yet labour-intensive and require a significant amount of funding which may not be accessible to everyone in the community (Nakre and Harikiran 2013). In addition, educating children in primary schools about tooth brushing provides short-term benefits but has its barriers and challenges (Chandio et al. 2022). This highlights the potential need for more such public engagement events and activities where parents and children can be educated together on maintaining oral hygiene outside of formally implemented oral health education programs.

Finally, a QR code was included in the activity sheet which was linked to a YouTube video [https://www.youtube.com/watch?v=XXQExhyfLDc&t=1s] on basic tooth anatomy that the attendees could watch later to consolidate their knowledge and understanding from the station. Upon completing the activities, attendees were awarded a sticker for their '*Discover Anatomy*' passport.

Since the station was busier than expected and most children required one-to-one attention, there were instances when some people had to wait or were asked to come back as there was not enough physical space to accommodate more than 12 people at one time. In future iterations of this station, either more space and/or more student volunteers could be used to ensure the ability to accommodate and manage larger numbers of participants.

#### 8.2.2.4 Gastrointestinal System

This station was the largest in physical area, to demonstrate the size and function of structures throughout the gastrointestinal system. The station was split into six subsections: (1) Mouth, (2) Oesophagus, (3) Stomach, (4) Small intestine, (5) Large intestine, and (6) Rectum and anal canal. Each participant would go through the subsections chronologically to mimic the journey of food through the gastrointestinal tract. The aims of the station were to:

- Highlight how there are many distinct structures that make up the gastrointestinal system
- Demonstrate the major functions of the structures of the gastrointestinal system.
- Explain how the function of each structure enables the next part of the digestive system to perform its own function.

Each group of participants would work through the subsections, each featuring a unique set of activities tailored to their specific topic. As the age of participants was variable, it was ensured that the activities were adequate for a diverse age range. For example, some terms were simplified to be easier to understand for younger audiences, such as 'esophagus' being termed the 'food pipe' where necessary.

##### 8.2.2.4.1 Mouth

The mouth section included SOMSO® [www.somso.de/en/] anatomical models of the head (Adam, Rouilly, Kent, UK) that showed the oral cavity and muscles of mastication (Fig. 8.5). Participants were asked to bite down whilst palpating these muscles to demonstrate their action, comparing how they felt to how they appeared on the model. Participants then built their favourite food out of modelling clay, which they were then asked to tear into smaller pieces so that they could be dropped down a cardboard tube into a box. Facilitators explained that this is compara-

**Fig. 8.5** Photograph from the 'Gastrointestinal/digestive system' station at the 'Discover Anatomy' event in October 2022 at the Life Science Centre. The image shows one of the SOMSO® anatomical models of the head used on the mouth section of the station

ble to how the mouth chews food so that it can be passed into the 'food pipe' (esophagus) where it can travel to the stomach.

#### 8.2.2.4.2 Esophagus

The oesophagus section was designed to demonstrate the action of swallowing and peristaltic contractions, using a pair of tights and soft balls. Facilitators explained how the oesophagus (the tights) is a muscular tube that receives food (the softball) from the mouth and can then squeeze the chewed food towards the stomach. This was demonstrated by pushing a softball through the tights into a box. The substation was kept light-hearted and fun and included facts that would be of interest to the layperson, such as describing how the esophagus is strong enough to push food toward the stomach even if a person were to be suspended upside down.

At the end of the esophagus activity, the participants were provided with a small workbook to complete the remainder of the sections.

#### 8.2.2.4.3 Stomach

The stomach section had a table with a question displayed: '*How much can the stomach hold?*'. Possible answers were displayed as different numbers of one-litre bottles labelled with cards A–D (A = one litre, B = two litres, C = three litres, D = four litres). Participants were asked to guess which they thought was the correct answer and to turn over each labelled card until they found the correct answer (D). This was often met with surprise from the participants and encouraged them to ask questions in relation to the answer.

#### 8.2.2.4.4 Small Intestine and Large Intestine

The small and large intestine subsections were question-based and fact-based, allowing participants to complete their workbook and work through this section at their own pace. This made the section more self-directed and reduced the staffing volume required. The small intestine section was unique in that it was located on a winding walkway already present in the venue, which worked well to represent the structure and course of the small intestine. The large intestine station was located on the final part of the walkway and led directly into the rectum and anal canal section.

Participants would walk along the walkway, stopping at each question/fact card and using the information found there to fill in their workbooks. The questions and/or facts were at varying degrees of difficulty, allowing participants to appreciate the complexity of the topic, whilst still being able to complete their workbook without the requirement of previous knowledge. A student volunteer was present along the walkway to help participants, answer any questions they had, or explain topics further.

#### 8.2.2.4.5 Rectum and Anal Canal

This final subsection was a fun and light-hearted end to the station, where attendees were given a plush 'smiling poop emoji' to throw into a miniature toilet, representing the end of the gastrointestinal tract. The student volunteer responsible for this subsection would explain to participants how they had followed the same journey as the food they had ingested. The fun ending to the station was designed to be memorable and create a sense of achievement for having followed the 'journey' of ingested food. Participants could then collect a sticker for their '*Discover Anatomy*' passport.

This was a multi-stage station covering a large region of the science centre, which created challenges at busy times of the day. Several student volunteers were present at the station, but having even more may have made the station flow more easily and ensured there was adequate support for attendees. However, the light-hearted and interactive nature of the activities meant the public easily engaged with and enjoyed the station.

**Fig. 8.6** Photograph from the 'Cardiovascular system' station at the 'Discover Anatomy' event in October 2022 at the Life Science Centre. The image shows two of the facilitators demonstrating an ultrasound of the neck using a Vscan Extend™ [GE HealthCare, GE HealthCare Technologies Inc., Chicago, IL, USA] portable ultrasound machine

#### 8.2.2.5 Cardiovascular System

Ultrasound is used by a whole range of medical and healthcare professionals in hospitals and some primary care centres (Lee and Decara 2020). It can be a powerful diagnostic tool when used alongside clinical examination. As ultrasound allows the visualisation of living anatomy in a safe and non-invasive manner, it is a great tool for public engagement activities. The aims of this station were to:

- Increase knowledge of the utility of this scanning modality.
- Demonstrate what to expect when having an ultrasound scan, and,
- Allow the public to see living anatomy and to look under the skin.

To achieve this aim, the station was divided into two sections. Firstly, there was a demonstration of an ultrasound showing the major vessels in the neck (common carotid artery and internal jugular vein). This was followed by assisting participants in using the portable ultrasound kit themselves to scan phantoms (jelly-filled tubs containing small objects, to simulate using ultrasound on the body).

There were two members of the anatomy teaching staff team on hand to do all the demonstrations of the neck (Fig. 8.6). The demonstration began with a short description of how ultrasound images are generated and an initial description of the anatomy of the neck using a plastic model. The three major structures discussed were the trachea, common carotid artery and internal jugular vein. Participants were encouraged to follow along with printed guides which included labelled images of the key ultrasound views obtained. The guide also described the differences between an artery and a vein, by showing the variability in the size of the internal jugular vein if compressed, as compared to when the Valsalva manoeuvre (forced exhalation while the airways are closed at the glottis) is performed. This was followed by a demonstration of the use of spectral Doppler ultrasound imaging. The spectral Doppler imaging functionality allows for the differentiation between arteries and veins due to the inherent differences in the direction of blood flow (Maulik 2023). The participants were encouraged to ask questions throughout.

After the demonstration, participants were encouraged to use the ultrasound probes to scan the phantoms. Participants were given an instructional sheet on how to place the probe on the phantom what to look for and what the object would look like once found. Additionally, student volunteers were on hand to guide participants through the activity. To complete this station, we had five Vscan Extend™ handheld portable ultrasound machines (GE HealthCare, GE HealthCare Technologies Inc., Chicago, IL, USA), and seven ultrasound phantoms that contained a variety of objects. This included a balloon to mimic the gallbladder, a piece of sponge to mimic muscle fibers, and a plastic tube to mimic a blood vessel. The handheld portable ultrasound machines were owned by the University of Sunderland, usually used for teaching purposes. The phantoms were created at a low cost by the technical team within the anatomy department at the University of Sunderland. The major costs associated with this station were printing and laminating the activity guides.

There were four individuals running the station at any given time (two academic staff and two students) to complete the demonstration. Although there had been a training session for the student volunteers, in hindsight it takes time to build confidence in ultrasound scanning. The students were understandably hesitant to do any live demonstrations on the day. In the future, this additional time needs to be factored into the student volunteer training timetable. Additionally, it was a labour-intensive session, so having more individuals facilitating the station would be beneficial. The portable ultrasound machines were battery-operated, meaning staff were required to manage battery charging and replacement throughout the day. Furthermore, there was a steady flow of participants throughout the day, however, if there had been particularly busy time points, staff were not prepared to deal with crowds. Therefore, in the future, it would be prudent to examine the flow of individuals through the event and consider any crowd control considerations.

#### 8.2.2.6 Respiratory System

This station involved the use of animal (sheep) organs to demonstrate and discuss structures of the respiratory system and their function. The activity took place in a self-contained room/space, with clear signage indicating what attendees could expect to see inside. An alternative activity was available outside the room for those who did not want to view the animal organs, allowing them to still complete a task to earn a sticker for their ‘passport’. The aims of the station were to:

- Demonstrate the major structures of the respiratory system and their function, including the relationship between the respiratory and cardiovascular systems; and
- Allow the public the opportunity to handle animal organs in a safe and educational manner.

This station was repeated at regular intervals throughout the day, rather than running constantly as with other stations at the event (the alternative activity outside the room was available throughout). The station timings were publicised on entry to the event venue and on signage around the station. At the station entrance, signage described the nature of the station and informed attendees of the suggested age requirement (minimum age of seven, suggested by the venue). On entry to the room, participants were provided with personal protective equipment (PPE), consisting of latex-free gloves and a plastic apron, and invited to sit at tables arranged around the room. Delivery of the station involved two stages, beginning with a demonstration of respiratory anatomy using sheep lungs and trachea (Darwin Biological, Shawbury, UK) as well as relevant SOMSO® [www.somso.de/en/] anatomical models (Adam, Rouilly, Kent, UK). A member of staff or a student volunteer delivered the demonstration, which utilised an overhead camera and large screens to ensure all participants were able to easily view the anatomy. The demonstration involved talking through each region of the respiratory tract and its function, following the passage of air from the oral cavity to the trachea, bronchi and lungs (Fig. 8.7a, b). A foot pump and hose were used to show the change in lung size and dimensions on inflation. The demonstration was interactive, involving questions to the audience, and participants were invited to come up to the demonstration table to palpate the lungs, helping them appreciate the spongy, air-filled texture. Care was taken to discuss the anatomy at an appropriate level for a lay audience with large numbers of children, whilst also including correct anatomical terminology. The demonstration lasted 10–12 minutes but was flexible depending on the size and interest of the audience. A shorter presentation was better suited to an audience consisting of younger participants. The second stage of the station allowed participants to interact with an animal organ specimen. Each table (five tables, up to six people per table) was provided with a sheep heart (Darwin Biological, Shawbury, UK) in a plastic tray, which had been dissected prior to the event to

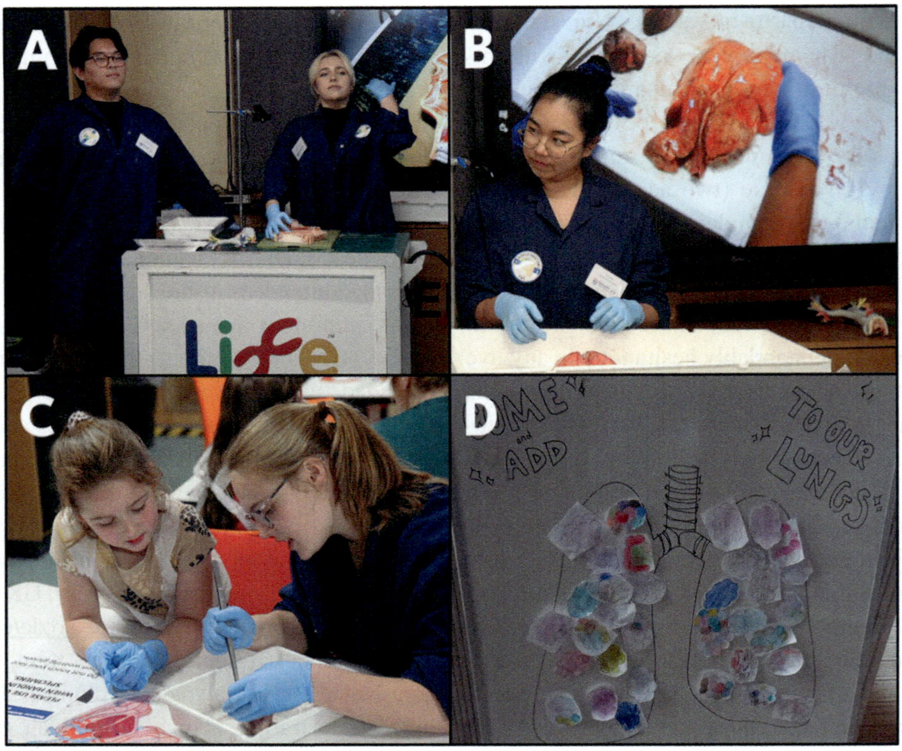


**Fig. 8.7** Photographs from the 'Respiratory system' station at the 'Discover Anatomy' event in October 2022 at the Life Science Centre. (**a**) Two student volunteers demonstrated the function of the upper airways to an audience using SOMSO® anatomical models. (**b**) A student volunteer demonstrated the anatomy of the lungs using sheep lungs with a young audience watching closely. (**c**) A student volunteer talking through a sheep heart specimen with a young attendee. (**d**) The completed collage of alveoli at the end of the event was an alternative craft activity for attendees at this station

allow viewing of internal structures. Instructional sheets placed on each table guided attendees on which structures to look for. Cardiac anatomy was chosen for this aspect of the station (despite the respiratory theme) due to the convenient size and ease of availability of the specimens. This also served to demonstrate the close relationship between respiratory and cardiac anatomy, including a discussion of pulmonary versus systemic circulation and the role of the lungs in the oxygenation of blood and the removal of carbon dioxide. Attendees were given between 15–20 minutes to explore their specimens, with a staff member and three student volunteers available to give guidance, answer questions and assist with identification of structures (Fig. 8.7c).

For any attendees who did not wish to view the animal organs, an alternative activity was available outside the space (Fig. 8.7d). Images of alveoli obtained from Servier Medical Art by Servier [smart.servier.com] were available for attendees to colour in, cut out and add to a board, creating a collage of alveoli throughout the day to represent lung tissue. A SOMSO® [www.somso.de/en/] anatomical model of the torso (Adam, Rouilly, Kent, UK) was also present, allowing a student volunteer to use this to talk through the respiratory anatomy. Space for this alternative activity was limited; in the future, it would be useful to have a larger alternative space to hold a more engaging separate activity for those unable to attend the specimen demonstration.

Due to the involvement of animal tissue, careful consideration was given to the safety and logistics of this station. This was a major advantage of holding the event at a science venue; as they had previously facilitated sessions using animal tissue, they were already aware of the equipment, space and disposal required. The following steps were taken to ensure the station was delivered in a safe and appropriate manner:

- PPE (e.g. a plastic apron and pair of gloves) was provided to all attendees, and they were advised on safe practices for handling the material; for example, not touching their face, tying their hair back and washing their hands at the end of the station.
- The space had wipe-clean floors and surfaces and had a sink for handwashing after handling the animal material. Antibacterial hand gel was also provided.
- The specimens were handled in plastic trays to further minimise contaminants and reduce the risk of them being dropped.
- The space for the station was not immediately visible to the public, allowing attendees to choose whether they wanted to see the animal organs or not. Attendees were able to leave the room at any time.
- Advice was given to attendees on what to do should they feel unwell, and the venue had first-aiders immediately available on-site if needed.
- Using tables and chairs allowed for limited attendees (up to 30 per demonstration) to prevent overcrowding and allow staff and student volunteers to adequately monitor the session.
- Organs were sourced from an educational company (Darwin Biological, Shawbury, UK) and were by-products of the food industry, meaning they were food-safe and therefore free from disease. Sheep organs were used rather than pig or cow to remove the risk of attendees being unable to engage for religious reasons.
- Specimens were dissected the day before the event, so no sharps were present at the venue.
- The venue was able to be air-conditioned and keep the space at a cool temperature. Specimens were stored on ice in an insulated cool box when not in use. An excess of heart specimens was brought to allow the organs to be swapped between sessions to keep them as fresh as possible throughout the event.
- The venue already had safe practices in place for the disposal of the material at the end of the day.

Signage and a clear announcement at the beginning of the demonstration informed attendees of the nature and source of the specimens, so they were aware they were sheep organs that were sourced as by-products from the food industry. The station began with a briefing outlining how to handle the specimens, and student volunteers were trained to support participants with this. Student volunteers were also given the opportunity to deliver the demonstration if they wished, which was an excellent opportunity for them to develop their teaching, public speaking and engagement skills. Many members of the public commented on their dynamic and engaging presentation style. Although this station required a large amount of preparation time and safety considerations, the station received excellent feedback, with participants appreciating the opportunity to both handle real organs and watch a demonstration to learn more about the anatomy. The station successfully created an air of excitement, and several attendees returned to view the station again later in the day.

#### 8.2.2.7 Musculoskeletal System

This station focused on the muscles of facial expression and how they work together to enable us to pull a variety of expressions, aiding in nonverbal communication, and helping to portray emotion. The station had three aims:

- Demonstrate that the skull consists of several bones fused together,
- Highlight how the muscles of facial expression enable different facial expressions; and
- Understand how muscles attach and relate to the underlying bones.

Each group or family was provided with a kit, containing an illustrated workbook, a model skull, a container of modelling clay in varying colours, and a plastic mirror. This activity was largely self-directed; the workbook acted as a guide that informed the 'artist' about each muscle, its skeletal attachments, morphology and function.

Two workbooks were created, one for children aged eight and below, and one suitable for anyone aged nine or above. The eight and below workbook used basic shapes in bright colours to represent the muscles upon an image of a cartoon skull, as well as simplified names, such as 'smiling muscle', to avoid overwhelming young children. The workbook for those aged nine and over made use of example photographs of the muscles of facial expression modelled on an anatomically correct resin skull. They also included the conventional Latin names for the muscles, as well as a phonetic guide to their pronunciation. This increased the complexity of the activity for older children and adults, meaning that the experience was tailored to the abilities and needs of the individual.

Following a similar thought process, younger children were provided with a simple polystyrene skull model whilst older children, adolescents and adults were provided with a SOMSO® [www.somso.de/en/] anatomical model of the human skull (Adam Rouilly, Kent, UK). The modelling clay provided was the same for all participants. The arguably arbitrary age range acted purely as a guide; some younger children who were particularly passionate about science requested an 'older children's kit' whilst some older participants preferred to work from the simpler version. Assuming younger children were supervised by a parent or guardian who understood that the anatomically correct skulls were expensive and, if dropped may hurt the child, this was accommodated.

Participants were encouraged to follow the steps detailed in their chosen workbook, to replicate several of the muscles of facial expression, using a different colours of modelling clay for each one (Fig. 8.8). Once the muscle had been created, individuals were encouraged to perform the movement facilitated by that muscle. Children then created these different facial expressions in their mirrors. Upon completing the workbook, all participants were awarded a sticker for their '*Discover Anatomy*' passport.

**Fig. 8.8** Photograph from the 'Musculoskeletal system' station at the 'Discover Anatomy' event in October 2022 at the Life Science Centre. The image shows a young attendee engaging in the modelling clay activity on a polystyrene model. The printed guides contain images of skulls that were generated from Servier Medical Art, provided by Servier, licensed under a Creative Commons Attribution 3.0 unported licence

The station was delivered by two academic staff and five student volunteers from the University of Sunderland, two in the morning and three in the afternoon. The students' assistance was invaluable as they were able to talk through their own recent learning experience of the muscles of facial expression with the public.

As a teaching station, this received incredibly positive feedback, primarily for its kinesthetic and hands-on approach to learning, as is best suited to young children (Garvis et al. 2023). Because of its location within the centre, and self-directed nature, this was a quieter station than most. This provided any younger or potentially overwhelmed children the opportunity to have some quiet time and recharge before attending any further highly stimulating stations. Additionally, for participants who were highly engaged and interested in science, the academics and student volunteers had the opportunity to discuss their careers and experiences, further inspiring the young people in attendance.

One limitation of this activity that was noted throughout the event was the modelling clay; inevitably, the colours pressed into each other at the end of the sessions, limiting their capacity for reuse. Additionally, it was difficult to remove all traces of the clay from the anatomically correct skulls increasing staff workload during the cleaning process after the event.

## 8.3 Student Involvement

### 8.3.1 Planning and Recruitment

A staff member from each university was allocated to organise the student recruitment for '*Discover Anatomy*'. Through discussion with the staff involved with event planning, it was decided that 40 students were required to support the '*Discover Anatomy*' full-day event, with 20 being requested from each university.

A description of the event and student expectations were written up and provided to students when they were recruited. Students were also provided with event- and station-specific training.

#### 8.3.1.1 University of Sunderland Student Recruitment

The recruitment process began in the 2022 UK summer break prior to the event. Students who had completed years one, two or three of the medical (MBBS) program were sent a 'save the date' message by email. It was important to send this information early, as the event was to be held on a Saturday near the start of the semester, meaning students were likely to have made alternative plans if they were not made aware of the event.

Students were contacted again in September, once the semester had begun and teaching resumed. It was decided that students in the first year would not be contacted as most would have limited anatomical knowledge at this stage, having been at university for less than a month. The email included a link to an online form to allow students to sign up to volunteer, which requested data such as name, student ID, course, year of study, contact details and availability. A response for each question was required for the form to be successfully submitted. In addition to the formal emails, announcements were made at the start of anatomy classes for second and third-year students.

#### 8.3.1.2 Newcastle University Student Recruitment

In a similar fashion to the University of Sunderland's student recruitment, students who had completed years one and two, and members of the Newcastle University Anatomical Students' Society, were contacted in August 2022 by email. The email contained a link to an online form requesting their name, email address and course details. A total of 35 completed forms were received, and a follow-up email was sent in September to these students to confirm their year group and availability. Around half of these students were still able to attend. The dental students were recruited separately via colleagues working in this department.

#### 8.3.1.3 Student Incentives and Rewards

Whilst there was the belief that students would volunteer because of their passion for anatomy and their desire to work with the public, it was important to ensure they would be rewarded for their efforts. It was made clear from the outset that all student volunteers would be provided with lunch, reimbursement for their travel, a goody bag and a certificate of participation from the Head of the Medical Schools of both universities. As a further incentive, all student volunteers were entered into a prize draw for a £50 GBP book voucher, with two winners being chosen at random across each of the Newcastle University and the University of Sunderland students.

Students were not required to support the event all day, being allocated to either morning or afternoon slots depending on their availability. Where possible, students from Sunderland University were allocated to stations that were being run by Sunderland staff and vice versa, therefore making students feel more comfortable as they would be working with staff whom they were already familiar with. The dentistry station only involved Newcastle University dental student volunteers, as the University of Sunderland does not have a dental school. Students were allocated to stations before the event, giving time for staff members to organise and deliver training on both station content and engagement skills beforehand. Where possible, students were placed in their friendship groups to further increase their enjoyment of the day. This was decided due to the role being voluntary for the students, making it important to ensure maximum comfort and enjoyment.

### 8.3.2 Training

Due to the scale of the event, it was essential that students were well-prepared and equipped to deliver the engagement activities to a sufficient standard. The students would be representing the universities in a public space, so training was important to ensure they could perform their roles confidently and professionally, maintaining and improving the public's perceptions of the institutions. As volunteers, many of the students had little or no experience with carrying out public engagement. Students were therefore provided with both generic public engagement training, and training and instruction on the specific station they would be involved with.

All students were provided with an asynchronous pre-recorded training video, which gave an overview of the event and explained their role on the day. This training also covered what was expected of them in terms of attitude and behaviour and included a short 'how to' guide on engaging the public. This guide included tips such as positioning (i.e. not standing behind tables and 'blocking' engagement opportunities), suggested questions to ask attendees, ways to tailor delivery depending on audience age, and communication tips. This initial training was important to both inform students of the wider context of the event and its aims, helping facilitate buy-in from the students, and to give them important basic knowledge on how best to engage the public.

Students were then contacted individually with further training specific to their allocated station. They were given any resources or materials for the station in advance, to allow them time to familiarise themselves with the station content. They were also given specific instructions for how the station would be delivered, and where necessary, given reminders of anatomical content. Delivery of this training was 'live' for all stations (rather than pre-recorded), to allow students opportunities to ask questions or make suggestions. As much station-specific training as possible took place online in advance of the event, to give students time to prepare. However, for some stations, where specific spaces or equipment were being used, it was necessary to deliver this training in person on the day of the event, before it opened to the public.

Ideally, all training would take place in advance of an event of this scale, but as all station setups took place on the morning of the event, there were limited opportunities to provide training at the venue in advance of the day itself. In hindsight, a venue-familiarisation session with the students would have been a useful exercise, to reduce stress and the amount of training required on the day of the event.

### 8.3.3 Delivery

Students who were volunteering in the morning slots were asked to arrive well in advance of the event opening time. On arrival, they were shown the location of a break room and toilets and were given a general briefing from both the anatomy team and venue staff, covering basic health and safety, and reminding them of the aims and expectations for the event. Students were then encouraged to familiarise themselves with their station and with the wider venue. They were able to observe staff delivering the station before approaching members of the public themselves.

When students volunteering in the afternoon arrived, there was a 30-min handover period where the morning students could brief the afternoon students on the station, and incoming volunteers could receive the health and safety briefing and be shown around the venue. This time was useful as it allowed the students to give incoming volunteers useful tips and techniques that they had picked up through delivering the station. Students were provided with lunch regardless of whether they volunteered in the morning or afternoon, and on departure were given a certificate and given a gift bag of small items such as stationery and badges. Students were also permitted free entry to the science centre, so could use the time before or after their volunteering shift to explore and enjoy the rest of the venue.

A member of staff circulated the venue as the event progressed to regularly check on the students, to ensure they were behaving professionally and that there were no issues or support needs. This member of staff was also responsible for ensuring all students had a break and that they received their lunch and other rewards.

### 8.3.4 Evaluation

The Contemporary Science Manager that we liaised with at the Life Science Centre throughout the event provided the anatomy staff with positive feedback, both on the overall event and on student involvement. Students were contacted to complete an informal evaluation of their experience of the event. The questions focused on recruitment, incentives, training and general running of the event and contained both Likert scale and open-ended questions. The responses were positive and provided useful feedback for future events. Students seemed to really enjoy the event, they felt well prepared and felt that the staff communicated well ahead of the event. They also felt that they were provided with more incentives than they were expecting. The main areas of improvement included better advertisement of recruitment, having an opportunity to choose the station they were delivering on, and having an opportunity to volunteer for the whole day.

## 8.4 Discussion and Reflection

Public engagement events generally pose a range of benefits to both the local community and to higher education institutions themselves. It was felt that the '*Discover Anatomy*' event harnessed many of these benefits, including improving the anatomical knowledge and health literacy of the local community and building a strong partnership between institutions and venues within the region. Feedback from the venue, public, staff and students was very positive.

### 8.4.1 Benefits of Student Involvement

The involvement of students enabled the event to be run on a much larger scale than if only anatomy staff were involved. With 520 attendees in total during the day, it was essential that we had enough workforce to support and facilitate the activities planned. Without the student volunteers, many visitors would not have been able to engage in the activities planned or they would have missed out on key information the stations were trying to teach.

Additionally, involving students contributed to a more diverse and enthusiastic delivery team. The demographics of people involved with delivering engagement and outreach events can contribute to the 'hidden curriculum' of these events, for example by creating an image or impression of the range of student demographics that may study courses such as medicine (Finn and Brown 2023). It is therefore important that delivery teams, where possible, represent the diversity of prospective students the event is designed to reach. At both Newcastle University and the University of Sunderland, approximately over 25% of the MBBS intake are from widening participation backgrounds (meaning they are from particular backgrounds that are traditionally underrepresented at UK medical schools, based on factors such as socioeconomic status, presence of disability or long term health condition, being a young carer, or being from a black, Asian or minoritised ethnic group), and involvement of these students in outreach would be of benefit. The demographic information for the students who volunteered is not available for publication (as ethical approval was not sought for research around this event) but would be an interesting area to explore further in the future. As the students volunteered, rather than were paid employees, it was felt that keen students with an interest in anatomy and public engagement were naturally more likely to be recruited. This resulted in an enthusiastic and positive delivery team.

Involving students in community and public engagement and outreach has benefits for both the students and for the young people they engage with (Cook et al. 2020). Allowing students to work with other young people creates excellent opportunities for role modelling for younger audiences. Being closer in age to many of the attendees, student volunteers are more relatable and can be seen as less intimidating than a university staff member. This can make it easier for younger participants to engage with the activities and creates a feeling that science and anatomy are accessible and available for all, regardless of age or background. Student volunteers are also well positioned to be able to talk to prospective pupils about their authentic experiences with studying at higher education, and honestly answer questions from both prospective students and their families about university life. This is of benefit to attendees by raising aspirations, making the university seem more accessible, and benefits the university by creating a positive image of the institution and increasing interest and therefore applications to study.

A further benefit of working with student volunteers during the event was the opportunity to develop positive staff-student relationships. Newcastle University is a large medical school, with upwards of 350 students per year, which can present challenges for staff in getting to know students well and forging good working relationships. The event allowed staff and students to work closely together, building rapport and enabling students to represent the medical school, improving their sense of belonging. This was an inadvertent, yet invaluable, benefit to the event. Building student engagement and belonging has proven to increase the academic and health outcomes of higher education students (Walton and Cohen 2011; Thomas 2012).

Staff and students were able to observe each other delivering the stations and teaching and engaging with the public, providing opportunities for professional development, and honing the skills of all involved. Although the University of Sunderland Medical School involves a smaller student cohort, staff and students still valued the event as an excellent opportunity to solidify staff-student relationships and create a sense of identity and pride for the students in representing their university.

Students also developed key transferable skills throughout the event. For example, they practised good communication skills (Fong Ha and Longnecker 2010), by talking with different members of the community, and they gained professionalism skills (Morreale et al. 2023) by working with participants of different demographics (Simons et al. 2023). There may also have been pedagogical benefits, particularly on stations where students were heavily involved with teaching anatomy to attendees (Peets et al. 2009), although this was not formally assessed.

### 8.4.2 Potential for Curriculum Integration

There is the potential for future public engagement events of this kind to be student-led, rather than simply facilitated by student volunteers. This is something that could be integrated into the anatomy curriculum (Nagesh et al. 2018). It is hypothesised that this would positively contribute to the student learning experience and the hidden curriculum (Neve and Collett 2018) through (1) building transferable skills relevant to their future career, (2) improving anatomical knowledge and retention, and (3) enriching the student experience within the programme.

#### 8.4.2.1 Building Transferable Skills Relevant to Their Future Career

Many students studying anatomy are training to be healthcare professionals. Undergraduate programs which contain anatomy education within their curriculum include medicine, dentistry, physician associates, physiotherapy, speech and language therapy, nursing and midwifery.

Interpersonal skills, such as communication, are essential for these professions. For example, communication skills are stated in 'Domain 3' of Good Medical Practice in the General Medical Council (GMC) guidance (GMC 2024). This skill is something that can be developed through public engagement activities (Simons et al. 2023). Other skills include leadership, organisation, time management, teamwork and problem-solving, and student involvement in public engagement and outreach activities has been shown to aid in the development of these skill sets (Adams et al. 2020; Karlsson and Janssens 2023).

Even for students of anatomy who do not intend to enter a healthcare career, these are highly transferable skills that will be of benefit to any future work or study. For example, students who wish to pursue a career in academia will need to know how to discuss complex research findings with a wide range of individuals. By leading and creating a public engagement event, opportunities are created for developing and honing these skills.

#### 8.4.2.2 Improving Anatomical Knowledge and Retention

Students could be asked to design an innovative public engagement activity for a specific target audience. 'Generating', 'Planning' and 'Producing' are higher cognitive processes as shown in the revised Bloom's Taxonomy framework (Krathwohl 2002). Therefore, the process of designing an educational anatomical activity could benefit students by encouraging them to approach the anatomical topic in a different way; this can help build understanding and enhance knowledge retention.

Additionally, during the delivery of the activities, they would explore several additional study methods such as teaching others (Peets et al. 2009) and the Feynman (Reyes et al. 2021) study techniques. Similar techniques, such as reciprocal peer teaching and near-peer teaching, have been proven to be effective and popular learning strategies in the anatomical sciences (Evans and Cuffe 2009; Krych et al. 2005; Manyama et al. 2016).

#### 8.4.2.3 Enriching the Student Experience Within the Programme

Student involvement in public engagement activities poses a range of benefits for students, higher education institutions and the local community (NCCPE 2018). Integrating public engagement activities within curricula helps foster a sense of belonging and purpose within the institution and local region, supporting students to build relationships with their peers, staff and local parties such as venues, charities and services. It also encourages the development of a range of transferable skills and the confidence to support students as they navigate through higher education (Murphy and Kelp 2023; Nagesh et al. 2018).

Additionally, there is an opportunity to provide interdisciplinary collaboration with students and staff from other universities, faculties, or departments (Lacey et al. 2023). For example, students studying hospitality-related programmes could work alongside anatomy students to support the development of public engagement events or activities. Experience of interprofessional working is an excellent learning opportunity.

### 8.4.3 Drawbacks of Student Involvement

There were some drawbacks to using student volunteers for the facilitation of the activities. For example, recruitment and training were staff- and time-intensive and there were numerous organisational and administrative duties involved. Ensuring the students were well managed and supported on the day created additional tasks for staff to manage as well as running their stations and/or coordinating the event.

Additionally, there were challenges around deciding how best to incentivise volunteering at the event, whilst also ensuring we recruited students who were genuinely interested in supporting it and who would be passionate and engaging on the day. We felt it important to adequately reward students for their time and effort, but it was a delicate balancing act to provide this whilst also avoiding a situation where students would apply simply to gain a reward or payment without having any interest in the event. We were unable to provide paid employment for student ambassadors for this event, but this may be an option for other institutions or events and would avoid some of the challenges around recruiting on a voluntary basis. Employing and paying students to be involved with the event comes with its own risks, as students may not have a particularly strong interest in anatomy or public engagement. Solely relying on volunteers meant a higher risk of not recruiting enough students but meant that those we did recruit were genuinely interested and passionate about the opportunity. There was a risk that some volunteers may not feel obliged to turn up on the day or engage as we would expect without being paid; fortunately, this was not a problem we faced in our event. Providing a reward (in the form of a gift bag of small items, such as stationary, and a signed certificate), as well as reimbursing travel costs and providing refreshments for volunteers, did create additional funding costs.

The event relied heavily on the students, and regardless of whether they were paid or not, there was a concern that some students may not behave as expected of a representative of the university. Fortunately, this was not the case for the student volunteers on the day but nevertheless requires careful attention in future planning of such events.

## 8.5 Conclusion

Based on this successful experience, more public engagement events delivered as student-staff collaborations are planned for the future. An interesting future project will aim to explore the integration of public engagement activities into anatomy curricula, helping to develop transferable skills and consolidate student anatomical knowledge through teaching a variety of audiences, mimicking skills required in the clinical area and patient-facing roles.

**Acknowledgements** The authors would like to thank all staff and student volunteers who helped design and deliver the World Anatomy Day events in 2022. The authors would also like to thank the Life Science Centre including Ben Rutherford-Orrock for their support in hosting the event. Most importantly, the events would not have been possible without the generous support of the event sponsors and funders which include Adam, Rouilly, the British Association of Clinical Anatomists (BACA), the Anatomical Society and the Newcastle University FMS Engage Tilly Hale Award. The latter was set up in memory of Tilly Hale, a woman diagnosed with primary biliary cholangitis (previously called primary biliary cirrhosis) who dedicated her life to supporting liver patients and highlighting the importance of patient involvement in research and healthcare.

## References

Adams JS, Nelson HM, Strong E, Wisco JJ, Klappa SG (2020) Mentors' experience in teaching basic science concepts in the community outreach program, anatomy academy: a phenomenological study. HAPS Educ 24(2):27–35

Arantes M, Arantes J, Ferreira MA (2018) Tools and resources for neuroanatomy education: a systematic review. BMC Med Educ 18:1–15

Arjomandi Rad A, Subbiah Ponniah H, Shah V, Nanchahal S, Vardanyan R, Miller G, Malawana J (2023) Leading transformation in medical education through extended reality. In: Biomedical visualisation: volume 16–digital visualisation in biomedical education. Springer, Cham, pp 161–173

Arnaldo B, Bonar SK (1996) Historical perspective: Andreas Vesalius. Spine 21(11):1388–1393

Auffret M (2024) Teaching neuroanatomy in the 19th century: Dr Auzoux's Papier-mâché anatomical models (S12. 003). Neurology 102(17):3600

Biglino G, Layton S, Hamer A, Milano EG, Caputo M, Wray J (2022) Interdisciplinarity and patient engagement: new representations of cardiovascular anatomy. Rev Cardiovasc Med 23(11):1–7

Burns ER (2002) Anatomy of a successful K–12 educational outreach program in the health sciences: eleven years' experience at one medical sciences campus. Anat Rec 269(4):181–193

Chandio N, Micheal S, Tadakmadla SK, Sohn W, Cartwright S, White R, Sanagavarapu P, Parmar JS, Arora A (2022) Barriers and enablers in the implementation and sustainability of toothbrushing programs in early childhood settings and primary schools: a systematic review. BMC Oral Health 22(1):242

Chorcora E, Bray A, Banks J (2023) A systematic review of widening participation: exploring the effectiveness of outreach programmes for students in second-level schools. Rev Educ 11(2):3406

Conway-Morris S (2004) Life's solution: inevitable humans in a lonely universe. Cambridge University Press

Cook DP, Steed K, Read C, Baysarowich R, Redway T, Robineau-Charette P, Carnegie J (2020) Science outreach: six examples of programs that enrich the learning environments of students and educators. HAPS Educator

Dewalt DA, Berkman ND, Sheridan S, Lohr KN, Pignone MP (2004) Literacy and health outcomes: a systematic review of the literature. J Gen Intern Med 19:1228–1239

Donald E, Dulohery K, Khamuani M, Miles H, Nott J, Patten D, Roberts A (2023) Putting the cart before the horse? Developing a blended anatomy curriculum supplemented by cadaveric anatomy. In: Biomedical visualisation: volume 16–digital visualisation in biomedical education. Springer, Cham, pp 15–35

Drake RL, Vogl W, Mitchell AWM, Gray H (2020) Gray's anatomy for students, 4th edn. Elsevier

Evans DJ, Cuffe T (2009) Near-peer teaching in anatomy: an approach for deeper learning. Anat Sci Educ 2(5):227–233

Finn GM (2014) Using body painting and other art-based approaches to teach anatomy. In: Teaching anatomy: a practical guide. Springer, pp 155–164

Finn GM (2019) Painting, pirouettes and proving: a toolkit of methods for engagement in anatomy. FASEB J 33(S1):21–23

Finn GM (2020) Using body painting and other art-based approaches for the teaching of anatomy and for public engagement. In: Teaching anatomy: a practical guide. Springer, pp 185–197

Finn GM, Brown ME (2023) The hidden curriculum of public engagement for creative methods of instruction. In: Teaching, research, innovation and public engagement. Springer, pp 139–152

Fong Ha J, Longnecker N (2010) Doctor-patient communication: a review. Ochsner J 10(1):38–43

Furco A (2010) The engaged campus: toward a comprehensive approach to public engagement. Br J Educ Stud 58(4):375–390

Garvis S, Barton G, Hai Le A (2023) Kinesthetic learning: a systematic review in early childhood education context. In: Kinesthetic learning in early childhood: exploring theory and practice for educators, 1st edn. Routledge, New York, pp 10–21

GMC (2024) Domain 2: patients, partnership and communication [Online]. https://www.gmc-uk.org/professional-standards/professional-standards-for-doctors/good-medical-practice/domain-2-patients-partnership-and-communication. Accessed 30 May 2024

González-Solá M, Rosario-Canales M (2015) Encouraging Puerto Rican high school students to pursue a STEM career through an anatomy & neurobiology outreach. FASEB J 29:693–696

Gröning F, Perez-Pachon L, Aliotta S, Collas C, Fertikova A, Koos N, Struck R, Curtis N, Crozier R (2021) 3D photogrammetry models of human specimens: A useful resource for teaching and public engagement. J Anat 239(4):940–941

Hall JE, Hall ME (2020) Guyton and Hall textbook of medical physiology e-book. Elsevier Health Sciences

Heaslip V, Hutchings M, Collins B, Crowley E, Eccles S, Hunt C, Tooth R, Wardrop A (2020) Situating the evidence for impact of outreach strategies: a systematic review for improving access to higher education. Widening Particip Lifelong Learn 22(1):25–54

Jędrzejewski Z, Loranger B, Clancy JA (2020) Virtual anatomy museum: facilitating public engagement through an interactive application. Biomed Vis 7:1–18

Johnson M, Danvers E, Hinton-Smith T, Atkinson K, Bowden G, Foster J, Garner K, Garrud P, Greaves S, Harris P, Hejmadi M, Hill D, Hughes G, Jackson L, O'Sullivan A, Otuama S, Brown PP, Philipson P, Ravenscroft S, Rhys M, Ritchie T, Talbot J, Walker D, Watson J, Williams M, Williams S (2019) Higher education outreach: examining key challenges for academics. Br J Educ Stud 67(4):469–491

Jozefowicz RF (1994) Neurophobia: the fear of neurology among medical students. Arch Neurol 51(4):328–329

Karlsson AW, Janssens A (2023) Patient and public involvement and engagement (PPIE) in healthcare education and thesis work: the first step towards PPIE knowledgeable healthcare professionals. BMJ Open 13(1):e067588

Keenan ID, Green E, Haagensen E, Hancock R, Scotcher KS, Swainson H, Swamy M, Walker S, Woodhouse L (2023) Pandemic-era digital education: insights from an undergraduate medical programme. Adv Exp Med Biol 1397:1–19

Khaleel I, Wimmer BC, Peterson GM, Zaidi STR, Roehrer E, Cummings E, Lee K (2020) Health information overload among health consumers: a scoping review. Patient Educ Couns 103(1):15–32

Krathwohl DR (2002) A revision of Bloom's taxonomy: an overview. Theory Pract 41(4):212–218

Krych AJ, March CN, Bryan RE, Peake BJ, Pawlina W, Carmichael SW (2005) Reciprocal peer teaching: students teaching students in the gross anatomy laboratory. Clin Anat 18(4):296–301

Kuzma JM, Wright W (2013) Using social networks as a catalyst for change in global higher education marketing and recruiting. Int J Contin Eng Educ Life Long Learn 23:53–66

Lacey MM, Capper-Parkin K, Schwartz-Narbonne R, Hargreaves K, Higham C, Duckett C, Forbes S, Rawlinson K (2023) University student-led public engagement event: increasing audience diversity and impact in a non-science space. Access Microbiol 5(8):acmi000534.v3

Lee L, Decara JM (2020) Point-of-care ultrasound. Curr Cardiol Rep 22:149

Manyama M, Stafford R, Mazyala E, Lukanima A, Magele N, Kidenya BR, Kimwaga E, Msuya S, Kauki J (2016)

Improving gross anatomy learning using reciprocal peer teaching. BMC Med Educ 16:1–12

Márquez S (2013) Anatomy education day: A global outreach. FASEB J 27(1_Meeting Abstracts):19.5

Márquez S, Mtui E, Curcio DF, Laitman JT (2019) The long arm of anatomy education outreach programs: from NYC to Tanzania and Brasil. FASEB J 33(S1):441–445

Martin AJ, Beska BJ, Wood G, Wyatt N, Codd A, Vance G, Burford B (2018) Widening interest, widening participation: factors influencing school students' aspirations to study medicine. BMC Med Educ 18

Maulik D (2023) Spectral doppler: basic principles and instrumentation. In: Doppler ultrasound in obstetrics and gynecology. Springer, pp 21–37

McMenamin PG (2023) The third dimension: 3D printed replicas and other alternatives to cadaver-based learning. In: Biomedical visualisation: volume 16–digital visualisation in biomedical education. Springer, pp 39–61

McRae C (1890) Fathers of biology. Percival, London

Meyer ER, Williams S, Conway M, Notebaert A (2018) Kids in the gross anatomy lab: how an outreach program in anatomy educates high school and undergraduate students about health care. HAPS Educ 22(3):262–267

Morreale MK, Balon R, Louie AK, Guerrero APS, Aggarwal R, Coverdale J, Beresin EV, Brenner AM (2023) The vital importance of professionalism in medical education. Acad Psychiatry 47:430–343

Murakami S, Mealey BL, Mariotti A, Chapple ILC (2018) Dental plaque-induced gingival conditions. J Periodontol 89:17–S27

Murphy KM, Kelp NC (2023) Undergraduate STEM students' science communication skills, science identity, and science self-efficacy influence their motivations and behaviors in STEM community engagement. J Microbiol Biol Educ 24(1):e00182–e00122

Nagesh NM, Giurca BC, Lishman S (2018) Innovating undergraduate pathology education through public engagement. Virchows Arch 472(5):853–863

Nakre PD, Harikiran AG (2013) Effectiveness of oral health education programs: a systematic review. J Int Soc Prev Community Dent 3(2):103–115

NCCPE (2018) Students [Online]. https://www.publicengagement.ac.uk/support-engagement/strategy-and-planning/edge-tool/dig-deeper-with-our-mini-edge-tools/students. Accessed 26 July 2023

Neve H, Collett T (2018) Empowering students with the hidden curriculum. Clin Teach 15(6):494–499

NHS England (2023) NHS equality, diversity, and inclusion improvement plan [Online]. https://www.england.nhs.uk/long-read/nhs-equality-diversity-and-inclusion-improvement-plan/. Accessed 15 Aug 2024

Nutbeam D (2008) The evolving concept of health literacy. Soc Sci Med 67(12):2072–2078

Oaks M, Franklin N, Bargerstock AB (2009) Situating outreach and engagement in the university: concepts, challenges, and opportunities. Contin High Educ Rev 73:224–233

Ortug G, Midi A, Elbizim DS, Karaot H, Yilik E, Uluisik IE (2021) Introducing children to anatomy: "getting to know our bodies: the first step toward becoming a scientist". Anat Sci Educ 14(2):232–240

Peets AD, Codderre S, Wright B, Jenkins D, Burak K, Leskosky S, McLaughlin K (2009) Involvement in teaching improves learning in medical students: a randomized cross-over study. BMC Med Educ 9(55)

Peres KG, Thomson WM, Chaffee BW, Peres MA, Birungi N, Do LG, Feldens CA, Fontana M, Marshall TA, Pitiphat W, Seow WK, Wagner Y, Wong HM, Rugg-Gunn AJ (2020) Oral health birth cohort studies: achievements, challenges, and potential. J Dent Res 99(12):1321–1331

Persons JE, Sipla JS (2023) The cure for neurophobia: an approach for progressive mastery of medical neuroscience. J Neuropathol Exp Neurol 82(8):745–750

Reyes EP, Blanco RMFL, Doroon DRL, Limana JLB, Torcende AMA (2021) Feynman technique as a heutagogical learning strategy for independent and remote learning. Recoletos Multidiscip Res J 9(2):1–13

Rosario MG, Gonzalez-Sola M, Hyder A, Medley A, Weber M (2019) Anatomage virtual dissection table: a supplemental learning aid for human anatomy education during an undergraduate outreach activity. FASEB J 33(S1):604–609

Ruth AA, Nesbitt A, Johnson LE (2023) Flexible, short-duration outreach sessions in the human anatomy laboratory provide authentic, humanistic experiences. Anat Sci Educ 16(2):280–290

Sanders K, Philp J (2019) Anatomy nights—a public engagement format bringing experts to the public. FASEB J 33(S1):21–21

Sanders KA, Philp JA, Jordan CY, Cale AS, Cunningham CL, Organ JM (2022) Anatomy nights: an international public engagement event increases audience knowledge of brain anatomy. PLoS One 17(6):e0267550

Simok AA, Hadie SNH, Ismail ZIM, Asari MA, Kasim F, Yusof NAM, Shamsuddin SA, Amin MSIM, Mukhtar SF (2022) Anatomy outreach through the world anatomy day celebration in Universiti Sains Malaysia. Educ Med J 14(4):113–120

Simons G, Birch R, Stocks J, Insch E, Rijckborst R, Neag G, McColm H, Romaniuk L, Wright C, Phillips BE, Jones SW, Pratt AG, Siebert S, Raza K, Falahee M (2023) The student patient alliance: development and formative evaluation of an initiative to support collaborations between patient and public involvement partners and doctoral students. BMC Rheumatol 7(1):36

Stearns SC, Nesse RM, Govindaraju DR, Ellison PT (2010) Evolutionary perspectives on health and medicine. Proc Natl Acad Sci USA 107(1):1691–1695

Taylor AM, Diggle P, Wessels Q (2018) What do the public know about anatomy? Anatomy education to the public and the implications. Anat Sci Educ 11(2):117–123

Thomas L (2012) Building student engagement and belonging in higher education at a time of change. Paul Hamlyn Foundation, 100(1–99)

Tubbs RS, Shoja MM, Loukas M (eds) (2016) Bergman's comprehensive encyclopedia of human anatomic variation. Wiley
Vargiu A (2014) Indicators for the evaluation of public engagement of higher education institutions. J Knowl Econ 5:562–584
Venkatesan M, Mohan H, Ryan JR, Schürch CM, Nolan GP, Frakes DH, Coskun AF (2021) Virtual and augmented reality for biomedical applications. Cell Rep Med 2(7):100348
Walton GM, Cohen GL (2011) A brief social-belonging intervention improves academic and health outcomes of minority students. Science 331(6023):1447–1451
Xiang J, Venkatesan S (2023) The role of Vesalius and his contemporaries in the transfiguration of human anatomical science. J Anat 242(2):124–131
Żytkowski A, Tubbs RS, Iwanaga J, Clarke E, Polguj M, Wysiadecki G (2021) Anatomical normality and variability: historical perspective and methodological considerations. Transl Res Anat 23:100105

**Kayleigh Scotcher** B.Sc., (Hons), M.Sc., S.F.H.E.A., was a Lecturer in Anatomy within the School of Medicine at Newcastle University in Newcastle upon Tyne, United Kingdom. She is currently undertaking a Ph.D at the University of Melbourne, Melbourne, Australia. Her research interests include exploring the autonomic innervation of the pelvic viscera and the development of innovative teaching approaches in anatomy education. She is also involved in outreach and public engagement.

**Emily Green** B.Sc., (Hons), M.Sc., F.H.E.A., is a Lecturer in Anatomy within the School of Medicine at Newcastle University in Newcastle upon Tyne, United Kingdom. Her research interests relate to student support and inclusivity within anatomy education, innovative approaches to anatomy education and the use of art-based learning and 3D software. She is currently undertaking a part-time Ph.D at Newcastle University in Medical Education, with a focus on inclusive anatomy education for neurodivergent students and students with specific learning difficulties.

**Thomas Butts** B.Sc., (Hons), D.Phil., F.H.E.A., is a Senior Lecturer in Neuroscience at the School of Medicine at the University of Sunderland, United Kingdom. His research interests principally include the development and evolution of the brain across the vertebrate phylogeny, particularly in relation to the control of neurogenesis.

**Evie M. Donald** B.Sc., (Hons), M.Sc., is a Lecturer in Anatomy within the School of Medicine, University of Sunderland in Sunderland, United Kingdom. Her research interests primarily centre around evolutionary human and primate anatomy.

**Kate Dulohery** B.Sc. (Hons), M.Sc., PhD, F.H.E.A, is a Senior Lecturer in Anatomy at the School of Medicine, University of Sunderland, United Kingdom. Her research interests include anatomical education including technology-enhanced learning and ultrasound.

**Munesh P. Khamuani** BDS, MSc, MMedSci, FHEA, is a Lecturer in Histology and Anatomy at the School of Medicine, University of Sunderland in Sunderland, United Kingdom. His research interests include the exploration of student perceptions and effectiveness of the educational interventions in anatomy education.

**Harry Miles** B.Sc., M.Sc., F.H.E.A., is a Lecturer in Anatomy at the School of Medicine, University of Sunderland in Sunderland, United Kingdom. His research interests include biomechanics, anatomical pedagogy and the effectiveness of educational videos and digital resources in anatomical education.

**Debs Patten** B.Sc., F.H.E.A., is a Professor of Anatomy at the School of Medicine, University of Sunderland in Sunderland, United Kingdom. Her research interests are focused on the utility of technology-enhanced learning and ultrasound in anatomical education.

**Alice E. M. Roberts** B.Sc., (Hons), M.Sc., is a Lecturer in Anatomy within the School of Medicine, University of Sunderland in Sunderland, United Kingdom. Her research interests predominantly relate to osteology and skeletal trauma. She is the widening participation and school outreach co-lead for the School of Medicine, University of Sunderland.

**Hannah Swainson** B.Sc., (Hons), M.B.B.S, F.H.E.A., is a Lecturer in Anatomy within the School of Medicine at Newcastle University in Newcastle upon Tyne, United Kingdom. Her research interests relate to clinical anatomy and the use of digital technology in teaching and assessment.

**Joanna Matthan** M.B.B.S., M.A., S.F.H.E.A., N.T.F., F.A.S. is Professor of Anatomy and Dean of Academic Affairs and Director of Education at Newcastle University Medicine Malaysia (NUMed). Her research interests relate to several aspects of anatomical/medical education and public engagement, and she serves on several societies in an honorary capacity.

GPSR Compliance

*The European Union's (EU) General Product Safety Regulation (GPSR) is a set of rules that requires consumer products to be safe and our obligations to ensure this.*

*If you have any concerns about our products, you can contact us on ProductSafety@springernature.com*

In case Publisher is established outside the EU, the EU authorized representative is:

Springer Nature Customer Service Center GmbH
Europaplatz 3
69115 Heidelberg, Germany

**Batch number: 10755744**

Printed by Printforce, the Netherlands